Advancing Technologies for Asian Business and Economics:

Information Management Developments

Dasho Karma Ura
Centre for Bhutan Studies, Bhutan

Patricia Ordóñez de Pablos
The University of Oviedo, Spain

Information Science
REFERENCE

Managing Director:	Lindsay Johnston
Senior Editorial Director:	Heather Probst
Book Production Manager:	Sean Woznicki
Development Manager:	Joel Gamon
Development Editor:	Heather Probst
Acquisitions Editor:	Erika Gallagher
Typesetter:	Russell A. Spangler
Cover Design:	Nick Newcomer, Lisandro Gonzalez

Published in the United States of America by
Information Science Reference (an imprint of IGI Global)
701 E. Chocolate Avenue
Hershey PA 17033
Tel: 717-533-8845
Fax: 717-533-8661
E-mail: cust@igi-global.com
Web site: http://www.igi-global.com

Library of Congress Cataloging-in-Publication Data

Advancing technologies for Asian business and economics: information management developments / Dasho Karma Ura and Patricia Ordonez de Pablos, editors.
p. cm.
Includes bibliographical references and index.
Summary: "This book offers the latest research available within the field of information management as it pertains to the Asian business market, promoting and coordinating developments in the field of Asian and Chinese studies, as well as presenting strategic roles of IT and management towards sustainable development"--Provided by publisher.
ISBN 978-1-4666-0276-2 (hardcover) -- ISBN 978-1-4666-0277-9 (ebook) -- ISBN 978-1-4666-0278-6 (print & perpetual access) 1. Information technology--Technological innovations--Asia. 2. Management information systems--Asia. 3. Database management--Asia. 4. Knowledge management--Asia. I. Ura, Dasho Karma, 1961- II. Ordsqez de Pablos, Patricia, 1975-
HD30.2.A347 2012
658.4'038095--dc23

2011049600

British Cataloguing in Publication Data
A Cataloguing in Publication record for this book is available from the British Library.

All work contributed to this book is new, previously-unpublished material. The views expressed in this book are those of the authors, but not necessarily of the publisher.

Table of Contents

Detailed Table of Contents

This paper examines the role of creative milieu and cultural heritages in the development of Chinese cultural industry. Through case study of Beijing's Nanluoguxiang, where several arts institutions and theaters concentrate, it depicts the birth and growth of a creative place for free artists in the institutional changing of Chinese cultural setting. Based on field survey data from artists, managers and visitors, it shows the spouting and growth of local creativities in a transitional economy needs atmosphere of tolerant and frequent social gathering, especially in spaces like cafes and bars. It also suggests that making a good use of local cultural and creative resources like heritages and folk customs enables the superiority of localization standing out in the wave of globalization. The key findings indicate that the newly-built creative parks might be useful to breed the creative products, but not necessary.

This paper indentifies a main barrier when doing business with China, the cultural gap, and provides the strategies that companies can use when entering the Chinese market. This empirical study examined 40 Australian organisations in their activities when entering the Chinese market. Alarmingly after 30 years of attempting to do business in China, companies are still not addressing the issue of cultural differences. Companies are also caught by surprises due to lack of preparation how large the cultural gap is between Australian and Chinese business culture. The findings of the study have important implications for businesses considering entry to China, and for Australian businesses already doing business in China. The strategies investigated include human resource strategies, dealing with Chinese staff, relationship building, getting outside support (employing consultants), learning about the culture, and adapting to the culture.

This paper analyzes the country-of-origin effects of Chinese brands going international. To approach this topic, the author uses a comparative case study, in which a Chinese manufacturer of white goods, Haier, and a German manufacturer, Miele, serve as examples. Fueled by "China trash" reports in EU and US media, the image of China as country of origin still has a negative connotation. The Chinese government and Chinese companies are working to change this image, since they both regard branding as important means for corporate internationalization. This paper argues that Chinese brands could achieve a similar development as did Japanese and Korean brands in the second half of the last century.

This paper focuses on the effects of the *Go Global* policy in Italy. The main type of Chinese investments, their strategic motivations, as well as the role of ethnic networks are analyzed. The phenomenon is still not significant in quantitative terms, but trends are impressive. Italy can provide an access to western markets and strategic logistic services, as well as to a wide array of distinctive skills/intangible assets in manufacturing industries. At the moment, there is a prevalence of greenfield initiatives, but acquisitions are rising sharply. In geographical terms, the locations chosen by Chinese investors favor areas that offer a wealth of distinctive skills (typically, but not only, district areas), but are not limited to industries in which Chinese ethnic groups are involved.

Trillions of dollars were evaporated worldwide as a result of the financial tsunami that shocked the world in 2007 (Soros, 2008). This paper argues that in a time of global recession, banks seeking to expand wealth management service in the People's Republic of China (P.R.C.) need to evaluate their entry strategies thoroughly to gain a competitive edge. Through applying the theory of planned behavior, a sequential mixed methods study was conducted post recession with the aim to identify wealth management service entry strategies into the P.R.C. A sample of 227 Taiwanese nationals based in the P.R.C. indicated three managerial implications consisting of the bank entry identity, deployment of expatriates, and brand-image position.

In past years, China recorded a fast sustainable economic growth with an estimated average GDP growth rate of 9.7% in the period of 1980-2008, turning China into the world's second largest economy. With

Preface

INTRODUCTION

Since 2008, the economies of many western countries (the United States, eurozone) have been suffering the dramatic consequences of the international financial crisis: continuous destruction of employment, high unemployment rates (especially among young people), the closing-down of small and medium companies, interventions in banks by central banks, growth of the risk premiums and intervention in some countries, like Portugal, Greece and Ireland. Spain suffers in a critical situation too, with more than 5 million unemployed people, the country is walking towards levels of recession and rumours about a possible intervention, as happened in other neighbouring European countries recently.

The earthquake and nuclear disaster that struck Japan in March 2011 had a negative impact on the Asian economy. Japan is the world's fourth largest economy (after the EU, U.S. and China), so its decline would drag down the global economy. This country also hires temporary workers from South Asian countries, who are now being laid off in droves. In November 2011 Japan made the decision to intervene in currency markets due to the appreciation of the yen which was putting Japan's economic recovery at risk.

In the *Asian Development Outlook Update 2011* published in September 2011, the Asian Development Bank trimmed its full year forecast to 7.5% from 7.8%. The 2012 projection is also lowered slightly to 7.5% from 7.7% previously. Some of these Asian economies observe how this inflation comes closer or exceeds the targets of the central banks at the time that they experienced a credit boom. Also there is the latent risk of rising prices of energy and food entering an inflationary spiral.

Undoubtedly the Asian economies are going to have an essential role in the economic recovery of the Euro area and USA. For this reason to know its economic policies, models of economic development, tendencies in management of companies and influence of the new information technologies will be of strategic interest. But not only to look at the top emergent markets but also to other countries that can provide surprises in the future, as are Bhutan, Laos, Mongolia, Myanmar and Vietnam.

In most academic journals on economics and business available in the libraries of European and North American universities there is little research published by authors from these countries. For example, Bhutan is developing the philosophy of the Gross National Happiness (GNH), an index measuring the happiness of the inhabitants of the country. Also Bhutan is implementing a tourist policy designed to attract quality tourism ("high value, low impact" policy) and at the same time trying to maintain its ancestral traditions and protect the beautiful natural environment of the country. Nevertheless, there is little knowledge about these issues in Europe and the United States. It would be very interesting to know about empirical researches on these topics and learn from their experiences.

The main emerging markets (China, India, Brazil, Indonesia, Philippines, Poland, Mexico, Russia, South Korea, South Africa and Turkey) are redeeming a critical role in the economic global recovery.

Some experts point out that India may become the fastest growing major economy in the world till 2012 and will be the 3rd largest by gross domestic product within the next two decades, behind China and the United States.

Nevertheless the help of the emerging Asian markets – the two most important ones are China and India --, to getting out of the recession in other economies, might be affected if the crisis in Europe continues, as pointed out recently Christine Lagarde, the president of the International Monetary Fund (IMF). The decrease of the demand from Europe and the United States begins to be a reason to worry for Asian countries.

THE BIRTH OF THE JOURNAL

In 2010, the *International Journal of Asian Business and Information Management* (IJABM) came to light. It is an academic journal published quarterly by IGI Global. The mission of the journal is to establish an effective channel of communication between academic and research institutions, policy makers, government agencies, and individuals concerned with the complexities of Asian business, information technologies, sustained development, and globalization.

The journal provides in-depth analysis on the globalization and sustained development of Asian commerce and management technologies with a clear international and interdisciplinary approach. A comprehensive resource for policy makers, government officials, academicians, and practitioners, this journal contains a wealth of valuable research on theories and practices which underpin successful business and management in Asia. *International Journal of Asian Business and Information Management* promotes and coordinates developments in the field of Asian studies, as well as presents strategic roles of information technologies and management towards sustainable development with special focus on practical aspects in Asia.

The Editor in Chief of *International Journal of Asian Business and Information Management* is Patricia Ordóñez de Pablos (The University of Oviedo, Spain). The Associate Editors are *Dasho* Karma Ura (Centre for Bhutan Studies, Bhutan), W.B. Lee (The Polytechnic University of Hong Kong, Hong Kong) and Jose Luis de Sales Marques (Institute of European Studies of Macau, Macau).

The Editorial Board of the *International Journal of Asian Business and Information Management* is formed by leading academics and experts mostly from China, Hong Kong, India, Japan, Philippines and Taiwan. The members of the Editorial Board were chosen following a review process of CVs submitted by those interested in joining the Editorial Board of the new journal in 2009. The Editorial Board members are Ilan Alon (Rollins College, USA), Siah Hwee Ang (The University of Auckland Business School, New Zealand),Rolando T. Dy (University of Asia and the Pacific, Philippines), David S. Hong (Taiwan Institute of Economic Research, Taiwan), Rabindra Ku Jena (Institute of Management Technology, India), Lluís Jovell, (Universidad Autónoma de Barcelona, Spain), Vicent Kong (Ministry of Foreign Affairs and Trade, New Zealand), J.P. Li (CEO China, China), Mingfang Li (California State University Northridge, USA), Jyhjong Lin (Ming Chuan University, China), Kai Meng Mok (University of Macau, Taipa), Maria Fernanda Pargana Ilheu (ChinaLogus, Portugal), Hong-Chi Shiau (Shih-Hsin University, Taiwan), Danchi Tan (National Cheng-Chi University, Taiwan), Jonathan Unger (Australian National University, Australia), Chihiro Watanabe (Tokyo Institute of Technology, Japan), Wei Ping Wu (Hong Kong Baptist University, Hong Kong), Qingrui Xu (Zhejiang University, China), Xi Zhang (Institute of Policy and Management, CAS, China), Xiaoyang Zhang (The Open University of Hong Kong, Hong

Kong), Jing-Yuan Zhao (Harbin Institute of Technology, China) and Wang Zhiwei (Ministry of Education of China, China). It is our duty to thank all of them for their generous work for this academic journal in 2010 and we hope they continue collaborating with us in the future.

The journal focuses on the role of business and information technologies in Asia and its implications on Asia region, Asia-European Union relations, Asia-Latin America relations, Asia-USA relations, business, business ethics, competition in a global economy, eco-tourism, entrepreneurship, globalization, education and cultural change, human resource management, information technologies, international trade, leadership, management, competitive, technological, social, and political environments, new technologies and new modes of e-learning, organization structure, planning processes, resource allocation, strategic decision processes and sustained development.

Currently the *International Journal of Asian Business and Information Management* is indexed in Bacon's Media Directory, DBLP, Google Scholar, The Index of Information Systems Journals, INSPEC, MediaFinder, The Standard Periodical Directory and Ulrich's Periodicals Directory.

During the first year of life of the academic journal numerous manuscripts have been received for review. In particular, the journal received more than 100 manuscripts, most of them written by researchers located in East Asia and South Asia, mainly from China and India. All submitted manuscripts were initially reviewed by the Editor in Chief or one of the Associate Editors. If the topic of the paper fits with the aims and scope of the journal and it offers added value to the field of research, then the paper is sent for review to two anonymous reviewers, to follow a double-blind peer review process. *The International Journal of Asian Business and Information Management* has a pool of reviewers with different areas of expertise. When a paper is submitted for review, our goal is to choose reviewers experts in the topic and also with a good knowledge of the economy, market or country where the paper is focused. After a double blind review process, only twenty papers were accepted for publication in 2010.

CONTENTS OF THE BOOK

In 2010 four issues of the International Journal of Asian Business and Information Management have been published. The book, *Advancing for Technologies Asian Business and Economics: Information Management Developments*, offers the collection of papers previously published and classifies them in the following sections: A) Chinese Cultural Values and Industry; B) Marketing in Asia; C) Economic Development and Investments and D) Managerial Issues in China. Now we present a summary of contents of the book chapters included in the four sections mentioned before.

Chinese Cultural Values and Industry

The Section "Chinese cultural values and industry" offers a collection of seven chapters that analyses the impact of Chinese cultural values on business in China and what companies need to learn to overcome the cultural gap when doing business in this country. Factors such as the Confucianism heritage, the dynamics of *guanxi* or social networks and the importance of reputation are key to understand the success factors of business in China.

The chapter "Local Milieu in Developing China's Cultural and Creative Industry: The Case of Nanluoguxiang in Beijing" by Jici Wang, Chun Zhang, Ching-Ning Wang, and Ping Chen, all from Peking University (China) studies the role of creative milieu and cultural heritages in the development of Chinese cultural industry. The chapter covers a case study of Beijing's Nanluoguxiang, a zone where several arts institutions and theatres are located. It suggests that the growth of local creativities in a transitional economy needs an atmosphere of tolerant and frequent social gathering, like cafes and bars. It also shows that the newly-built creative parks might be useful to breed the creative products, but are not necessary.

The chapter, "Australian Businesses in China: Strategies to Overcome the Cultural Gap," by Mona Chung and Jane Menzies, both from Deakin University, (Australia) points a main barrier when doing business with China: the cultural gap. It studies the strategies that companies can use when entering the Chinese market. With this aim, the authors develop an empirical study with 40 Australian organisations that penetrated the Chinese market. These strategies refer to human resource strategies, dealing with Chinese staff, relationship building, learning and adapting to local culture, among others.

The chapter, "Confucianism Values for Transformational Leadership Development in the Chinese Context," is written by Balakrishnan Muniapan and Ee Yaw Seng both from Winburne University of Technology (Malaysia). Transformational leadership has received a great attention in the leadership literature during the past two decades. The chapter examines the importance of Confucianism and its impact in transformational leadership development. It focuses on the Analects or *Lunyu* - a collection of sayings by Confucius and dialogs with his disciples). The Analects provide the essence of Confucianism to explore the relevant transformational leadership concepts. According to the authors of the chapter, Confucianism offers an inside-out approach to transformational leadership development, which is self leadership first, before leadership of others.

The chapter, "Study on Cultural Commercialization and Public Consumption: The Case of China," is written by Yuhong Li, Jingyuan Zhao and Han Weixi, all from Beijing Union University (China). The authors analyze the industrialization process of Chinese culture, probes into the cultural consumption demand and structure in contemporary China, and analyze the cultural consumption phenomenon of the Chinese public. It offers advice on the destined cultural commercialization and the development trend of cultural consumption in China.

Valerie (C. Y.) Zhu (Xi'an Jiaotong University, China) and Linyan Sun (Xi'an Jiaotong University, China) are the authors of the chapter, "The Interactive Relationship between Corporate Cultures, Leadership Style and Knowledge Management." The chapter studies the dynamic relationship and effect mechanism between corporate culture, leadership styles and knowledge management in organizations. It shows that corporate culture can serve as a propeller or inhibitor in the knowledge management strategies. For example a knowledge-sharing organizational culture has a positive effect on knowledge management while hierarchical culture has a negative impact. Leadership style has an impact (both direct and indirect) on knowledge management through the formation and components of organizational culture.

The next chapter is, "Interactive Strategies for Cultural Heritage Tourism and Macau City Development: The Tourists' Perspectives," written by Aliana M. W. Leong (Macau University of Science and Technology, China) and Jacky X. Li (Macau University of Science and Technology, China). It analyzes the cultural heritage resources of Macau and the development of this city. The research analyzes the condition of Macanese cultural heritage tourism and the challenges and opportunities for its future development. Using data from a survey of visitors to Macau, the chapter provides understanding of the tourists' behaviour, awareness and evaluation of scenic spots that are part of the cultural heritage tour.

The last chapter of this section is "Embracing *Guanxi*: The Literature Review" by Jilong Zhang and Nattavud Pimpa both from RMIT University (Australia). It studies the basic differences between Chinese and Western approaches to business dealings, to identify the factors that make it difficult for Western managerial practises to develop and use *guanxi* or social connections. It identifies the differences in the perception of trust in both societies and how this variable leads to cultural clashes.

Marketing in Asia

The section "Marketing in Asia" analyses marketing issues in China. In particular it has two papers that focus respectively in the Malaysian market and the effect of Chinese versus German brands.

The section starts with the chapter "A Study of Customer Satisfaction of an Automotive Company in Malaysia" written by Noor Azlina Fakiruddin (International Islamic University Malaysia (IIUM), Malaysia), Hebat Allah A. Wagih (International Islamic University Malaysia (IIUM), Malaysia), and Waleed F. Faris (International Islamic University Malaysia (IIUM), Malaysia). It covers an analysis of the current level of customer satisfaction on the services offered by an automotive outlet in Malaysia. It is clear that customer satisfaction is one of the most important factors in the achievement of a long term competitive advantage for firms and organizations. Customer satisfaction is a dependent variable on how well the product performs. The authors of the chapter develop a study to understand how companies and organizations can improve the customer level of satisfaction through providing continuous efforts to improve their level of service and product productivity.

The other chapter in this section is "Creating a Chinese 'Miele'? Asymmetries in the Consumer Perception of Chinese and German Brands," written by Cora F. Jungbluth (Heidelberg University, Germany). It examines the country-of-origin effects of Chinese brands going international and offers a comparative case study (Haier, a Chinese manufacturer of white goods, and Miele, a German manufacturer). The image of China as country of origin may still has a negative connotation. Chinese government as well as Chinese companies are working hard to change this image. The chapter discusses if Chinese brands could change this negative image and how they could do it.

Economic Development and Investments

The third section of the book ("Economic development and investments") offers a collection of eight chapters, analyzing key topics on investments from entry strategies in the Chinese market to Chinese foreign direct investments.

The first chapter of this section is "Chinese Investments in Italy: Is the Wave Arriving?" by Francesca Spigarelli (University of Macerata, Italy). It chapter analyzes the effects of the Go Global policy in Italy and focus on main type of Chinese investments, their strategic motivations, as well as the role of ethnic networks. Why does Chinese investment choose the Italian market? This country can provide an access to western markets and strategic logistic services industries.

The chapter, "Entry Strategies into the P.R.C. for Post Recession Wealth Management Service," is written by Andrew Chan (City University of Hong Kong, China), and Carlson Chan (City University of Hong Kong, China). It states that in the current economic situation, banks seeking to expand wealth management service in the People's Republic of China need to evaluate their entry strategies thoroughly to gain a competitive edge. The authors developed an empirical study with a sample of 227 Taiwanese nationals based in the People's Republic of China which suggested three managerial implications: bank entry identity, deployment of expatriates, and brand-image position.

Fernanda Ilhéu from ISEG/Technical Universal of Lisbon in Portugal is the author of the chapter, "Chinese Outward Foreign Direct Investment: In Search of a New Theory." It addresses the role of China as the most attractive developing country for FDI flows, both short and long term. During the current global financial crisis, China was able to resist better than other major world economies, clearly profiting from this downturn to implement policies to reduce its economic imbalances, such as the gap between Chinese FDI and OFDI, which is now progressively narrowing. The chapter also explores whether Chinese OFDI can be explained by current theories or if a new theory is required.

Bradley J. Koch from Grand Valley State University (USA) is the author of the next chapter titled, "No Strategy is an Island to Itself: China First-Mover and Other FDI Strategies' Interaction Effects." It examines the first-mover strategy for foreign investment to determine under what conditions an important effect exists when it is combined with other foreign investment strategies like partner selection, geographical market focus, joint-venture control, and resource commitment strategies. Through a survey data from Sichuan, the results suggest the existence of significant interaction effects that can delete a first-mover advantage, create a first-mover effect that previously didn't exist, or even modify the direction of this effect.

The chapter, "Study on the New Tendency of the International Industry Shifts and Chinese Enterprise Implementation Tactics," by Jing Sun (Shenyang University, China), and Mingyou Wang (Shenyang University, China) addresses multinational corporation international industry shift new tendency and analyzes how China uses the international industrial shift's present situation and the superiority and proposes the development countermeasure.

The chapter, "Blue Ocean Strategy: A Vehicle for Entrepreneurship Development and Economic Growth in Rural Areas," by Vinay Sharma (Indian Institute of Technology, India), Piyush Seth (Sahara Arts and Management Academy, India), and Shujauddin Niyazi (Sahara Arts and Management Academy, India) analyzes the role of entrepreneurship in economic development of a society with the perspective of Blue Ocean strategy used by organizations around the world and in particular in Indian rural markets. The chapter also includes examples of organizations going towards uncontested rural markets and is suggestive of the crux and the core definition of entrepreneurship in correlation to the core philosophy of Blue Ocean strategy.

The chapter, "Study on Environmental Tax: A Case of China," is written by Huifeng Li (Beijing Union University, China), and Xuanwei Zhang (Beijing Union University, China). The authors propose that to solve problems related to ecological destruction and environmental pollution, the implementation of environmental tax is an effective measure. The chapter analyses the theoretical foundation of environmental taxes, international practice and experience in environmental tax system. It also describes the status quo of the Chinese environmental tax system, and proposes the future strategy for building this system.

The last chapter of the, "Economic Development and Investments" Section is "Study on Chinese Low Carbon Economic Model," written by Huifeng Li (Beijing Union University, China), and Xiaofang Wang (Beijing Union University, China). Global warming is a major concerns of nations and societies as it impacts global ecological balance, social and economic development, energy security, ecological safety, water and food safety In this sense, low carbon economy is a new economic development model based on low energy consumption, low pollution, and low-emission based. This chapter focuses on the status quo and limits of the low carbon economy in China as well as the design and implementation strategy of this model.

Managerial Issues in China

Finally the last section of the book, "Managerial Issues in China", provides interesting insights on the management of companies in China, analyzing both success and failures in this economy. The section is formed by 3 chapters.

The first book chapter of this section is, "Mutual Guarantee Systems of Small and Medium-Sized Enterprises in China: An Application of Self-Organization Theory." The authors Yi Yang (Beijing Union

University, China), Hong-yan Yan (Beijing Union University, China), and Ze-yun Yang (Beijing Union University, China) argue that mutual guarantee systems (MGS) of small and medium enterprises (SMEs) represent a complex system mainly serving the SMEs, including the mutual guarantee institutions formed by SMEs, financial institutions, industry associations and credit re-guarantee institutions. The authors address the self-organization mechanisms and the self-organization characteristics of openness, far-from-equilibrium, nonlinearity and fluctuation of mutual guarantee systems of SMEs. It also provides suggestions to improve the operation efficiency of these systems in China.

The second chapter of this section is "A Trilogy of Unfortunate Events in China: Reflecting on the Management of Crises," written by Zhang A. Long (China University of Geosciences, China), William Crandall (University of North Carolina at Pembroke, USA), and John Parnell (University of North Carolina at Pembroke, USA). The chapter analyzes three organizational crises that recently occurred in China; the SK-II cosmetic incident, the Sharon Stone comment on the May 2008 earthquake in China, and the melamine milk contamination crisis. It presents an overview of each company and the specific crisis and outlines reflections on the management of these crises and their cultural context.

Finally the last chapter of the section is "Implementation Concerns of Scholar-Practitioners: A Pilot Study of the Link between Research and Practice," by Claretha Hughes (University of Arkansas, USA), Jai Wang (Texas A&M University, USA), Wei Zheng (Northern Illinois University, USA), and Laird McLean (McLean Global Consulting, USA). It shows the results of a case study with insights from seven scholar-practitioners to help both human resource development professionals and researchers understand implementation concerns of scholar-practitioners. The empirical evidence show scholar-practitioners' perceptions of their roles, the link between research and practice, and actions they took to overcome barriers and challenges related to integrating research and practice.

COMING CHALLENGES OF THE JOURNAL

One of the goals of the journal for the next years is to increase the interest of researchers and experts in economy, business, information technologies and tourism from Bhutan, Cambodia, Laos, Mongolia, Myanmar, Nepal, Tibet and Vietnam, among others, so that they submit their manuscripts with the outcomes of their recent research activities to the *International Journal of Asian Business and Information Management*. Additionally, we would also like to increase the attention and interest from researchers, universities and research centres in Central Asia, such as Afghanistan, Kazakhstan, Kyrgyzstan, Tajikistan and others. Related to this goal is the organization of Special Issues focused on topics like Buddhism, regional economics and business; Central Asian countries, geopolitics and business; climate change, policies and economies in Asia; Confucianism and its impact on the economies of China and Japan and other Asian countries; corporate social responsibility and business in Asia; impact of Asian economies in global business; information technologies and its role in the preservation on cultural heritage in Asia; information technologies and educational development in Asian countries; Institutional transitions and strategic responses in Asian economies; Innovation and technology parks in Asia; and sustained development in Himalayan countries, among others.

Special issues focused on specific countries, like Bhutan, Mongolia, Tibet and Afghanistan will be scheduled too, addressing the role of information technologies in the economic development of these countries. For example, Afghanistan, one of the world's poorest countries, has a post war economy. When foreign spending dries up, the country may lead to economic depression. Recently the World Bank stated that the current gross domestic product (GDP) growth of 9% a year in this Central Asian country will fall to 5-6%. Rapid population growth clearly indicates that, measured by GDP per person, this land locked country will hardly get any richer in the near future. Could information technologies based business help to the economic development of this country that suffered wars for more than 30 years? If so, what specific policies and actions could contribute to this development?. Another example is a land locked country too: Lao People's Democratic Republic. According to the World Bank, its economy grew an average 6.4% per annum over the last decade and poverty declined from 46% in 1993 to 27% in 2008. This small country has undertaken reforms that undertaking reforms that are helping to embark it on an increasingly sustainable development path. What is the role of information technologies in this strategic process?

Furthermore we will look for top academics and experts from these countries to join the Editorial Board of the Journal and advice on topics of interest related to their country of origin as well as on potential titles for Special Issues. The journal seeks to increase the current pool of reviewers with academics from these areas of Asia too.

Additionally the *International Journal of Asian Business and Information Management* wants to increase its collaboration with academic and business conferences, summits and workshops focused on Asia to promote the journal and publish best papers from these events.

Dasho Karma Ura
Centre for Bhutan Studies, Bhutan

Patricia Ordóñez de Pablos
The University of Oviedo, Spain December 2011

Section 1
Chinese Cultural Values and Industry

Chapter 1
Local Milieu in Developing China's Cultural and Creative Industry:
The Case of Nanluoguxiang in Beijing

Jici Wang
Peking University, China

Chun Zhang
Peking University, China

Ching-Ning Wang
Peking University, China

Ping Chen
Peking University, China

ABSTRACT

This paper examines the role of creative milieu and cultural heritages in the development of Chinese cultural industry. Through case study of Beijing's Nanluoguxiang, where several arts institutions and theaters concentrate, it depicts the birth and growth of a creative place for free artists in the institutional changing of Chinese cultural setting. Based on field survey data from artists, managers and visitors, it shows the spouting and growth of local creativities in a transitional economy needs atmosphere of tolerant and frequent social gathering, especially in spaces like cafes and bars. It also suggests that making a good use of local cultural and creative resources like heritages and folk customs enables the superiority of localization standing out in the wave of globalization. The key findings indicate that the newly-built creative parks might be useful to breed the creative products, but not necessary.

DOI: 10.4018/978-1-4666-0276-2.ch001

INTRODUCTION

Since the end of last century the set of cultural and creative industries have presented increasingly important employers, export earners, and competitive strengths in developed countries. This poses a formidable challenge to developing countries. Since cities in China used to compete for business with tax incentives and well-trained, docile and low-cost labor forces, the imperative for a change is undeniable.

Along with the on-going debate on the importance of cultural and creative industries for the economic development of cities, this discourse has been enhanced by the milieu concept and cluster concept. More and more studies look at creative cities of developed countries. However, there is a lack of literature on the milieu of cultural and creative industries in developing countries. This paper is intended to fill this gap.

In the background of the fever of building "creative parks" in China, this paper focuses on Beijing's old Nanluoguxiang where the artists are clustering. By an in-depth interview and survey with free artists and bar managers, the paper argues that the role of local space and the milieu of creative and cultural industries can be significant in the creative processes. While the local tolerant atmosphere has been released by effective institutional reform, talent will enhance the creative milieu, the related groups such as the local fans of art may spark off positive feedback to the locality, and the place can be a catalyst to creative activities. The creative infrastructure like cafes and bars plays an important role. From this point of view, the newly-built creative parks might be useful to breed the creative products, but they are not necessary. The key for the development of local creative activities is to attract the creative talents to be there.

POLICY BACKGROUND OF THE CASE OF NANLUOGUXIANG

China's cultural undertaking was treated as a political propagandistic tool but separated from economy for three decades after 1949. The traditional cultural goods and services fell in stagnancy during the overwhelming "modernization" throughout the country, especially in cities. As the economic reform proceeding, the state art agencies have gradually gotten rid of a purely academic style, starting to promote art products to the market. Since the 16th National Congress of CPC in 2002, to meet requirement for China's entry into the WTO, China has formally encouraged non-government sectors to participate in cultural undertakings[1]. In the Tenth Five-year Plan of National Economic and Social Development of China in 2005, "improving culture industrial policy, strengthening the management of market of culture products and services, promoting the development of culture-related industry" was addressed. Since deepening reform of cultural system, general art troupes, publishing houses, newspapers and magazines on culture, art, life and popular science, bookstores, movie studios, theaters, TV play producers, cultural intermediary organizations have gradually restructured into market businesses. Large part of the cultural undertakings has undergone the change of commercialization, which makes it distinctive to study the milieu breeding of the cultural and creative activities. The contemporary change of Nanluoguxiang in Beijing reflects this situation.

Nanluoguxiang - South Gong and Drum Lane - is in the Dongcheng district of Beijing in the old part of town. This lane used to be the geometry centre of the city in Yuan dynasty, and has a history of about 740 years. Nanluoguxiang is intersected by 16 branch parallel alleys (Hutong) on both sides, showing the pattern of "fish-bone" dating from Yuan dynasty. Among the heritage protection areas of China, Nanluoguxiang is the one which reserves the most complete urban pattern, possesses the most history heritage and story, and has the richest cultural connotation. In the recent years, more than 30 various cafes and bars are springing out along this lane, which makes Nanluoguxiang a place of interest for tourists and a place of leisure for local residents.

Along the lane and in the area nearby, there are several advanced arts research institutions and good theaters, including The Central Academy of Drama (CAD), Beijing Art Association, Chinese Drama Theater (CDT), Beijing Seven-Color Children's Theater (BBCCT), Yi-Fu Theater, Trial Theater and so on. The actors, directors, script writers and other artists usually come to the cafes and bars (Figure 1).

Around the area of Nanluoguxiang independent studios are established and dedicated to creating contemporary Chinese theatre productions outside the old official cultural system. They have a potential to break the barriers among traditions, modern theatre and different disciplines, integrate with Chinese operas and folk art and established its own style. The studios also focus on the development and promotion of young Chinese play writers.

Since 1990s the activities of the cultural and creative industries in China have appeared in the three types of locations. The first type is the rural villages' adjacent to big cities, such the painter village in Songzhuang[2], Beijing. The second type is the location of the abandoned old factories, such as Factory 798 factory[3] dated back to 1950s in Beijing, the 1930s' food mechanical factory in the #200 Road Taikang[4] and the flour factory in the #50 Road Muoganshan[5] in Shanghai, as well as the standard factory building of 1980s fostering lots of 3D animation talents in Shenzhen, which rebuilt by the digital media company named Global Digital Creations Holdings. The third type is the new developing zones such as Digital and Entertainment Industrial Park in Hangzhou. The case of Nanluoguxiang analyzed in this paper, however, is totally different from the three types discussed above. It is embedded in the sedimentation of Chinese long history.

Because the development of cultural and creative industries has been written in the Eleventh Five-year Development Program of the country, there is a bandwagon of creative industries in various cities. They are more or less in the trap of "old wine in the new bottle", or in the Chinese saying - "wear on new shoes and take the old road". The government starts to build parks and provide preferential policies. In big cities like Beijing, Shanghai, Shenzhen, and smaller ones like Suzhou and Zhuhai, many plots of land are now encircled as creative parks with attractive names. For example, more than twenty bases for the industry of animation are established in China by different sector authorities, while many cities claimed to develop digital arts and digital entertainment industry.

The current fever in developing creative industries in China is extremely similar to the fever in high-tech industries in the mid-and-late 1980s. Many local governments only replant from developed countries the name of the set of industries, such as "high-tech", "creative industries", and their sub-sectors' names, such as new material industry and animation industry, while ignoring the position of their activities in the value chain. In other word, those low-end activities might have same or similar sector names as the one of high-end activities. In addition, many local governments intend to drive the high-tech or creative industries by real estate development. They set up special management administration office to build industrial parks and rule the enterprises in such parks. This property-driven development achieved a success at some degree in 1990s, since it provided land necessary for enterprises from outsiders and/or deregulation policy environment for innovative start-ups. But this model could not be repeated to promote the real creative activities of cultural and creative industry in the new century.

It is necessary to address the importance of the local milieu for contextualized knowledge creations. The creative milieu happens only when developing the high-end creative activities. However, the so-called creative parks could become the venue to attract the low-value-added activities of creative sector, such as the one in outsourcing activities of animation processing which needs low-cost skilled labor of art. Cultural and creative industries may not emerge in some green places which are good looking but not attractive for talent people.

COLLECTVE CREATIVITY AND CREATIVE MILIEU

Creativity or creative capacity was used to be seen as inherited and learned characteristics attached to the individual. The creative process would also be seen as something which occurs in isolation in the mind of the individual. For many, solitude is a prerequisite for the concentration that creativity requires. At the same time, however, it is clear that new ideas are built upon a capital of experiences gathered through interaction with one's surroundings, instead of developing in an empty space for long without stimulation and new outside impulses. Consensus of promoting creative activities increasingly has focused on creativity as a collective social process rather than as a process taking place within the minds of individuals.

Cultural and creative industries have a high proclivity to cluster in particular locations, especially in big cities, and their products have close connections with local cultural context. The rationale behind agglomeration is the localization economies and in some areas urbanization economies. Dominant creative agglomerations in countries closely match the dominant urban areas.

Based on the recognition of collective creativity, existing theory on geographical agglomeration of the creative industries in cities has deepened the understanding of the role of place and its milieu as necessary condition for creative activities. The milieu of creativity should be seen primarily as places and groupings that have attracted competencies within specialized disciplines (Tornqvist, 2004). A milieu may offer meeting places and opportunities for communication. Of even greater importance is that they may attract intellectual elite which places specific demands on its neighborhood. A creative milieu is a place that contains the necessary requirements in terms of 'hard' and 'soft' infrastructure to generate a flow of ideas and inventions (Landry, 2000). Research shows that creative setting or creative milieu is that there is always more important in any place for regional innovation than we would think (Porter & Stern, 2001).

Becker focused on creativity in the visual arts but argued that even the visual arts, often perceived as a predominately individualized process, are, in reality, a collective creative process (Becker, 1982). Others have reinforced this emphasis of collective creativity. And creativity is not due to natural gifts but is a product of the social, economic and educational background or environment of the individual. Although some others argued that the emergence of the creative industries was having a negative impact on individual creativity.

Hall (2000) addressed lessons from history, wrote the book Cities in Civilization (Hall, 1998), and summarized many scholars' studies on creative milieu. He strongly supports the issue about milieu seem to have some things in common, including special kinds of ability that take a long time to develop, as happened in Vienna around the year 1900; but these special kinds of competence can be attracted to certain magnetic places — the Vienna case, again. They need communication between individuals and between different areas of competence, so there must be a certain density of communication, which seems to require a rich, old-fashioned, dense, even overcrowded traditional kind of city (Tornqvist, 1983).

In periods of intense creativity historically, creative people have acquired a set of common characteristics, a kind of accumulated culture and style of life. As the French geographer Philippe Aydalot refer to the idea that a creative milieu has four key features: information transmitted among people, knowledge or the storage of information, competence in certain activities, and creation of something new out of these three activities (.Aydalot & Keeble, 1988).

Anderson argues that such a creative milieu has six critical prerequisites: a sound financial basis, but without tight regulation; basic original knowledge and competence; an imbalance between experienced need and actual opportunities; a diverse milieu; good internal and external possibilities for personal transport and communication; and structural instability — a genuine uncertainty about the future within general scientific and technical environment (Andersson, 1985). The "creative milieu" will brings about the integration in a single field of the various socially specialized sub-fields, particular markets which are completely separate in social and even geographical space, in which the different fractions of the dominant class can find products adjusted to their tastes, whether in the theatre, in painting, fashion, or decoration" (Bourdieu & Johnson, 1993)

Previous discussion echoes Drake's argument that place or creative milieu has four different types featurcs .These can be categorized as: a resource of visual raw materials and stimuli, locality-based intensive social and cultural networks, a brand based on reputation and tradition, and locality-specific communities of creative workers (Drake, 2003). But Scott emphasis on local production system, training or attracting a relevant labor force, appropriate programming of urban space and ensuring that all the different elements involved work more or less in harmony with one another (Scott, 2005).

Much of the literature has spurred an impressive amount of empirical research. For example, entertainment industry in Hollywood (Scott, 2005; Christopherson & Storper, 1986), natural history film-making in Bristol (Bassett; Griffiths; Smith, 2002), and Toronto's film clusters (Vang & Chaminade, 2007), etc.

Scott focuses on the way in which the attributes of particular places have the ability to promote creativity within clusters of creative economic activity. He refers to the ways in which creative industries can benefit from a local community's "characteristic styles, sensibilities and themes" (Scott, 2000, p. 329). Bourdieu (1993) adopts the concept of the "creative field" to refer to the way in which localized agglomerations of creative economic activity behave as catalysts for higher rates of creativity and innovation. Furthermore he said "it is the field of production, understood as the system of objective relations between these agents or institutions and as the site of the struggles for the monopoly of the power to consecrate, in which the value of works of art and belief in that value are continuously generated."(Bourdieu & Johnson, 1993, p. 78).

Furthermore, understanding the operations, the location decisions and the linkages of different creative sub-sectors and how they relate to the local economy, will help identify their potential contribution in underpinning economic growth and how they support cultural/creative life or communities and drive the overall transition to a "creative city" (Hall, 2000).

DADA AND RESEARCH METHODOLOGY

In the survey of Nanluoguxiang conducted in summer 2006 the data is collect mainly from qualitative interviews, supplemented with a small sampled questionnaire. The interviewees include 11 artists and students, 19 managers and employees from cafés and bars (Table1). The interview is semi-structured, and is designed to collect data on: 1) why do you choose Nanluoguxiang as a starting point of the art career? 2) Where and with whom the original ideas usually emerge, and what condition supported the ideas? 3) What is the role of the local environment? 4) What daily activities are considered to be relevant to creativity? 5) What formal and informal organization do you join, and how do these organizations help you? The interview is recorded, transcribed, and coded.

The small sampled questionnaire is designed in a structured format, and hand-out to the customers of cafés and bars. The sample size is 100 and 70% are valid, which including 41 Chinese customers and 29 foreign customers. Among these

Table 1. Details of the artists, bar managers and employees in the interviewing (Source: Filed survey.)

Number	Profession description	Gender	Age	Graduate university/institution / Home town (HW)
1	Logistics manager	M	40	The Central Academy of Drama(CAD)
2	Instructor & director	M	40	Department of performances of CAD
3	Stage costumes	F	33	Department of Stage art of CAD
4	DV maker	M	42	Heze Institute of the Arts and Music
5	Filmmaker	M	35	Department of performances of CAD
6	Examinee of CAD	F	19	Beijing Institute of Etiquette
7	Art photographer	M	45	College of the Arts of BNU
8	Free-lance writer	M	50	Chinese literature Department of BNU
9	Graduate student	M	23	Drama Literature Department of CAD
10	Undergraduate	F	21	Department of performances of CAD
11	Undergraduate	F	21	Department of performances of CAD
12	Community official in charge of economy	M	37	Veterans from army
13	Association President of Nanluoguxiang	M	40	Veterans from army
14	Community official in charge of local planning	F	25	PLA Institute of International Relations
15	Investor & Local People	M	26	Beijing Union University
16	Waiter in Bar	M	22	(HW)Chongqing City
17	Manager of Cafes	F	33	Zi-Yang Opera Mission,Si Chuan provice
18	Gaffer Waiter of Cafes	M	26	(HW) Baoding City, Herbei province
19	Investor of Bar	M	40	Department of performances of CAD
20	Waitress of Cafes	F	19	(HW)Fengyang City, Anhui province
21	Wife of the copartner of Bar	F	30	Dianmen comprehensive high school
22	Manager of Bars	M	33	Department of Art of Central University for Nationalities
23	Friend of Managers of Bar	M	36	No.201 (Art) Middle school
24	Waitress of Bars	F	18	(HW)Jilin City, Jilin province
25	Sister of Managers of Bar	F	28	Hua Xis Girls School
26	Manager of Bar	M	39	(HW)Local resident
27	Waitress of Bar	F	21	(HW)Anshan City, Liaoning province
28	Bartenders of Bar	M	25	No.179 (hotel management) middle school
29	Waitress of Bars	F	18	(HW)Chengdu City, Sichuan province
30	Gaffer Waiter of Bar	M	25	Beijing Chinatown cuisine technical
31	The Copartner of Bar	F	24	Beijing Modern Vocational School
32	Manager of Cafe	M	36	Dance Arts Theater in Heilongjiang Province

samples, half of them are employed in art field, with high level education and rather young (Table 2).

THE GENESIS OF CREATIVE MILIEU AT NANLUOGUXIANG

Lane NanLougu emerged as a creative art area when the state monopoly on art market broke in 1993. Initially creative activities sprouted from the art research and education institutes close to the Lane, say, at the "wall corner" of those agencies, and an increasing number of art fans came nearby. The creative activities which first appeared at the "wall corner" or "the very neighborhood" relied on these art agencies' hardware facilities, human capital, and qualification authorization, so that they could lower the cost to find opportunities in the market. Figure 2 shows the location of the state-owned research and education institutions and infrastructures related to art and the early emerged private clubs in the investigated region. At the beginning, when the people left arts faculty - "Danwei" - and rent nearby houses in the Nanluoguxiang in order to seek more opportunities, the distance should be close enough to "Be around whenever you call". Many artists who were interviewed referred this.

Table 2. Personal information of the questionnaire samples

Item	Details	Number	Ratio(%)
Gender	Male	26	43.3
	female	34	56.7
	Total	70	100
Age (year)	15-24	17	27.9
	25-35	31	50.8
	35-45	11	18.0
	>45	2	3.3
	Total	61	100.0
Education level	Undergraduate or higher	46	67.6
	College	8	11.8
	Senior high	6	8.8
	Junior high	1	1.5
	Elementary school	0	0.0
	Below elementary	7	10.3
	Total	68	100.0
Profession	Art	32	50.0
	IT	6	9.4
	Service	20	31.3
	Retail	8	12.5
	Finance	4	6.3
	Teacher	4	6.3
	Other	12	18.8
	Student	10	15.6
	Total	64	100.0

Source: Questionnaire done in summer 2006.

"...At that period, everyone wanted to make a fortune, but didn't dare to leave far - it was strictly controlled. (The institution) wouldn't let you go out. Once caught, you'd be kicked out. So, everyone had a pager. In case your boss called you, you had to run back as quickly as you can. The apartment I rented was less than 100m from the gate of CAD. Even that I was bolder than most people at that time - they didn't even dare to do that... (Interviewee #5, film maker)"

Lane NanLougu became a seedbed for creative activities' germination, because it provides the shelter for the underground part-time spin-offs from the arts institutions, and also features the historical tradition of culture. On the one hand, as the stagewise scenery for movie producing, Nanluoguxiang lets talented artists display their gift at a relatively low cost without the need of a huge amount of funds supported by the original institution where they worked.

"...Our films mainly sold to foreigners. They showed special interests to people and things inside Siheyuan[6]. It cost little to rent these yard. They would be very satisfied for ¥100 for a morning (Interviewee #4, DV maker)."

On the other hand, it's a best place for writers to conceive a script, and also a natural arena for actors to touch the history (e.g., inside the *Siheyuan*). The gesture and mood are influenced unconsciously.

"...Hutong and Siheyuan are the ground for conceiving people and stories happening here – they are a window to history, and a link between time and space, without which no such feeling would exist anymore." (Interviewee #8, Free-lance writer)

The local fans of art also play an important role in Nanluoguxiang. They have a certain artistic taste and skills. They include the students trying to get into art institutions nearby, the friends and partners of professional artists, and the enthusiasts who devote themselves to art. They boost the creative milieu in four aspects: 1) they are demanding audiences, who can be both fans and critics. 2) They provide prompt and creative raw materials, which is important to the creative idea inspiration. 3) They are subsidiary and cheap labors, and form art pool. 4) They press and challenge the professional artists for their pursuit of excellence constantly.

"...Sometimes, (the producer) to low the cost, they may found amateur actors. So, we had to perform better. Anyway, we spent so many years on the stage......and would not lose the face." (Interviewee #2, Instructor and director)

Long history and culture, many movie stars and directors' contributions establish a unique brand of Nanluoguxiang. The creative products worth a higher price here in the cultural market than other place, which make the artist can devote themselves to their career without warring about the spending.

Many junior actors and directors draw the fame and share the intangible assets here. Interviewees mention that their initial creative activities earn 'the first bucket of gold' with the accumulation of brand of place.

"...The script I wrote here could sell at ¥200000, but only worth less than ¥10000 if it was written in somewhere else. These famous directors survived by their name, and we rely on the name of Nanluoguxiang." (Interviewee #8, Free-lance writer)

CAFES AND BARS AS MEETING PLACES IN THE PHYSICAL MILIEU

The value of meeting places in the physical milieu will be examined below. In geographic as well as institutional milieu, there is a need for meeting places, facilities where meetings may take place more or less by chance. It is notable that these meeting places often exist outside the formal institutions and organizational forms, in contexts where professional and commercial competition is not able to block the free exchange of information.

The American sociologist Oldenburg identifies third places, or "great good places," as the public places on neutral ground where people can gather and interact. In contrast to first places (home) and second places (work), third places allow people to put aside their concerns and simply enjoy the company and conversation around them (Oldenburg, 1991).

In the case of Nanluoguxiang, cafes and bars play the role of breeding ground and became creative spaces. At the beginning, the birth of cafes and bars satisfy the need of anytime contacts and exchanges inside the "art coterie". The artists need get together to exchange and discuss, thus collective creativities take place to grasp the fleeting inspiration. So it is necessary to find a reasonable place to accommodate and continue these exchanges which benefit the creative activities, the first bar - PassBy - appeared under this circumstance. Initially, the artists only have trading talks in a fixed cafes and bars considering the need of privacy. However there isn't connection between the cafes and bars.

"In our coterie, the life is rather irregular. Drink and talk for the whole night... ...Later on, in order to avoid disturbing the neighborhood, we rent a house nearby to consume ourselves. Therefore, the cafe born." (Interviewee #32, Managers of Cafe)

As the creative activities took place more and more frequently, the cafes and bars of the secret sites of contacts and exchanges gradually opened to the public. The "brain storm" of the artists is more popular in collective creation in which the script-writer, director, actor, stage, costume charge and so on are all take part. At the same time, entering the cafes and bars - an extensive interaction space supplementing their classroom - was the neccssary cxperience to become a future star for those late comers. The spill-over effect was notable and prevalent, unconscious over-hearing might be beneficial to inspiration of new ideas.

At this stage, the creative activities were not limit to the "art coterie" in one or two cafes and bars. With the commercialization in art field, the transaction on cultural and creative products didn't depend on the connection of the managers of the cafes and bars. Besides, the managers of the cafes and bars shift their role form the organizer of club gathering to lay out the environment thoughtful maintaining the creative art milieu, which including the elaborate decoration, dedicate food and drinking and the scrutinized service.

In Nanluoguxiang, while the cafes and bars fostered the creative milieu as a space, it also could be creations themselves. Visual creative production can be accumulated and transfer to customs in form of layout, background music and demonstrate style. Therefore, the cafes and bars became a window to talking with history and former artists. Like as book galleries, art galleries and museums, the creative products are preserved, performed and re-produced through visual elements in these cafes and bars, which illume and induce later creative inspiration. The cafes and bars also brought the elegant art down of the altar, and make it accessible to the public. Everyone could be an artist here and the creative production process is not mystic. The customs in the cafes and bars usually get on the process of creation unconsciously with fun.

As the number of the cafes and bars kept increasing, based on relation of the alumni association of CAD, the managers got together and NanLuogu association was founded in 2005. By 2007, there are already more than 40 cafes, bars, tea house, book shop, photograph workshop and also special dressing shop join the association.

The interview reviewed that there were more cooperation than competition among the cafes and bars. Through diversification to avoid cut-throat competition, the creative milieu was protected. Take the club parties for example, NanLuogu association encourage differentiation on topic or time.

"...Zha-Zha Café called together the poets every Monday night, the Luogu Cave restaurant organized English corner every Tuesday and Thursday night, Sandglass Café hold a mini-music promulgate meeting every Friday and Xiaoxin Bar hold the photographers salon every Sunday... (Interviewee #12, Community official in charge of the local economy)"

However, the creative activity needs quite scrutinized environment, not only the artistic milieu, high level taste and tangible element, but also relatively cheap cost for the local artists and fans. NanLuogu Association tries its best to maintain the delightful creative milieu, but it seems not effective. It is facing the challenges: 1) Managers of cafes and bars are increasingly seeking profit but neglecting quality in operation. 2) Some artists are involved in the flubdub but lack of real creativity. 3) House lords ask for more and more rent. 4) Higher government official proposed unpractical policy, such as change it to a commercial street.

"The rent is rising dramatic. Last year it's ¥20000 a year, this year doubled, and may be triple again next year. Many of the first owners have disappeared. They can not stay on. It's absolutely no good." (Interviewee #26, Manager of Bar)

CONCLUSION AND DISCUSSION

The cultural and creative industries have been largely recognized as two forms. One is digital creative industries in which new digital technologies alter the way that cultural goods and services flow around the globe. Much remains to be done to develop the digital creative industries into a global net and this can be achieved through a range of innovative new partnerships, capacity-building and tighter piracy control. The fragmentation of global production and distribution systems of digital cultural and creative industries affect local development strategies. Within the global system of digital creative industries, the ways in which the enterprises of developing countries integrate into global value chains affects upgrading, but this is out of our paper here.

The other form is the traditional arts and crafts, intangible heritage such as dance and ritual, as well as tangible cultural heritage goods. Besides the "creative park" model, this form can be an alternative way to develop cultural and creative industry in inner cities through the use of local historical and cultural resources. All the resources such as natural heritages, cultural heritages, and folk customs heritages could be transformed into the elements of creative industry. Making a good use of local cultural and creative resources enables local characteristics and the superiority of localization to stand out in the wave of globalization.

To sum up, the spouting and growth of local cultural and creative activities in a transitional economy needs the local tolerant atmosphere released by effective institutional reform. Under the institutional reform towards encouraging the commercialization of cultural undertaking in China, cultural and creative infrastructure like cafes and bars plays an important role. In the case of Nanluoguxiang with the state-owned art academies and institutions nearby, bars and cafes are developing very fast and act as the planting pots for the rooting and flowering of creative seeds. The art education and research in the art academy is important in awakening and developing the individual's creative capacity. Yet conversations and meetings allow genuinely creative personalities to bring together and formulate ideas that may lead to creativity. The cafes and bars themselves become the product of creativity and inspire the production of new creative activities. As the cultural and creative activities contributed by the artists' group extend to art fans, the artistic and creative concepts gradually penetrate into the daily life of the people through the interface of bars and cafes.

This paper did some research on the role of the historical heritage and the creative infrastructure of those cafes and bars in Nanluoguxiang. But it didn't provide the details of those creative people inside the cafes and bars: how they interact in the bars and cafes and what products they create. The future research of this area is to investigate the creation process of creative idea and creative products itself.

REFERENCES

Adorno, T. (1991). *The Culture Industry: Selected Essays on Mass Culture*. London: Routledge Press.

Andersson, A. E. (1985). Creativity and Regional Development. *Papers / Regional Science Association. Regional Science Association. Meeting, 56*(1), 5–20. doi:10.1007/BF01887900

Aydalot, P., & Keeble, D. (Eds.). (1988). *High Technology Industry and Innovative Environments: The European Experience*. London: Routledge Press.

Bassett, K., Griffiths, R., & Smith, I. (2002). Cultural Industries, Cultural Clusters and the City: the Example of Natural History Film-making in Bristol. *Geoforum, 33*(2), 165–177. doi:10.1016/S0016-7185(01)00032-X

Becker, G. (1998). Human Capital and Poverty. *Religion and Liberty, 8*(1), 5–7.

Bourdieu, P. (1993). *The field of cultural production: Essays on art and literature*. New York: Columbia University Press.

Christopherson, S., & Storper, M. (1986). The City as Studio; the World as Back Lot: the Impact of Vertical Disintegration on the Location of the Motion Picture Industry. *Environment and Planning. D, Society & Space, 4*(3), 305–320. doi:10.1068/d040305

Drake, G. (2003). This Place Gives Me Space: Place and Creativity in the Creative Industries. *Geoforum, 34*(4), 511–524. doi:10.1016/S0016-7185(03)00029-0

Hall, P. (2000). Creative Cities and Economic Development. *Urban Studies (Edinburgh, Scotland), 37*(4), 639–649. doi:10.1080/00420980050003946

Landry, C. (2000). *The creative city: A toolkit for urban innovators*. London: Earthscan Press.

Oldenburg, R. (1999). *The Great Good Place: Cafes, Coffee Shops, Bookstores, Bars, Hair Salons, and Other Hangouts at the Heart of a Community* (3rd ed.). New York: Marlowe & Company.

Porter, M., & Scott, S. (2001). Innovation: Location Matters. *MIT Sloan Management Review, 42*(2), 28–36.

Scott, A. (2000). *The Cultural Economy of Cities: Essays on the Geography of Image-Producing Industries*. London: SAGE Publications.

Scott, A. (2005). *On Hollywood: The Place, The Industry*. Princeton: Princeton University Press.

Törnqvist, G. (1983). Creativity and the Renewal of Regional Life. In Buttimer, A. (Ed.), *Creativity and Context: A Seminar Report* (pp. 91–112). Lund, Sweden: Gleerup Press.

Törnqvist, G. (2004). Creativity in Time and Space. *Geografiska Annaler, 86*(4), 227–243. doi:10.1111/j.0435-3684.2004.00165.x

Vang, J., & Chaminade, C. (2007, March). Global-local Linkages, Spillovers and Cultural Clusters: Theoretical and Empirical Insights from an Exploratory Study of Toronto's Film Cluster. *CIRCLE Electronic* (Working Paper Series). Retrieved from http://www.circle.lu.se

ENDNOTES

1. Jointly issued by the Communist Party of China (CPC) Central Committee and the State Council, the document indicates that China's top leadership has decided to embrace marketization to invigorate the country's cultural undertakings. Only organizations that offer public cultural service such as libraries, museums, cultural and scientific centers, and Party periodicals, news agencies, radio and TV stations, major news websites, politics periodicals, a few publishing houses for political and public welfare purposes, major social science research institutes, art troupes showing national features will still enjoy state support.
2. Before the 1990's, Songzhuang was nowhere but a small town in Tongzhou District, Beijing suburb. The first art community in Songzhuang came into being in the early ninety's in a village called Fu Yuan Men close to the Old Summer Palace. After some time they dissolved due to the political shock and the conflict between their unconventional

way of living in groups and the restriction of census registration system, and some of the core artists moved into Songzhuang later. Nowadays, with artists living next door, "Songzhuang" has become a substitute for the largest contemporary art community in China.

3. "798" is located in the Dashanzi area, to the northeast of central Beijing. It is the site of state-owned factories including Factory 798, which originally produced electronics. Beginning in 2002, artists and cultural organizations began to divide, rent out, and re-make the factory spaces, gradually developing them into galleries, art centers, artists' studios, design companies, restaurants, and bars. It became a "Soho-esque" area of international character, replete with "loft living," attracting attention from all around. Bringing together contemporary art, architecture, and culture with a historically interesting location and an urban lifestyle, "798" has evolved into a cultural concept, of interest to experts and normal folk alike, influential on our concepts of both urban culture and living space.

4. Shanghai's Art Street there is art shops and various art studios. In the International artists' factory where creative designers, photographers, home decoration and retail shops are nestled together in one building.

5. Shanghai Chunming Roving Factory was built in 1930s. It was renamed Shanghai Chunming Metropolitan Industrial Park in 2002 and Chunming Art Industrial Park in 2004 by Shanghai Municipal Economic Committee. Within a few years the Park has introduced more than 130 artists who come from over 10 provinces and 16 foreign countries and regions, as well as the galleries, plane design institutes, architect firms, movie and television production firms, environmental artistic design institutes and artwork (jewelry) design institutes and so on.

6. Siheyuan is a classical architecture style of residential housing of Beijing citizens dated from the Ming Dynasty. The narrow street between the Siheyuan are called Hutong. A single implementation of Si He Yuan comprises of garden surrounded by four houses and protected by high walls. Siheyuan with Hutong which have been very familiar for people, having a history of more than 700 years now suffer from the urban renovation and economic development of China, and they are slowly disappearing from Beijing.

This work was previously published in International Journal of Asian Business and information Management, Volume 1, Issue 1, edited by Dasho Karma Ura and Patricia Ordóñez de Pablos, pp. 10-22, copyright 2010 by IGI Publishing (an imprint of IGI Global).

Chapter 2
Australian Businesses in China: Strategies to Overcome the Cultural Gap

Mona Chung
Deakin University, Australia

Jane Menzies
Deakin University, Australia

ABSTRACT

This paper indentifies a main barrier when doing business with China, the cultural gap, and provides the strategies that companies can use when entering the Chinese market. This empirical study examined 40 Australian organisations in their activities when entering the Chinese market. Alarmingly after 30 years of attempting to do business in China, companies are still not addressing the issue of cultural differences. Companies are also caught by surprises due to lack of preparation how large the cultural gap is between Australian and Chinese business culture. The findings of the study have important implications for businesses considering entry to China, and for Australian businesses already doing business in China. The strategies investigated include human resource strategies, dealing with Chinese staff, relationship building, getting outside support (employing consultants), learning about the culture, and adapting to the culture.

INTRODUCTION

Since the open door policy in 1979 China has seen an influx of capital in the form of foreign direct investment (FDI) (Yeung, 2000). In 2005 and 2006, China received 70 billion worth of FDI (Davies, 2007), and is becoming one of the world's principle destinations (Bellabona & Spigarelli, 2007). According to the Department of Foreign Affairs and Trade (DFAT), China has become Australia's 21st largest investment destination with $3 billion being invested by Australian businesses in 2006 (DFAT, 2007). However, China is not a straight forward market to invest in (Chung, 2008). Among many influencing factors, the cultural differences have been a challenging area for foreign investors.

DOI: 10.4018/978-1-4666-0276-2.ch002

Previous research by Chung (2008) has identified that a key reason for business failure is not recognising the cultural differences. This paper addresses the questions:

1. *What is the cultural gap that Australian businesses experience when entering and operating in China?* and
2. *What are the strategies that Australian businesses can use to overcome the cultural gap*?

This paper will first investigate the cultural distance by examining literature in the area, and then examine the strategies Australian businesses use to decrease the gap, through an empirical investigation.

EXAMINING THE CULTURAL GAP

Cultural gap as a term has been used widely in describing the distance between elements of two different cultures. In this paper we discuss the lack of understanding of Chinese culture by Australian businesses when dealing with Chinese and hence the gap of knowledge. Psychic distance was initially introduced by Nordstrom and Vahlne (1992) as the "factors preventing or disturbing the flow of information between potential or actually suppliers or customers". Psychic distance is defined as "a firm's degree of uncertainty about a foreign market resulting from cultural differences and other business difficulties that present barriers to learning about the market and operating there" (O'Grady & Lane, 1996, p. 330). The psychic distance concept has been operationalised in the following way by O'Grady and Lane (1996): this includes the level of economic development in importing countries, the difference in the level of economic development in host countries, the level of education in importing countries, the difference in the level of education, difference in business languages, difference in culture and local language, and existence of previous trading channels (O'Grady & Lane, 1996). Psychic distance is influenced by differences in the culture and language of the home and target countries (Kogut & Singh, 1988). Therefore the larger the cultural differences between two cultures the larger the psychic distance.

Researchers have suggested that entering countries that are "psychically close" (meaning have smaller psychic distance) reduces the level of uncertainty firm's face in the new market (Johanson & Vahlne, 1992); and for countries that are "psychically close" it is easier for companies to learn (Kogut & Singh, 1988). When investing in foreign countries, companies perceive less risk when the psychic distance is smaller and greater risk when the psychic distance is larger. Therefore companies tend to start their international experience in countries where the psychic distance is close first and then venture into those which are larger.

THE DISTANCE BETWEEN AUSTRALIAN AND CHINESE CULTURE

Hofstede (1984, p. 21) defines culture as "the collective programming of the mind which distinguishes the members of one human group from another. Given this definition, the programming of the mind determines the differences of behaviours which are generally formulated over a long period of time under the influence of cultures. Previous research by Chung (2008) has identified there is a large cultural gap between Australian business and Chinese business.

China has a distinctive cultural difference that they are described by the Chinese government as "a market economy with socialistic characteristics". It is the only one in which both past and present worlds exist. Its characteristics of a centrally planned resources allocation combined with a degree of free market is unique to any western understanding of markets.

Johanson and Vahlne (1977) postulated that when establishing international operations, firms need accurate market knowledge, which come from the direct experience in the foreign market and an understanding of its internal relationships, rather than more factual, general market information that is easily transmitted without the need for experience in interpreting it (Johannson and Valhne, 1977). For doing business with China, reading books and articles alone were simply not enough to learn about real cultural differences of the Chinese market.

Guanxi and Face – Two Most Different Chinese Characteristics

In terms of central values of Chinese culture, such as trust, reciprocity, respect for age, authority and time (Wang, Zhang, & Goodfellow, 1998). Two important values of Chinese culture are: *guanxi* and *face*. Both are difficult to explain and understand. They are specific Chinese characters with specific Chinese values attached.

Guanxi as a concept has not been well understood by majority of the Western business people let alone mastering it (Chung, 2008). *Guanxi* commonly comprises relationships, networks, and the development of trust within these relationships (Wang et al., 1998) and accordingly networks of associates and friends are important. *Guanxi* allows for the sidestepping of formal regulatory or contractual opportunities to seize opportunities. It is an essential element in the Chinese business culture. Many mis-believe that Guanxi can be "purchased" through people. Guanxi is built not bought (Chung, 2008). To build *guanxi*, one needs to invest time and money into the relationship, and once established it may remain long term (Linong, 2006). For example, a good guanxi with government can assist with obtaining investment and planning permits. Building *guanxi* with local levels of government and senior members is a useful strategy when doing business with China (Tian, 2006). The development of *guanxi* and relationships among individuals creates obligations for the continued exchange of favours. Australian business culture also relies on relationships, but outward displays of favours to connections or family members would be considered to be nepotism, and inappropriate conduct. Ability to recognise such cultural differences between Australian and Chinese businesses is the first step however not nearly enough. Capacity in building Guanxi (relationships) in China for long-term is one of the required elements in doing business with China.

Face is an important element in doing business with Chinese (Chung, 2008), and can be a difficult concept for western business people to understand. Chung (2008) suggested "mutual respect is suggested because, in practice, not only does one try to maintain his or her own face, but one also tries to preserve the other party's face through self-respect and respect for others. Respect is not about giving in; but it is about parties not losing face during negotiations. Face concerns one's status and respect received (Linong, 2006). Face of officials ranks their level and influence, business people their wealth and their influence, professionals their intelligence and reputation, which implies that people need to give respect, where respect is due, and means addressing people according to their relevant rank (Linong, 2006). People's words give them face, so people who go back on their words or break promises lose the face and confidence of others. Other face losing activities can include showing anger, frustration, disrespect, being dismissive or mocking.

Hofestede's Cultural Dimensions

Hofstede's (1984) dimensions are the most used constructs in cross-cultural studies. Chinese culture is often characterised as having high power distance, which may have resulted from centralised political power and Confucian doctrines (Linong, 2006), and Australian culture tends to be low power distance (Hofstede, 1984). Uncertainty avoidance "is the degree to which people feel threatened by uncertain situations and have created beliefs and institutions that try to avoid these" (Linong, 2006, p. 537). Chinese are characterised as having high uncertainty avoidance as Chung (2008) explains: "Hofstede defines uncertainty as a situation in which anything can happen and one has no idea what might happen, and says that uncertainty avoidance is not the same as risk avoidance, which can be a common misunderstanding. The level of uncertainty in society in many areas, such as technology, law, and religion, forces us to live on the brink of an uncertain future. To cope with uncertainty, people constantly manage anxiety." Chinese find it is easier to cope with a less clear legal system than Australians. Chinese culture can also be defined as collective, where people feel a sense of belonging and loyalty to the group, and group interests override individual interests, where Australian culture is categorised as individualistic (Hofstede, 1984). Chinese culture is high on masculinity that success is highly regarded. At the same time there are some feminine characters within the Chinese culture. Chinese are more caring for others and place an importance on the quality of life (Linong, 2006). So given these differences there is a culture gap in regard to Australian business people dealing with Chinese business people, consumers and suppliers.

Communication and Language

Communication in Chinese culture is considered to be high-context (Hall, 1976) and diffused (Trompenaars, 1993) and Australian culture is considered to be low-context and specific. "People from a high-context and diffused culture expect listeners to follow the train of thought of the speaker and understand the hidden agenda behind the words. People from a low-context and specific culture will ask specific questions straight to the point and expect direct answers. People from a low-context and specific culture will not be able to read behind words to look for hidden agendas or interpret words differently from how they are presented" (Chung, 2008).

Language is an important part of communication. Mandarin is very different from English and other major western languages. To start there are no alphabets, to master Chinese, a large number of characters need to be memorised. For Australians, it is difficult to learn and pronounce, especially for adults who already have developed vocal cords. The Australian education system has been criticised as being poor in teaching Chinese. The likelihood that Chinese business people speak English is dependent on whether they have western education (Linong, 2006). Hence, the language gap is a key impediment to Australians conducting business with Chinese counterparts and is a fundamental barrier to overcome (Wang et al., 1998). Table 1 summarises the differences between Australian and Chinese culture.

THE STUDY ON AUSTRALIAN BUSINESSES IN CHINA

To investigate the differences between Australian and Chinese culture in business relationships, we interviewed managers of 40 Australian businesses who had activity or operations in China. The study was conducted between August to December 2007. The research was supported by Australia China Business Council (ACBC), Victoria Branch. Invitations were emailed to Victorian members. Non-respondents were followed up with phone calls to encourage participation. Research team members' personal contacts were also used in selecting companies who were not members of ACBC. The snowballing method of interview participant recruitment was also used (Burgess, 1987), where interviewees were asked to identify potential interviewees from other companies.

Upon acceptance of participation, interviewees were explained the aims and background of the study. Interviewees were assured confidentiality and anonymity. A time was then made to conduct an interview. Interviews mainly took place at the interviewees' offices. Interviews were an hour

Table 1. Cultural differences between Australia and China: understanding the gap

Australia	Cultural Elements	China
Low	Power Distance	High
Individualistic, Immediate Family	Social Orientation	Collective, Extended Family is Important
Low	Uncertainty Avoidance	High
Masculine	Goal Behaviour	Feminine moving to Masculine
Low Context (Hall, 1976)	Communication	High Context (Hall, 1976)
English	Language	Mandarin and Dialects
Predominantly Christian	Religion	Atheist, Muslin, Budhist, Daoism, Taoism, Christian, Confucianism
99%	Literacy Levels	81.5%
English	Business Language	Standard Chinese/Mandarin/Cantonese
Reciprocity is important, but not to a large degree	Reciprocity	Renqing feelings of reciprocity, those who do not return a favour will lose face.
Relationships are more business like, and pragmatic	Relationships	Relationships are important, need to be developed overtime to establish trust and rapport
Westerners tend to face conflict straight away.	Conflict	Chinese will tend to avoid conflict

on average which were taped and transcribed by a transcriber. One participant wished not to have their interviews taped, and instead copious notes were taken. A total of 43 interviews were conducted in 40 Australian Companies. A thematic analysis was conducted, where themes were identified by the researchers, and illustrative quotes provided.

THE SAMPLE

This research was extensive as a range of industries and sectors were covered: Manufacturing (10); Business and Property Services (5), Building, Construction and Engineering (3), Education (4), Finance and Insurance (4), Agriculture, Forestry and Fishing (3), and Government, Administration and Defence (3). The remainder of the firms were scattered across various industries such as, Mining (2), Legal (1), Publishing (1), Accommodation, Restaurants and Cafes (1), Information Technology (1), Medical (1), Transportation and Logistics (1).

The modes of entry represented in the study include being Project/Client Based (9), and some with representative offices. Other modes included Joint Ventures (10), Wholly Owned Foreign Entity (WOFE) (7), using agents to source customers, Agency Agreements and Partnerships (4), others did Exporting (3), Supplier Relationships (2), (1) others had a representative office (1), or provided Exporting Assistance (1), Memorandum of Understanding (MOU) (1), a strategic alliance (partnership) (1) and licensing (1).

In terms of locations, major cities were evidently more popular. (Shanghai (23), Beijing (15), Guangzhou (10), Shenzhen (5), and Tianjin (4)) This was reasoned by interviewees as being closer to facilities and resources.

RESULTS AND FINDINGS

A major issue identified by respondents in this research was the cultural differences between Australia and China. Companies found themselves caught by surprise on this issue which suggests that they are ill-prepared for dealing with cultural differences. Coming from the perspective of Australian businesses, the cultural differences between the two countries are vast that the Chinese culture is referred to as an 'alien culture' (Chung, 2008) by Australian expatriates. Our research established that Australian organisations do not pay enough attention to cultural issues as they should do. Although some companies conduct basic training, such as a couple of hours prior to an assignment and expatriates may read a little about China and even learn a few words of Chinese, they do not really extend their way to understand some of the deeper issues such as the behavioural differences that were underpinned by the cultural differences. The cultural understandings of the way to do business are the most dominant issue here, because they are inherent within any business relationship and Chinese business relies heavily on relationships.

The result of not addressing the cultural issue means some fail badly and some find it hard to make an inroad into the market place. For example, one company who attempted opening a pizza shop in Shanghai failed because they did not understand the cultural preferences of Chinese consumers and their taste preferences, in addition to choosing the wrong location of a food hall for their business based on the knowledge they gained in Australia. They made no attempt of understanding their customers or doing any research prior to their market entry. The business ended up losing approximately $2 million from 14 private investors in Victoria. Prior research of a depth understanding of Chinese consumers, and a more thought out decision in terms of location would be fundamental to this project.

As mentioned previously a prominent feature of the cultural differences in doing business with Chinese is the relationship factor. Relationship building to bridge the cultural divide was one strategy identified by participants however is a slow process. Respondents identified that business relationships were not as pragmatic or to the point as business relationships in Australia, that is they take time. However failure to address this issue may seriously impact market entry and the ability to establish clients in China, as the following quote suggests:

It can affect the way you do business. Although you're selling the same product in the same industry, you might have to ... to do business with these people it is first contingent on firstly establishing a rapport/relationship with them. No one walks in and says "this is my product" and walks out with a purchase order. And, the cultural differences that exist in various parts of China can impact on how you go about establishing those relationships or the sorts of things you would do to build relationships, some things will work better in other areas than others.

Not only developing relationships takes time, once established they also need to be maintained. The costs associated with doing this, such as spending money on trips, lunches, and dinners over a long period of time need to be factored into a market entry budget. Participants in the current study suggested that it would take a number of business trips before relationships were established, and that those trips needed to be maintained. Participants also suggested that it was not appropriate to only take hospitality that Chinese offered, but also to match it, which can be seen as an investment into business relationships with Chinese:

And that's how I guess we work, is the level of reciprocity. Whatever we get we reciprocate. And sometimes we push the envelope and then they'll follow up. And particularly Chinese government departments are incredibly hospitable. They have huge budgets, compared to ours, for official entertaining and things like that. And so they do offer things, and I think people forget that sometimes. They think, 'Oh well, that's just the Chinese. I'll take it." but when they come out here they give them a plate of sandwiches. It's just ridiculous.

So we focus on that, and I think because we focus on that, it certainly gives us a bit more leverage.

Eating and drinking is one special cultural characteristic in doing business with Chinese (Chung, 2008). Eating and drinking gives business people an opportunity to socialise, get to know one another and work together rather than satisfying hunger alone. It is important in the life of Chinese business culture, and is used as a major business tool. Important deals and can be made during a banquet. Trust can be built at a dinner table. Negotiations can be continued at lunch. Food, to the Chinese, is a business tool rather than a form of bribery. Food is the opportunity to be social. Food is eaten and shared to demonstrate personal quality, and used as an effective tool in massaging and creating business relationships. Drinking goes with it (Chung & Gong, 2009). Participants in the current study, commented on the usefulness of drinking sessions as relationship building exercises, the following quote is indicative of this:

Gan Bei. That's the biggest cultural factor for me, Gan Bei is to drink a whole glass and get drunk. It's horrible beer. That's the key cultural factor for me.

Therefore the best strategy at building relationships with Chinese will benefit organisations to succeed in China. One company uses a consultant of mainland Chinese background (Chung & Smith, 2007) who had extensive experience working with both Chinese and Australian individuals. Other strategies identified included using western educated Chinese individuals and staff of similar background to bridge the relationship building with Chinese companies. Also having Western individuals who have extensive Chinese business experience is important. The following quote is indicative of this:

There are some important cultural things that really do matter and that's about having experienced people and maximizing the number of Chinese nationals in your business. The country president, who's an Australian, he has being there for 25 years in China not working all the time for us, so he established China experience.

However, it is critical to make the point of distinguishing overseas Chinese between those of mainland background and those of other country background (Chung & Smith, 2007). Some firms did err on the cautious side of using those overseas Chinese who were brought up in Western countries:

Their managing director in China is a Chinese born fella who at a relatively young age moved to Canada and then subsequently to the US where he became the employee of this company. Now they've put him in as Managing Director in China because he's Chinese but I've discussed him with our managing director who was fully Chinese and he's of the view that he's much more American than Chinese. He's lost it. So a Western education is a good thing if it enables a Chinese person to understand how a Western company needs to be managed and controlled but it can go too far. They can lose too much of that cultural connection. This guy is not well liked at all by the workforce; he just doesn't do things in a Chinese way. He's much more American than Chinese.

Chung and Smith (2007) were the first to distinguish mainland Chinese from overseas Chinese. In the past companies recruited overseas Chinese and mistakenly believed they are familiar with the mainland Chinese and the market. This strategy had proven to be ineffective (Chung, 2008). This is because individuals' minds are programmed by the surrounding culture that overseas Chinese, such as the American-Chinese mentioned above, are products of the culture they were brought up in. The above mentioned American-Chinese is culturally a Canadian rather than a Chinese.

Participants also commented on setting up a Chinese style business, rather than using Australian models of business operation, as a way to deal with the cultural divide as suggested by the respondent below:

It depends on the nature of the business you want to set up, but for us, we set out with the intent of creating a Chinese business, bringing the best of what we have from Australia into that business. We never set out to create a foreign business in China. There is a big difference between how you approach the two. A lot of foreign businesses will have a foreign general manager, a foreign CFO, will have a Chinese team that they all speak English. We actually have a Chinese General manager a Chinese CFO. We've had some ex-pats in our business, but of 250-300 people there now, there are probably only 3 or 4 Australians in that business.

Large corporations especially those well established ones do tend to take an ethnocentric approach to their own organisational culture which is often forced on to joint venture partners, Chinese staff, managers and customers when they enter the Chinese market (Chung, 2008).

Participants also commented on the engagement strategies with Chinese staff and how they had to show them what to do, or engage in cooperative strategies and gain consensus from their workers, or find flexible staff who would appropriately listen to what was told to them. This suggested that participants would communicate with their staff in a friendly manner, and get them to agree on things through a process of consensus, which required participation, and openness, rather than simply telling people what to do. Here the key to managing Chinese staff is found against the literature by suggesting Chinese should respond well to autocratic style of management (Chung, 2008). The difficulty may be that Chinese staff will not express their disagreement openly in the culture that conflict is to be avoided. Nevertheless, procedures may only be followed when expatriate managers are present (Chung, 2008).

The answer and solution to these cultural difficulties is suggested by a successful organisation is the engagement of bi-cultural consultants in order to effectively bridge the culture gap (Chung, 2008), as the following quote suggests:

...My first two trips were not nearly as productive due to cultural and communication difficulties. XXX set a busy schedule and demonstrated excellent communication, organisational and cultural skills with top results. We are now moving ahead fast in China. Without XXX, the project would have been abandoned by now...

They also suggested that spending time learning about the culture first was important, and to continue to learn about it as time goes on; as one participant suggested that they keep on learning as they go along: "*I always leave with questions... each time I go I try to learn more about the local culture.*" Other important strategies were to engage in ways to combine the cultures between Australians and Chinese. A bridging cultural course was also suggested as a way to do this, to combine the cultures together. Other strategies identified included being subtle with Chinese individuals, and try to understand the deeper issues surrounding their behaviour.

Once individuals had learnt about the culture, they believed that they needed to adapt to Chinese culture and interact it, and view Chinese culture through their eyes. This suggested that people should not be ethnocentric about their views held of Chinese culture and individuals. It was also suggested to behave like Chinese when working with them. One participant elaborated on the pragmatic nature of people he does business with:

So they go out and get stuff done and then figure out later how they're going to do it. Whereas we sit down and plan it and do it. I actually ran that past a few Chinese guys, and they said, "That's the way we do things." I'm going to sell paint! Righto! Find out, blah, blah, blah and then go back and say, "Oh yeah, I can get you paint." But there is a disconnect between... Then they give you an order for some paint. We haven't discussed the price and all the details. The pressure comes back on me to provide the price. And I say, "Why do you do it?" "Well there is money to be made. Get going." So the culture thing is like this. That's my personal experience that I have found. You've just got to accept that that's the way they are."

Another participant stated that they engaged in bribery, or what they would perceive to be bribery to achieve successful business deals, as this was viewed as being part of the culture, but pointed out that it was more about relationship building, such as taking people out to dinner. Table 2 summarises the strategies that were articulated by the interviewees.

Table 2. Strategies to deal with the culture gap

Strategies mentioned by participants
Human Resource Strategies
• Use Chinese nationals in the business
• Use western educated Chinese staff with mainland background
• Use people with China experience
• Setup a business that is Chinese in its orientation as opposed to Western in Orientation
Dealing with Chinese Staff
• Have to show Chinese staff how to do things
• Use diagrams, pictures, drawings, numbers and mobile phone converters to get the message across
• Engage in cooperative strategies to gain consensus from Chinese workers
• Find flexible Staff who will listen to you
Relationship Building
• Make Relationship building a slow process
• Drinking sessions are relationship building
• Engage in face to face interaction with Chinese
• Try not to get things done by email
Getting Outside Support
• Work on strategies to combine cultures
• Bridging cultural course
• Employee a bi-cultural consultant
• Use translators and interpreters
Learn about the Culture
• Learn from their culture
• Be subtle
• Learn about it as time goes on
• Understand the deeper issues
• Have language training
Adapt to Chinese Culture
• Do not use ethnocentricity and see through their eyes what Chinese culture is.
• Conduct negotiation through Chinese eyes
• Behave like Chinese

DISCUSSION AND IMPLICATIONS

Previous research has indicated that there are many strategies that can be used to bridge the cultural divide. Researchers have been seeking to develop frameworks that assist in overcoming obstacles to help cultures working together (Bond, 2003). Other researchers have identified using Chinese nationals as strategies to deal with cultural issues. Chung and Smith (2007) have pointed out that using overseas Chinese (those who grow up outside mainland China) may be problematic that they lack of acceptance of Chinese nationals of these people, and they often have a different value set then Chinese (Gomez, 1999; Tan, 2000). The only possible successful method of the HR strategy is to use bi-cultural consultant/staff (Chung, 2008). This is referring to those who have sound understanding of both Chinese and Australian business cultures and reducing the cultural gap. The final HR strategy was to use expatriates, but it was emphasized by participants that those expatriates need to have extensive China experience. Individuals also commented on the need to establish a Chinese business, using Chinese models of business, rather than Australian models. It was commented by participants that Australian models of businesses could not be easily replicated in China and just work. It was suggested that the business model needed to be converted to the Chinese environmental context.

When dealing with Chinese nationals a range of strategies was suggested which included showing staff how to do things rather than telling them. This may be as a result of Australian managers not knowing Chinese language and Chinese workers knowing little English, Chinese using indirect language styles (Hall, 1976), and Chinese staff preserving face when interacting with western managers (Linong, 2006). Showing staff how to do things maybe a form of visual learning rather than verbal learning, which maybe more concrete, especially where language is a barrier. The findings also indicated that it was important to use cooperative strategies to engage Chinese staff in work. This is consistent with past researchers who have pointed out that cooperative strategies resulted in better trust, and commitment, and assisted them in working together (Chen & Tjsovold, 2007), which is similar to the findings presented in this paper. Other strategies included finding flexible staff that would listen and implement what managers had to say.

Relationships are important in Chinese-Western business activities and much literature discusses this concept. Establishing relationships and building relationships was emphasized to be a slow process. Of course, relationships are not exclusive to the Chinese, although, along with other Asian business people, they pay more attention to relationships for reasons related to Chinese culture. As Mu (1995) said: "...the three most important factors in doing business with the Chinese are relationships, relationships and relationships".

Mu (1995) proposed a two-level and one-level interaction theory to suggest that Westerners are capable of a two-level relationship, business/professional and personal, while Chinese are only capable of a one-level relationship, namely personal. Chung (2008) disagrees with that theory. On the surface the Chinese do give the impression of mixing personal and business relationships. This confusion perhaps is caused by other characteristics of their culture, such as the emphasis on trust in business transactions. Without a trustworthy personal relationship having been established, business is unlikely to proceed (Beamer & Varner, 2001). However, the concept of a relationship is closely related to cultural factors, making this a complex issue for non-Chinese.

The use of Chinese consultants was seen as an important element of understanding Chinese culture and assisting with successfully doing business there. Previous research in the field has found that it is important to engage the help of Chinese advisory consultants when conducting business in China (Menzies, Chung, & Orr, 2008). This was found to be more important because firms often had no strategic plans to conduct business in China. Without plans the necessary homework and learning about Chinese culture is difficult to achieve.

CONCLUSION

In summary it can be seen that Chinese culture is significantly different to Australian culture, and there exists a culture gap, which cause impediments to Australians doing business with China. Some of these differences were highlighted in the literature research. Empirical evidence was used to develop strategies on how to reduce the culture gap. The strategies were categorised under human resource strategies, dealing with Chinese staff, relationship building, getting outside support (employing consultants), learning about the culture, and adapting to the culture. Further research in the area could further develop these strategies, and empirically examine whether they do in fact have an influence on reducing the cultural gap. From a practical perspective future work could develop training and cross-cultural training programs for business executives that seek to further reduce this gap, considering the importance of Australia's international trade with China.

REFERENCES

Beamer, L., & Varner, I. (2001). *Intercultural Communication in the Global Workplace*. New York: McGraw-Hill Higher Education.

Bellabona, P., & Spigarelli, F. (2007). Moving from Open Door to Go Global: China Goes on the World Stage. *International Journal of Chinese Culture and Management*, *1*(1), 93–107. doi:10.1504/IJCCM.2007.016170

Bond, M. H. (2003). Cross-cultural social psychology and the real world of culturally diverse teams and dyads. In D. Tjosvold & K. Leung (Eds.), *Cross-cultural management: Foundation and Future* (pp 45-58). Aldershot, Hampshire: Ashgate Publishing.

Burgess, R. (1987). *In the field: An introduction to field research*. London: George Allen and Unwin.

Chen, N., & Tjsovold, D. (2007). Guanxi and leader member relationships between American managers and Chinese employees: open-minded dialogue as mediator. *Asia Pacific Journal of Management*, *24*(2), 171–189. doi:10.1007/s10490-006-9029-9

Chung, M. (2006). *The Impact of Cultural Difference: An Australian Business in China.* Unpublished Doctoral Thesis, Monash University, Melbourne.

Chung, M. (2008). *Shanghaied: Why Foster's Could Not Survive China*. Melbourne, Australia: Heidelberg Press.

Chung, M., & Gong, W. Y. (2009). *How Food Is Used as an Effective Business Tool in China?* Paper presented at the Symposium on Food Law & Policy, Melbourne, Australia.

Chung, M., & Smith, W. (2007). Overseas Chinese as Expatriate Managers in China – Is Their Recruitment a Solution to Cross-Cultural Management Problems for Multinationals Operating in China? *International Journal of Diversity in Organisations, Communities and Nations, 7*.

Davies, K. (2007). China's Investment Watch. Organisation for Economic Cooperation and Development. *The OECD Observer. Organisation for Economic Co-Operation and Development*, *260*, 19.

Department of Foreign Affairs and Trade. (2007). *People's Republic of China Country Brief - August 2006*. Retrieved February 19, 2008 from http://www.dfat.gov.au/geo/china/cb_index.html

Department of Foreign Affairs and Trade. (2008). *Australia-An Overview*. Retrieved June 27, 2008 from http://www.dfat.gov.au/aib/overview.html

Eisenstadt, S. N. (1973). *Tradition, Change and Modernity*. New York: John Wiley & Sons Inc.

Fisher, G., Hughes, R., Griffin, R. W., & Pustay, M. W. (2006). *International Business: Managing in the Asia-Pacific*. Frenchs Forest, NSW: Pearson Education.

Gomez, E. T. (1999). *Chinese business in Malaysia: Accumulation, accommodation and ascendance.* Richmond, Surrey: Curzon Press.

Goodall, K., Li, N., & Warner, M. (2007). Expatriate managers in China: The influence of Chinese culture on cross cultural management. *Journal of General Management*, *32*(2), 57–76.

Hall, E. T. (1976). How cultures collide. *Psychology Today*, 67–74.

Hofstede, G. (1984). *Cultures Consequences* (abridged ed.). Thousand Oaks, CA: Sage.

Johanson, J., & Vahlne, J. E. (1977). The internationalization process of the firm - Four Swedish case studies. *Journal of Management Studies*, *12*(3), 305–322. doi:10.1111/j.1467-6486.1975.tb00514.x

Johanson, J., & Vahlne, J. E. (1992). Management of foreign market entry. *Scandinavian International Business Review*, *1*(3), 9–27. doi:10.1016/0962-9262(92)90008-T

Kogut, B., & Singh, H. (1988). The effect of national culture on the choice of entry mode. *Journal of International Business Studies*, *19*(3), 411–432. doi:10.1057/palgrave.jibs.8490394

Linong, Z. (2006). *China Business: Environment, Momentum, Strategies, Prospects*. Singapore: Pearson Prentice Hall.

Menzies, J., Chung, M., & Orr, S. (2008). *Doing Business with China: How Australian Companies Make their Decisions when Entering the Chinese Market*. Melbourne, Australia: Deakin Prime.

Mu, D. P. (1995). Culture and Business: Interacting Effectively to Achieve Mutual Goals. *East Asian Executive Reports*, *17*(6), 6.

Neuborne, E. (2003). Bridging the culture gap. *Sales & Marketing Management, 155*(7).

Nordstrom, K. A., & Vahlne, J. (1992, April 22-25). *Is the globe shrinking? Psychic distance and the establishment of Swedish sales subsidiaries during the last 100 years*. Paper presented at the International Trade and Finance Association's Annual Conference, Laredo, Texas.

O'Grady, S., & Lane, H. (1996). The psychic distance paradox. *Journal of International Business Studies*, *27*(2), 309–333. doi:10.1057/palgrave.jibs.8490137

Oberg, K. (1979). Culture Shock and the problem of adjustment in new cultural environments. In E. C. Smith & L. F. Luce (Eds.), *Towards internationalism: Readings in cross-cultural communication* (pp. 185–186). New York: Newbury House Publishers.

Ralston, D. A., Yu, K.-C., Wang, X., Terpstra, R. H., Gustafson, D. J., & Wei, H. (1996). The cosmopolitan Chinese manager: Findings of a study on managerial values across the six regions of China. *Journal of International Management*, *2*, 79–109.

Tan, E. K.-B. (2000). Success amidst prejudice: Guanxi networks in Chinese businesses in Indonesia and Malaysia. *Journal of Asian Business*, *16*(1), 65–84.

Tian, X. (2006). *Managing international business in China*. New York: Cambridge University Press.

Trompenaars, F. (1993). *Riding the Waves of Culture: Understanding Cultural Diversity in Business*. London: Nicholas Brealey Publishing Ltd.

Wang, Y., Zhang, X. S., & Goodfellow, R. (1998). *Business culture in China*. Singapore: Butterworth Heinemen.

Webber, R. H. (1969). Convergence or divergence. *The Columbia Journal of World Business*, *4*(3), 75–83.

Yeung, H. W. C. (2000). Local politics and foreign ventures in China's transitional economy: the political economy of Singaporean investments in China. *Political Geography*, *19*(7), 809–840. doi:10.1016/S0962-6298(00)00027-5

Youzhou, Z., Jiesheng, F., Wong, M. Y., & Stewart, S. (1996). The 'catfish effect' of the private sector on the economy of The People's Republic of China. *Journal of Enterprising Culture*, *4*, 331–349. doi:10.1142/S0218495896000198

This work was previously published in International Journal of Asian Business and Information Management, Volume 1, Issue 1, edited by Dasho Karma Ura and Patricia Ordóñez de Pablos, pp. 42-53, copyright 2010 by IGI Publishing (an imprint of IGI Global).

Chapter 3
Confucianism Values for Transformational Leadership Development in the Chinese Context

Balakrishnan Muniapan
Swinburne University of Technology - Sarawak Campus, Malaysia

Ee Yaw Seng
Swinburne University of Technology - Sarawak Campus, Malaysia

ABSTRACT

Transformational leadership has received a great significance in the leadership literature during the past two decades. However, this research is dominated by Western traditions and scant attention is paid to Asian traditions. Therefore, in this paper, the authors explore the relevance of Confucianism, an Asian tradition, and its significance in transformational leadership development. For this purpose, the authors have selected the Analects, which provide the essence of Confucianism to explore the relevant transformational leadership concepts. This paper is based on hermeneutics, a qualitative research methodology that involves study, understanding and interpretation of ancient text. Confucianism provides an inside-out approach to transformational leadership development, which is self leadership first, before leadership of others. This inside-out perspective is also similar to other Asian traditions, such as ancient Indian wisdom. The prospects of analysis of Confucianism in other areas of leadership and management can be considered for future research.

1. INTRODUCTION

Leadership has a long history and has been the focus of research in numerous fields, from psychology, sociology, organizational theory, politics and management. Leadership is also one of the most relevant and an important aspect of management whether in work organizations, political parties, societies and countries. It is also one of the most researched management subjects and an interesting topic of discussions around the world. Ancient civilizations throughout the history had methods by which they were led and managed. Every king

DOI: 10.4018/978-1-4666-0276-2.ch003

or queen who ruled ancient kingdoms had his or her own leadership values to transform their people.

The western tradition and influence on leadership theories and concepts have dominated the world for over two centuries due to colonization and the widespread use of English in many countries and the readily available of leadership literatures in English language. However the studies on Asian tradition on leadership especially the application of leadership ancient wisdoms from China and India are beginning to grow due to the economic growth of both nations (Muniapan & Dass, 2008; Muniapan & Dass, 2009; Muniapan & Satpathy, 2010).

A careful analysis reveals that many of the recently popularized western leadership theories and concepts have been in practice in Asia for centuries. These practices however were not in the context business organization but in the state or political governance in those days (Muniapan, 2010). Governance in the contemporary context has both state and on-state sponsored dimensions. Sharma (2001) argues that for a leadership system, to be effective, it has to be rooted in the cultural soil of the country, where it is practiced. Many communities and countries in the world are now trying to discover and explore their own system of leadership and management, which includes financial management, human resource management, customer relationship management, corporate social responsibility and also corporate governance.

Maruyama (1994) was one of the authors who explored the Asian tradition of leadership beginning with wide understanding of Japanese leadership. Besides Japanese, the interest in exploring other Asian tradition especially from the Indian context of leadership has seen a tremendous growth over the past two decades. In the Indian context, the studies to explore ancient Indian wisdoms in leadership were made by scholars such as Chakraborty (1993, 1995, 1999), Chakraborty and Chakraborty (2008), Sharma (1996, 1998, 1999, 2002, 2003), Krishnan (2001, 2003), Kejriwala and Krishnan (2004), Satpathy (2006, 2007), Muniapan (2006, 2007, 2009), Muniapan and Dass (2008, 2009), Satpathy and Muniapan (2008), Muniapan and Satpathy (2010), Roka (2006), Parashar (2008) and others.

In the Chinese context, ancient wisdoms play an important role and relevance for leadership development. In China, Confucianism together with Buddhism and Taoism are among the great schools of thought. Confucianism has deeply influenced the lives of Chinese people for more than 2500 years. All the teachings of Confucius (Confucianism) have a strong emphasis on high standard of moral cultivation through human activity and experience.

2. THE BROAD PURPOSE OF THIS PAPER

From the perspective of the authors, there is a lack of studies on transformational leadership values from the Chinese wisdom. Most of the researches are in general leadership and not in-depth studies. Even this paper is also not an in-depth analysis of the Confucianism and transformational leadership as it only attempts to explore the relevance of Confucianism in transformational leadership. With growth of the Chinese economy, studies in Chinese leadership need to grow based on the Chinese cultural roots and upbringing. A Chinese grows up in a system, where family ties and sense of belongingness gets top priority and with this type of background, he or she may not be able to adjust or fit into the job environment practicing Western leadership based on individualistic, direct, low power distance and contractual. China is high in power distance as employees acknowledge and respect the authority of the leader based on his legitimate power and they seldom bypass the chain of command. Besides high power distance, China has a low uncertainty avoidance, which means leaders have a propensity for low risk aversions and employees exhibit little aggressiveness in businesses. China also has low masculinity and low individualism, which means group, family and society, are more important than the individuals (Hofstede, 1991, cited in Muniapan & Dass, 2009). As a result, it is imperative that the Chinese should focus their effort on transformational leadership development from their own Confucianism cultural roots. Therefore the broad purpose of the paper is an attempt to provide the relevance of Confucianism in transformational leadership development in the Chinese business and management.

3. METHODOLOGY

This paper is based on a qualitative research methodology called hermeneutics. Hermeneutics is related to the name of the Greek god Hermes in his role as the interpreter of the messages of the gods. In the current context, hermeneutics can be described as the interpretation and understanding of ancient literatures. It is also used in contemporary philosophy to denote the study of theories and methods of the interpretation of all texts and systems of meaning. A hermeneutic is defined as a specific system or method for interpretation, or a specific theory of interpretation (Muniapan, 2010). Hermeneutics is widely applied in many field of social science such as philosophy, religion and theology, law, sociology and also international relations. Besides social science, hermeneutics is also used in management and organization such in the study of organizational culture, information systems, accounting and international management (Noorderhaven, 2000).

The principles of Confucianism are found in the *Analects* or selected sayings of Confucius which were compiled over a period of time. Though the analects have been credited to Confucius, they also include a number of traditional proverbs and sayings. There are old and new interpretations of the Analects of Confucius. The main purpose of the old interpretations was to make the text easy to understand. On the other hand, the new interpretations were influenced by the philosophies of Taoism and Buddhism and are called by the term "Neo-Confucianism" (http://www.shvoong.com/books/1710809-analects-confucius).

For the purpose of this paper, the *Analects* were analyzed based on four stages; namely identification, investigation, interpretation and integration. The identification stage involves searching for the relevant direct and indirect leadership sayings from the text; this will be followed by a detailed investigation of the sayings in terms of context and context. The next stage involves interpretation by providing the meaning and the relevance of the sayings in the modern context. The final stage is the integration of the sayings involves adopting, modifying the lessons from the sayings and provide commentary from the perspectives of transformational leadership.

4. TRANSFORMATIONAL LEADERSHIP

Today, many approaches and theories of leadership are being developed to advance the academic and practical application of this crucial organizational element. All of these varied approaches and theories offer contributions to the on-going dialogue and debate on leadership. Leadership has also commanded the attention of several academics from a variety of disciplines, such as political science, organizational psychology, management studies, education, sociology and anthropology. Definitions of leadership are also numerous, varying greatly between and within disciplines. In fact, there are many definitions of leadership as there are persons who have attempted to define the concept.

From the organizational management context, House (2004) defines leadership as the ability of an individual to influence, motivate and enable others to contribute toward the effectiveness and success of the organizations of which they are members. Leadership is also the ability of an individual to set rules for others and lead from the front. It is an attitude that influences the environment around us. Other leadership scholars such as Jaques and Clement (1994) defined it as that process in which one person sets the purpose or direction for one or more other persons and gets them to move along together with him or her and with each other in that direction with competence and full commitment. Another famous author on leadership, James MacGregor Burns (1978) wrote that leadership is one of the most observed and least understood phenomena on earth and that we know a lot about leaders while knowing very little about leadership. He defined leadership as inducing followers to pursue common or at least joint purposes that represent the values and motivations of both leaders and followers.

As for transformational leadership, according to Burns (1978) involves a process in which leaders and followers work together, mutually influencing or even transforming each other to advance towards higher standards of moral and motivation for mutually desired outcomes. He proposed that leadership process occurs in one of two ways; either transactional or transformational. Transformational leadership is superior to mere transactional leadership, which relies on a mutual exchange of valued outcomes like monetary incentives and promotions for hard work. Transformational leaders not only recognize and exploit existing needs of potential followers but also look for potential motives and higher needs in followers. In this manner, they completely engage their followers and help them to achieve their fullest potential (Burns 1978 cited in Mulla & Krishnan, 2009).

Transactional leadership, on the other hand is a set of leadership behavior that emphasizes exchanges or bargains between the leader and the follower, and focuses on how the current needs of the followers can be fulfilled (Maher, 1997; Muniapan, 2010). The exchanges can be economical, political, or psychological in nature. Psychological exchanges are the primary characteristics that distinguish transactional from transformational leadership. Transactional leadership behavior includes contingent reward, which involves an interaction between the leader and follower based on exchange of resources and management by exception, in which leaders intervene only when problems emerge (Bass & Avolio, 1994). Most leaders engage in both transformational and transactional leadership; however they do so in differing proportions.

Transformational leaders adopt a long-term perspective. Rather than focusing solely on current needs of their followers or themselves, they also focus on future needs; rather than being concerned only with short-term problems and opportunities facing the organization, they also concern themselves with long-term issues; rather than viewing intra-and extra organizational factors as discrete, they view them from a holistic orientation (Bhargava, 2003).

According to Bass and Avolio (1994), also cited in Krishnan et al. (2004), Muniapan (2007) and Muniapan (2010), transformational leadership consists of four primary dimensions. They are as follows:

a) **Inspirational Motivation (IM):** This dimension is reflected by behaviors that provide meaning and challenge to followers' work. It includes behaviors that articulate clear expectations and demonstrate commitment to overall organizational goals, and arouse a team spirit through enthusiasm and optimism. Krishnan (2000) asserts that inspirational leadership also involves envisioning a desired future state, making followers see that vision, and showing followers how to get to that state. Envisioning is translating intentions into realities by communicating that vision to others to gain their support as the right vision attracts commitment, energizes people, creates meaning and establishes a standard of excellence. Vision inspires followers to transcend the outcome and getting people to commit voluntarily and completely something worthwhile.

b) **Idealized Influenced (II):** Idealized influence is described as behavior that results in follower admiration, respect and trust. Idealized influence involves risk sharing on the part of leaders, a consideration of follower needs or personal needs, and ethical and moral conduct. Idealized influence also refers to the leader's charisma. Krishnan (2000) defines charisma as a form of social authority that derives its legitimacy not from rules, positions, or traditions, but rather from faith in the leader's exemplary character. He further asserts that charismatic leader is seen different from an ordinary person and treated as endowed with supernatural, superhuman or at least exceptional power and qualities. Only charismatic leaders, with their sense of vision and empowering behavior could address the higher order needs of followers. Charismatic leadership is characterized by followers trust in the correctness of the leader's belief, unquestioning acceptance of the leader, affection for the leaders, willingness to obey the leader, and emotional involvement in the vision and mission of the organization.

c) **Intellectual Stimulation (IS):** Leaders who demonstrate this type of transformational leadership solicit new ideas and creative solutions for problems from their followers and encourage novel approaches for performing work. Krishnan (2000) asserts intellectual stimulation arouses in the followers the awareness of the problems and how they may be solved, and stirs the imagination and generates thoughts and insights. The intellectual stimulation provided by the leader forces the followers to rethink some of the ideas that they never questioned before.

d) **Individualized Consideration (IC):** This is reflected by leaders listening attentively to the opinion and feedback of their followers and pay special attention to the followers needs for achievement and growth. Krishnan (2000) refers individualized consideration as the developmental orientation of the leaders towards the followers. The transformational leader gives personal attention to the followers who seem neglected, treat each follower individually, and help each follower to get what they want. These leaders have empathy or the capacity to sense intuitively the thoughts and feelings of others.

Bass (1985), on the other hand found that transactional leadership consist of two factors. They are as follows:

a) **Contingent Reward (CR):** This refers to leaders who reward followers for their effort, support and doing what needs to be done by clarifying the followers' roles and task requirements to meet their personal goals and the organizational missions.

b) **Management by Exception (ME):** This dimension of a transactional leader refers to leaders taking corrective action only when followers deviate from expectations or fail to meet goals.

A number of researches have suggested that transformational leaders, in general, motivate followers to perform at higher levels and to exert greater effort than do transactional leaders (Bass & Avolio, 1994). Transformational leaders motivate others to transcend self-interest so as to benefit the group as a whole. Transformational leaders create vision and direction for their followers around a common mission and gave them a sense of purpose.

Table 1 gives comparisons between two-leadership style based on findings by Bass and Avolio (1994, also cited in Krishnan et al., 2004; Muniapan, 2007, 2010).

Table 1. Comparisons between leadership styles

Leadership Quality	Transactional Approach	Transformational Approach
Time Orientation	Short term	Long Term
Communication	Vertical, Downward	Multidirectional
Focus	Financial Goals	Customer (Internal and External)
Reward Systems	Organizational, Extrinsic	Personal, Intrinsic
Source of Power	Legitimate, Reward	Referent, Expert
Decision Making	Centralized, downward	Dispersed, Upward
Employees	Liability, Cost	Asset
Coordination Mechanism	Rules and Regulations	Goals and Value Congruence
Compliance Mechanism	Directive	Rational Explanation
Attitude Towards Change	Avoidable, Resistant, Status Quo	Inevitable, Embrace
Guiding Mechanism	Profit	Vision and Values
Control	Rigid Conformity	Self Control
Perspective	Internal	External
Task Design	Compartmentalized Individual	Enriched, Groups

The transformational leader, unlike the transactional leader, does not use people as a means to get his organizational objectives fulfilled. In fact, he is intent on the development of the people, who in turn get inspired by him and try to emulate him. The transformational leader is able to guide the destinies of many, because he changes the very visions and perspectives of the people around him. People get so influenced by him that they are naturally transformed. Ross and Offermann (1997) found that higher levels of need for change, self-confidence, and dominance predicted transformational leadership.

Tichy and Devenna (1986) in their studies on leadership asserted that transformational leaders possess the following characteristics.

a) Transformational leaders tend to develop organizations that are entrepreneurial, innovative and flexible. Their nature enables them to lead others in such an environment where change is the only factor that is constant.
b) Leaders who adopt transformational style are risk taker in the face of adversity.
c) Even though transformational leaders possess great authority and power in the circumstance, they are sincere and willing to motivate their followers. Wherever possible and appropriate, they will empower their followers.
d) Transformational leaders develop a set of essential values for their followers to attain.
e) Using their own attitudes and behavior and experience from the past, this type of leaders are able to tackle problems that are even complex, ambiguous and even radical.
f) Transformational leaders have visions and the ability to connect the situation while communicating them to the followers.

Krishnan (1990) in his studies on transformational leadership and Vedanta philosophy highlighted some of the following characteristics of the transformational leaders:

- They identify themselves as change agents.
- They are courageous individuals.
- They believe in people.
- They are value-driven.

- They are life-long learners.
- They have the ability to deal with complexity, ambiguity and uncertainty.
- They are visionaries.

Transformational leaders are considered to be a breed of individuals of certain flair. They have the knowledge, attitude and ability to influence their followers to change and grow to a level that is not expected before. Simic (1999) described this as the qualitative or mental change of an organization. The above dimensions are characteristics of transformational leadership have a lot of similarities with Confucianism tradition in the ancient China with contextual differences. In a nutshell Confucianism promotes transformational leadership as the Confucianism values are congruent to transformational leadership. The next section of this paper will explore Confucius and Confucianism values before highlighting its relevance to transformational leadership.

5. CONFUCIUS AND CONFUCIANISM VALUES

Confucius or K'ung Fu Tzu (551-479 BCE) was an early Chinese philosopher, thinker, political figure and an educator. He lived during the time known in Chinese history called the *Warring States* period (475-221 BCE) where several states fought against each other to gain control over what is now China (McDevitt, 2007). As an educated man from Eastern Zhou, he travelled among the states offering advice to feudal rulers but more importantly, he shone as a teacher to any students who were willing to learn for a fee (Theodore de Bary & Bloom, 1999). His teachings and sayings, which were preserved in the *Analects*, were based on the principles of how people should live their lives and interact with others and the kinds of society and government people should participate in. The *Analects* was compiled by disciples or the disciples of disciples of Confucius and it is considered the foundation of Confucianism. From China, Confucianism later spreads to other East Asian countries such as Hong Kong, Taiwan, Macau, Japan, Korea, Vietnam and Singapore. It forms the guidelines and manuals for proper mode of conduct in many Chinese families and societies and affected the management of numerous Chinese businesses and organizations in the East Asian region.

Confucianism is considered as being more familistic and collectivistic orientated. This is also consistent with the popular Hofstede (1991) four dimensions studies on collectivism in the Asian context. Collectivism refers to the notion that each individual does not exist alone, but as a part or a member of a family or group (whole). Collectivism, unlike individualism, holds the group as the primary and the standard of moral value. According to Kirkbride and Tang (1992), Confucianism urges individuals to build on collectivity, controlling of personal desires, avoiding self interest for the well being of the group, restraining conflict and ensure harmony.

As for the Confucianism values, Miles (2006) pointed out that in order for a stable and lasting social order, there must be civilized and cultured values which were developed through human wisdom, not imposing of strict rules and laws enforced on individuals. Confucian values does not emphasize on formalized rules as this will destroy the strong human relationships which are one of the foundations in Confucianism (Bond, 1992).

Tu (1998a) listed the five virtues as humanity/benevolence (*ren*), righteousness (*yi*), propriety (*li*), wisdom (*zhi*) and trustworthiness (*xin*) which are highly relevant for transformational leadership development. The elements of humanity and benevolence (*ren*) are key values in Confucian teaching. Showing concerns and care to other human beings is important for the development, progress and well-being of a harmonious society. Secondly, another Confucian value concerns the belief of righteousness (*yi*). Weber (2009) defines righteousness as 'what is the ethically best action'.

This is in contrast with actions that lead to self-interest. An individual for whatever reasons should try to follow the path that lead to the greater good. In summary, individuals should aim to do the right thing for the right reason. Humanity, benevolence and righteousness are important elements in the development of leadership charisma. Charismatic leaders have great influence on their followers and this is consistent with the idealized influence dimension of transformational leadership.

The third value is propriety (*li*) which is defined as 'the quality of being socially or morally acceptable' (Collins Cobuild, 2009). In Confucianism, the notion refers to the way things should be done or the right practice (Weber, 2009). In business, adopting the right practices is regarded as utmost important. This value is based on morality which leads to the introduction of authority. For Confucians, the father is the authority in a family whereas the king is the authority in a society and in leadership the authorities should set good examples for others. This is also an important motivational element in leadership as whatever action the leaders performs, common people will follow; and whatever standards the leaders sets by their exemplary acts, all everybody pursues as it is the heads which provides the directions and motivations to act. This is consistent with the inspirational motivation dimensions of transformational leadership.

Another Confucianism value is wisdom (*zhi*). To Confucius, one should not stop acquiring new knowledge. Life is a continuous process of learning and self-improvement and one should not stop to educate until death. This is not only carried out through learning but practice as well. Through education and acquisition of skills with mentoring by benevolent individuals are best suited to be leaders. Firms that are advocating Confucian values in Asia tend to value training highly. Confucius believed that education is fundamental in the development of just leaders (McDevitt, 2007). In Confucianism, effective leaders view them as a lifelong learners and strongly committed to learning from self and others as promoted in the modern day concept of learning organization. Therefore in Confucianism effective leaders are expected to stimulate the followers intellectually for their own growth and organizational growth. This is consistent with the intellectual stimulation dimension of transformational leadership.

The next Confucian value covers action based on trustworthiness (*xin*), honesty or fidelity. Trustworthiness is an important moral value for transformational leaders. A trustworthy leader is someone in whom the people can place their trust and rest assured that the trust will not be betrayed. In the Chinese it is said that a leader without trust and honor is a dead man in this world. He is despised and ignored in the community. In every community the concept of 'face' is an important element in social interaction. The 'face' determines status and the role of an individual in a society. A trustworthy leader will fulfill the promises given to people and will give individual consideration to meet their people's needs. This is consistent with the individual consideration dimension of transformational leadership.

In addition to the above, Weber (2009) wrote that business relationships built on trustworthiness and honesty is the foundation of ethical commerce. Apart from the values, there are five basic social relationships which can be applied by transformational leaders in the organizational context and business relationship. The five basic social relationships are: between father and son, elder brother and younger brother, ruler and ruled, husband and wife and friend and friend. Among the five, only the last relationship is equal in nature (Tu, 1998b, cited in Slote & DeVos). These social relationships takes various forms in an organizational context and transformational leaders at times will be in the roles of father, elder brother, ruler, husband or wife and also as a friend.

In the Confucianism values the relationship is governed by trust and being faithful. Thus, building trustworthiness among various stakeholders is crucial. In Confucianism, there are two "golden rules" that govern all types of social interpersonal relationships (Tu, 1998a). The first deal with consideration and reciprocity and the rule is interpreted as: "Do not do to others what we would not want others to do to us". Thus, this develops on trust among stakeholders and anything good for one party might not be so for the other. The other rule is: "In order to establish ourselves, we must help others to establish themselves; in order to enlarge ourselves; we must help other to enlarge them". So, Confucianism teaches that adopting the collective and not competition is the way to go. This brings us to another Confucian value which is related to the building of social and economic network relationships (guan xi). In the teaching of Confucianism, this is highly valued and is also good for business development as a transformational leader develops business network through his or her charisma. Relationship and trust (xin yong) are above formal contractual relationship (Micklethwait, 1996; Weidenbaum, 1996a, 1996b; Yeung & Tung, 1996).

6. CONFUCIANISM AND VALUE BASED TRANSFORMATIONAL LEADERSHIP

The Confucianism values provide relevance for transformational leadership development in the Chinese context. The above leadership values need to be integrated into the transformational leadership development trainings in Chinese organization as it is not only relevant to the Chinese culture but also universal.

One of the pillars found in Confucianism is the upholding of harmony, commonality and integrity. This forms the notion of building and creating social relationship by focusing on common and collective goals and objectives hence avoiding factionalism. Cheung and Chan (2008) reported that in certain organizations in Hong Kong, collectivism and harmony are translated into regular and frequent contacts or meetings with mass personnel or staff members in the organization. In addition and in relation to offering a harmonious environment, there are organizations that encourage every employee a shareholder through offering them opportunity to buy and own the organization's shares. Therefore the value based Confucianism transformational leadership advocates a collective approach and interest. In contrast, western transformational leadership values focuses more on individualism.

Confucian style of leadership favors the hierarchy of authority and structured relationships. This is defined in the parent-child, the ruler-subject and superior-subordinate relationships. The relationships are sort of reciprocal in nature in which the child would serve their parents or the ruled being submissive to their rulers and in return, they would treat their children or the ruled with kindness (Greenhalgh, 1994; J. Lee, 1996; Wong, 1985; Xing, 1995). According to Bond (1992), having a hierarchy of authority in itself is insufficient as authority needs to be delegated and the subordinates who are made responsible must receive training. This leads to the idea that organizations practicing Confucian values are to uphold more centralized decision making process. The leaders are expected to make the best decisions considering the contributions by the subordinates. This is in relation to one of the Confucianism virtues, which is respect. Likewise, Confucianism transformational leadership values treasures equality among the personnel in the organization with respect and care. The preference is based on the ideas of cooperation, coexistence and mutual support.

Wong (1985) reports that Chinese businesses that adopt Confucianism practice non-aggressive stance on management style. They placed trust and friendship highly. This is consistent with the Confucianism transformational leadership values which emphasized that "in order to establish ourselves, we must help others to establish themselves; in order to enlarge ourselves, we must help others to enlarge themselves" (Tu, 1998a). Moreover, negotiation and mediation are used to settle any disputes that arise between management and employees, between managers and employees and also among employees within the organization. Similarly in the western transformational leadership approach, technique such as the use of charisma to transform followers is adopted (Ardichvili, 2001; Bass, 2002).

Bond (1992) asserted that in organizations where Confucianism is the corporate ideology, there are less formalized and detailed rules concerning how organizations are to operate. An example of such rules may include checking of medical certificates for those who are on medical leave by the managers. Formalized and written agreements are taken as being destructive as individuals are supposed to possess a certain degree of trustworthiness or fidelity in them while the superiors are expected to have confidence in them. Confucius seldom discussed matters related to law and legislation as they are depicted as "punishments". Hence in organizations where Confucianism is upheld, the superiors' or colleagues' verbal agreements are noted as binding where written rules are just regarded as for formality.

Low (2008) asserted that the Golden Rule of "not to do unto others what one does not want others to do unto oneself" is an important transformational leadership values in managing the organizational stakeholders relationship. This principle is very crucial not only for leadership success but for creating and maintaining harmonious employment relations in the industry. Therefore, the transformational leaders are to move away from being self-centered and become less self-centered and recognize their responsibilities to their stakeholders and make decisions that reflect these responsibilities.

7. CONCLUSION

In this paper the relevance of Confucianism for value based transformational leadership has been explored. In summary Confucianism promotes an inside-out approach to transformational leadership. This means leaders must first strengthen their values to rightness of their actions. The leader needs to straighten his or her character and continuously refine it through practice. This requires commitment to self discipline and self restrain. Walking the talk and talking the walk is also one of the key requirements for effective leadership, as it is often said that a man who is full of words, but not deeds is like a garden full of weeds.

The philosophy of Confucius has in one way or another affected and influenced the behaviors and thoughts of the Chinese people. Even though the environment has changed due to globalization and advancement of technology, the effects of Confucianism cannot be ignored as it has permeated the Chinese society in many parts of Asia for the last 2,500 years. These Confucianism vales have a lot to contribute to the development of transformational leadership. The prospects of analysis of Confucianism in other areas of leadership and management such as strategic management, organizational behavior, human resource management and employment relations can be considered for future research.

ACKNOWLEDGMENT

The authors are thankful to their colleague Professor Peter Yu Kien Hong for his feedback and suggestion to the original version of this paper.

REFERENCES

Ardichvili, A. (2001). Leadership styles of Russian entrepreneurs and manager. *Journal of Development Entrepreneurship*, *6*(2), 169–187.

Bass, B. M. (1985). *Leadership and Performance Beyond Expectation*. New York: Free Press.

Bass, B. M. (1990). From transactional to transformational leadership: learning to share the vision. *Organizational Dynamics*, *18*(3), 19–31. doi:10.1016/0090-2616(90)90061-S

Bass, B. M. (2002). Cognitive, social, and emotional intelligence of transformational leaders. In Riggio, R. E., Murphy, S. E., & Pirozzolo, F. J. (Eds.), *Multiple Intelligences and Leadership* (pp. 105–118). Mahwah, NJ: Lawrence Erlbaum.

Bass, B. M., & Avolio, B. J. (1994). *Improving Organizational Effectiveness through Transformational Leadership*. Thousand Oaks, CA: Sage.

Bass, B. M., Waldman, D. A., Avolio, B. J., & Bebb, M. (1987). Transformational leadership and the falling dominoes effect. *Group and Organization Studies*, *12*(1), 73–87. doi:10.1177/105960118701200106

Bhargava, S. (2003). *Transformational leadership: Value-based management for Indian organizations*. New Delhi, India: Response Books.

Bond, M. H. (1992). *Beyond the Chinese face: Insights from Psychology* (4th ed.). New York: Oxford University Press.

Burns, J. M. (1978). *Leadership*. New York: Harper.

Chakraborty, S. K. (1993). *Managerial Transformation by Values: A Corporate Pilgrimage*. New Delhi, India: Sage.

Chakraborty, S. K. (1995). Wisdom leadership: leading self by the SELF. *Journal of Human Values*, *1*, 205–220. doi:10.1177/097168589500100205

Chakraborty, S. K. (1999). *Wisdom Leadership: Dialogues and Reflections*. New Delhi, India: Wheeler Publishing.

Chakraborty, S. K., & Chakraborty, D. (2008). Spirituality. In *Management: Means Or End?*Oxford, UK: Oxford University Press.

Collins Cobuild. (2009). *Collins Cobuild Advanced Dictionary.*

Greenhalgh, S. (1994). Re-orientalizing the Chinese family firm. *American Ethnologist*, *21*(4), 746–775. doi:10.1525/ae.1994.21.4.02a00050

Hofstede, G. (1991). *Culture and Organisations: Software of the Mind; Intercultural Cooperation and its Importance for Survival*. London: McGraw-Hill.

House, R. J. (2004). *Culture, Leadership, and Organizations: The GLOBE Study of 62 Societies*. Thousand Oaks, CA: Sage.

Jaques, E., & Clement, S. D. (1994). *Executive Leadership: a practical guide to managing complexity*. Cambridge, MA: Carson-Hall & Co. Publishers.

Kejriwal, A., & Krishnan, V. R. (2004). The Impact of Vedic Worldview and Gunas on Transformational Leadership. *Vikalpa*, *29*(1), 29–40.

Kreitner, R. (1995). *Management* (6th ed.). Boston: Houghton Mifflin Company.

Krishnan, A., Muniapan, B., Lew, T. Y., & Kong, E. E. F. (2004, December 13-15). *Exploring the Extent of Transformational Leadership in the Context of Miri Entrepreneurs.* Paper presented at the ASEMAL 4 Conference, Penang, Malaysia.

Krishnan, V.R. (1990, January 11). Transformational Leadership and Vedanta Philosophy. *Economics Times,* p. 7.

Krishnan, V. R. (2000, September). Training Programs on Leadership: Do They Really makes a Difference? In *Proceedings of the Seminar on "Role of HR: A New Agenda",* Delhi, India. Retrieved from http://www.geocities.com/rkvenkat/

Krishnan, V. R. (2001). Can Indian worldview facilitate the emergence of transformational leaders? *Management and Labour Studies*, *26*, 237–244.

Krishnan, V. R. (2003). Modernization without demolishing cultural roots: the role of transformational leadership. In J. Gifford & G. Zezulka-Mailoux (Eds.), *Culture and theState, Volume IV (Alternative Interventions)* (pp. 164–173). Edmonton, Alberta, Canada: Canada Research Chairs Humanities Studio, University of Alberta.

Krishnan, V.R., & Mulla, R. Z. (2009). Do Karma-Yogis make better leaders? Exploring the relationship between the leader's Karma-Yoga and transformational leadership *Journal of Human Values, 15* (No 2, 2009), 167-183.

Lee, J. (1996). Culture and management – A study of small Chinese family business in Singapore. *Journal of Small Business Management, 34*(3), 63–67.

Low, P. (2008). Confucian ethics & social responsibility – The Golden Rule & responsibility to the stakeholders. *Ethics & critical thinking journal, 4*, 46-54.

Maher, K.J. (1997). *Gender – Related Stereotypes of Transformational and Transactional Leadership.*

Maruyama, M. (1994). *Mindscapes in Management: Use of Individual Differences in Multi-Cultural Management.*

McDevitt, R. (2007). Confucianism – Understanding and Applying The *Analects* of Confucius. *Education About Asia, 12*(1).

Micklethwait, J. (1996). The limits of family values. *Economist, 338*(7956), S10–S18.

Miles, L. (2006). The Application of Anglo-American Corporate Practices in Societies Influenced by Confucian Values. *Business and Society Review, 111*(3), 305–321. doi:10.1111/j.1467-8594.2006.00274.x

Muniapan, B. (2006, December 19-21). *Can the Bhagavad-Gita be Used as a Manual for Management Development of Indian Managers Worldwide?* Paper presented at the 5th Asia Academy of Management Conference, Asian Management: Convergence and Divergence, Tokyo.

Muniapan, B. (2007). Transformational Leadership Style Demonstrated by Sri Rama in Valmiki Ramayana. *International Journal of Indian Culture and Business Management, 1*(1), 104–115. doi:10.1504/IJICBM.2007.014473

Muniapan, B. (2008). Perspectives and Reflections on Management Education in Malaysia. *International Journal of Management in Education, 2*(1), 77–87. doi:10.1504/IJMIE.2008.016232

Muniapan, B. (2009, July 5-8). *The Bhagavad-Gita on Leadership for Good Governance and Sustainable Development.* Paper presented at the 15th Annual International Sustainable Development Research Conference. "Taking up the Global Challenge: Analyzing the Implementation of Innovations and Governance for Sustainable Development", Utrecht, The Netherlands.

Muniapan, B. (2010). Perplexity, Management and Business in India. In Lowe, S. (Ed.), *Managing in a Changing Times: A Guide to Perplexed Manager* (pp. 317–346). Thousand Oaks, CA: Sage.

Muniapan, B., & Dass, M. (2008). Corporate Social Responsibility: A Philosophical Approach from an Ancient Indian Perspective. *International Journal of Indian Culture and Business Management, 1*(4), 408–420. doi:10.1504/IJICBM.2008.018622

Muniapan, B., & Dass, M. (2009). An Indian Leadership Perspective from Literature Works of Poet Kannadasan. *International Journal of Indian Culture and Business Management, 2*(4), 326–340. doi:10.1504/IJICBM.2009.023552

Noorderhaven, N. (2000). *Hermeneutic Methodology and International Management Research* (Occasional Paper 2000/2). Uppsala, Sweden: Department of Business Studies, Uppsala University.

Parashar, S. P. (2008). Winning over equals: insights from Bhagwad Geeta (Krishna–Arjuna framework). *International Journal of Indian Culture and Business Management, 1*(3), 354–359. doi:10.1504/IJICBM.2008.017790

Roka, P. (2006). *Bhagavad Gita on Effective Leadership: Timeless Wisdom for Leaders.* Bloomington, IN: iUniverse.

Ross, S. M., & Offermann, L. R. (1997). Transformational leaders: Measurement of personality attributes and work group performance. *Personality and Social Psychology Bulletin*, *23*(10), 1078–1086. doi:10.1177/01461672972310008

Satpathy, B. (2006). Transformational Leadership in the Bhagavad-Gita. *The Journal of Indian Management & Strategy.*

Satpathy, B. (2007). Hierarchy of Values in the Bhagavad-Gita. *Journal of SETU*, 15-27.

Satpathy, B., & Muniapan, B. (2008). The Knowledge of "Self" from the Bhagavad-Gita and its Significance for Human Capital Development. *Asian Social Science*, *4*(10), 143–150.

Sharma, G. D. (2001). *Management and the Indian Ethos*. New Delhi, India: Rupa & Company.

Sharma, S. (1996). *Management in New Age: Western Windows Eastern Doors*. New Delhi, India: New Age International Publishers.

Sharma, S. (1998). Enlightened leadership in Indian ethos: the way of the theory K. *Management and Change*, *2*, 93–104.

Sharma, S. (1999). Corporate Gita: lessons for management, administration and leadership. *Journal of Human Values*, *5*, 103–123. doi:10.1177/097168589900500203

Sharma, S. (2002). Corporate leadership model: an Indian model for corporate development and ethical leadership. In Pareek, U., Osman-Gani, A. M., & Rao, T. V. (Eds.), *Human ResourceDevelopment in Asia: Trends and Challenges* (pp. 291–296). New Delhi, India: Oxford and IBH.

Sharma, S. (2003). Towards corporate VEDA: Indian ethos and corporate development. *Journal of Human Values*, *9*, 163–172. doi:10.1177/097168580300900207

Simic, I. (1998). Transformational Leadership – The key to successful management of transformational organizational changes. *The Scientific Journal FACTA UNIVERSITATIS*, *1*(6), 49–55.

Slote, W. H., & DeVos, G. A. (Eds.). (1998). *Confucianism and the family.*

Theodore de Bary, W. (1960). *Sources of Chinese Tradition* (1st ed.). New York: Columbia University Press.

Theodore de Bary, W., & Bloom, I. (1999). *Sources of Chinese Tradition* (2nd ed.). New York: Columbia University Press.

Tichy, N., & Devanna, M. A. (1986). *The Transformational Leader*. New York: John Wiley.

Tu, W. (1998a). Joining east and west. *Harvard International Review*, *20*(3), 44–49.

Tu, W. (1998b). Probing the "three bonds" and "five relationships" in Confucian humanism.

Weber, J. (2009). Using exemplary business practices to identify Buddhist and Confucian ethical value systems. *Business and Society Review*, *114*(4), 511–540. doi:10.1111/j.1467-8594.2009.00352.x

Weidenbaum, M. (1996a). The Chinese family business enterprise. *California Management Review*, *38*(4), 141–156.

Weidenbaum, M. (1996b). The bamboo network. *Executive Speeches*, *11*(2), 5–6.

Wong, S. L. (1985). The Chinese family firm: A model. *The British Journal of Sociology*, *36*(1), 58–72. doi:10.2307/590402

Xing, F. (1995). The Chinese cultural system: Implications for cross-cultural management. *Sam Advanced Management Journal*, *60*(1), 14–20.

This work was previously published in International Journal of Asian Business and information Management, Volume 1, Issue 3, edited by Dasho Karma Ura and Patricia Ordóñez de Pablos, pp. 10-22, copyright 2010 by IGI Publishing (an imprint of IGI Global).

Chapter 4
Study on Cultural Commercialization and Public Consumption:
The Case of China

Yuhong Li
Beijing Union University, China

Jingyuan Zhao
Beijing Union University, China

Han Weixi
Beijing Union University, China

ABSTRACT

This paper demonstrates the commercializing trend and phenomenon of the contemporary culture, arguing that as an intangible commodity, culture has special and limited commercialization that it's only a superficial homogeneity brought by cultural commercialization. Cultural commercialization is the release of cultural intensity and connotation. This paper illustrates the industrialization process of Chinese culture, probes into the cultural consumption demand and structure in contemporary China, and analyzes the cultural consumption phenomenon of the Chinese public. This paper draws a conclusion on the destined cultural commercialization and the development trend of cultural consumption in China.

1. INTRODUCTION

There are many concepts of cultural economy mainly defined by experts, scholars, international authorities, and the governments that promote cultural economy. This is a consensus held by international organizations as the United Nations Conference on Trade and Development (UNCTAD) and the United Nations Educational, Scientific and Cultural Organization (UNESCO), because to the concept of culture there are different understandings of extensive social ideologies and

DOI: 10.4018/978-1-4666-0276-2.ch004

spiritual fields when viewed from the development of society and history.

From culture a lot of economic forms, including "creative industry", "content industry", "copyright industry", "leisure industry", "experience industry", "economy of attention", and "cultural trade" etc., are arising. This shows the dominant position of cultural economy in the future's economic development and marks the coming of cultural economy age. In this new age the traditional economics is challenged and people are re-defining various economic factors and their interactive forms, re-establishing the forms of economic organization and production organization, re-allocating resources, re-constructing value system and assessment criteria, and re-organizing industrial chain. The development of cultural economy wants the theoretical and practical thinking in a systematic way, such as the relationship between culture/creativity and commodity/commerce, the degree of cultural industrialization, the risks challenging cultural economy, and the realistic problems brought by the global culture and economy (Hartley, 2007).

Both agricultural economy and industrial economy are based on the occupancy and consumption of natural resources. However, countries of today's world are gradually separating from the economic growth mode with land and machine as the carrier and the consumption of natural resources and low labor pay as the cost. The economic mode dominated by culture, with an intellectual basis of creativity, innovation, and originality, has released more powerful energy than mechanical production and more fortune than tangible resources. Culture is now creating a new economic structure and mode (Li & Lu, 2007). The financial crisis that exploded in 2007 and lasted to today is rightly the puzzle and problem of new economic form that makes finance, IT, education, and cultural creativity as the new economic growth points. Although there are increasing doubts upon the intangible industry that causes economic bubbles, a fact that cannot be denied is that human economy is now progressing toward the age of operating culture and wisdom.

Industrial society brings us economic prosperity, and makes people absorb great satisfaction in the supply of sufficient materials, till such materials become overmany Technical progress and improved production efficiency lowered the price of tangible commodities. In 1950s, the major developed countries have realized that they have stepped into the post-industrial time. They became aware of the knowledge economy brought by rapidly developing computer and network technology. In 1980s, some countries chose to adjust and transform economy actively, for example America reformed the traditional industries of clothes and household appliance, and named the smokeless industries including computer, software, network, education, finance, and video publication as "rising industry". The United Kingdom (UK) and Australia in sequence positioned cultural industries or creative industries as the core industries to be developed by the governments. During this period cultural contents and value began to penetrate into all the industries of the society. By adjusting and changing economic structure, the consumption and production supply of these countries have had essential changes. The desire of people to get satisfied in the consumption of food and clothes for "survival" is gradually becoming lower than their desire to get satisfied in the consumption of education, culture, entertainment, and sport for "development". "Every day the world's creative economy creates 22-billion dollars, which is still increasing by a ratio of 5%." In some countries, the increase is faster. America has a speed of 14%, while UK has a speed of 12% (Li, 2008).

Viewing the whole world, the infinite cultural products, marketing, and service of developed countries have drawn the focus of the world's people, and prevailed in the world in the form of a huge tide of intellectual economy. The developed countries' cultural industries have by their unique orientation, field, and style been rapidly developing and produced huge economic benefits. As the developed countries now are still the representative of the advanced politics, economy, and culture of the world, we may say that the world of the 21st century has entered the age of cultural economy (Zou & Xu, 2006).

2. CULTURAL COMMERCIALIZATION

2.1 Culture and Commerce

Today culture has contributed a part of influence and offerings to human life scarcely seen in its whole history. The most important influential factor is cultural commercialization. The governments began to realize that culture could bring human beings great spiritual enjoyment as well as creating unprecedented huge fortune. Today's culture is endowed with multiple meanings of material and spirit, ideology and commercial value etc.

Traditional scholars and schools are inclined to believe that culture as a thing of grace, solemnity, and ideology shall be far away from commerce. Many countries take prudent steps when dealing with cultural commerce. For example French government adopts measures as trade barriers to block America cultural commerce represented by films of Hollywood, potato crisps of Mcdonald's, and chips of Intel. They worried that such American culture will influence and transform French culture. Japan also adopts the approaches in education and diplomacy to alter foreign culture and paint it with Japanese features (Hartley, 2007). In China the cultural resources are almost under the sole control of the government. Recently China government has undertaken extensive and diversified construction of cultural industrialization and cultural system reform. However, in China only attempts are made on connecting cultural system and rules with commercialization. No clear development logic and plan has been developed yet. However, cultural commercialization has been a trend of the world's economic development that cannot be blocked. Culture breaks off the barriers among regions and nations, and builds a digital media platform with its diversified contents and rich expression forms for it to go to the world's corners quickly. Cultural commercialization has greatly promoted the publicity and development of culture, and become an economic pillar industry and powerful productivity that drives forward social progress (Santos, 2008).

"Culture" in cultural economy will have the property of intangible commodity, be the object of commercial trade, and be the carrier of commercialization? Is that true? Cultural commercialization quickens cultural merging and consolidating around the world, but it may also floods everywhere to lower the public culture's value, overthrow traditional culture, invade national culture, shake or make chaos in national ideology, and impact the basic principles and faiths of human beings in cultural philosophy, cultural aesthetics, cultural ethics, and cultural history.

Creativity contained in culture is an expression of human wisdom. It is the precious fortune of human beings that can be used to create huge commercial value. Commercialization activates the space of creativity and blurs more the more the boundaries between reality and vision, foretime and presence, art and commonness, and fact and story. The truth of life is that the life is now being virtualized and coming to a virtual and visional world. Cultural creativity cultivates new people, and creates "culture" in cultural economy. Different from other commodity forms, cultural commercialization has made choices. Some parts of culture can be sold, for example cultural forms and expression techniques. But some parts of culture cannot be commercialized, for example national spirit and national ideology.

Cultural commercialization puts forward two major subjects. The first is to choose and determine which parts of culture can be sold, for example cultural forms and expression techniques, and which parts of culture cannot be commercialized, for example national spirit and national ideology. The commonweal nature of culture decides that partial management on culture shall be implemented by public or state-owned entities. But the commercial nature of culture promotes culture to be operated at the market. The second is that the development of cultural commodities shall take into account of moral criteria, namely such development shall meet human beings' cultural aesthetics, cultural ethics, and cultural morality, and the misuse of human weakness shall be avoided.

2.2 Contemporary Features of Cultural Commercialization

2.2.1 Diversification of Cultural Commercialization

The completely new publicity approaches and media forms as news release, global premiere, worldwide talent show, live show, and 24-hour air reporting etc. brought the world into the same cultural atmosphere wherein different nations, countries, and races are living. A regional popular fashion may cause a global tide which the world's people follow. Such regional fashion may even become a dominant culture of the world. Visual impact brings profound feelings to the soul of people, who are submerged in a sea of information, without time to stop for thinking and facing all the things around calmly. It seems that the whole world is drawn to the same direction which the puzzled people follow in blindness. People at remote places are unwilling to keep their quiet and simple life, and the urban places are filled with the so-called internationalized symbols and elements. Cultural phenomena induce cultural thinking that influences the social living style, the communication style between humans, and even the relationship between humans, and between human and nature.

Cultural commercialization brings no cultural homogeneity to the contemporary youths. On the contrary, their choices are diversified when they are receiving massive information and media publications. As they are of different countries, races, ages, and genders, they have their own understandings and definitions when to absorb various cultural elements. They pursue their own feelings of life. The youths seize the key tide of the times, namely they use the rich materials of the world to exercise their imagination and write stories and create original life. Cultural tendency may have no limitation of country or region, as possibility the young people of both America and China like Michael Jackson and Hua Mulan. However, such cultural tendency means not cultural homogeneity. But on the contrary it means the recognition diversification of penetrated cultures. This recognition will certainly stimulate more penetration of cultures and the prosperity of diversified cultures. In living style the youths are more willing to break off the homogeneity, despise the traditional creeds for life, living style, and belief of life, and parade their own personality.

2.2.2 Network Virtual Nature of Cultural Commercialization

The age of cultural economy owns the storytelling ability better than any time of the past. The invention of computer communication technology and network blurs the boundary of real life. Yesterday's life may be written into a blog story today. Both real and virtual network options are available for the community, house, medical treatment, car, and shopping in life. Everything of the reality, including psychological activities and even thinking can be re-produced or exaggerated at network. A student may choose not to attend his graduation ceremony but make his person appear at the group photo by using a Photoshop technique. People's imagination is enlarged without limit in real life. Everything true may turn out to be false. Truth and falseness are no more the focus of discussion, for people's real life is to create a virtual world. Never in any times before people had ever lived in so rich a tiered space that they jump to and fro between virtual and real worlds.

Virtual life has become the part of some people, especially the youths.

What the virtual life will bring to human beings after creations and interests of life, paradise in dreams, and the amusement to play with life? Will the traditional cultural principles such as ethics, morality, and faithfulness also become unreal? Will people lose mutual trust and find their emotions sleepy in the virtual world? Besides, the digital network media brings the commercial secrets, trademarks and signs, industrial designs, and writings and inventions to extensive regions and countries very fast. People know not the origin and source of such information. Intellectual inventions and creations are disclosed and knowledge publicized so soon that, is it necessary to protect the interests and rights of the inventors? (Zhang, Yin & Li, 2007).

2.3.3 Tradition and History of Cultural Commercialization

Cultural commercialization made people stretch their arms to the most precious cultural fortune of human society, namely traditional culture and historical relics. The problem of intellectual property rights concerning culture stands for more urgent solution in some countries with lasting history and tradition. In such countries their national traditional culture and historical relics are publicized, embezzled, and copied around the world without the protection and limit from any intellectual property rights (Xu, Bi & Qi, 2007). The cultural creativity industry presents the worldwide problem of intellectual property rights, namely by what style the world receives and publicizes the culture of different countries and nations, and by what approach the world acknowledges the traditional, primitive and sub-cultural value. The prosperity of cultural creativity industry endows intellectual property rights with more functions in addition to the existing commercial value, namely such rights can be used to protect the cultural niches of the world and to respect and inherit the unique cultures of different nations (Jiang, 2007).

As the Aborigines culture of Canada, folk art of Mexico, handcraft of American Indian, and the traditional cultures of the world's other countries and nations, the traditional culture and historical relics of China contains huge interests that have strategic significance and value for the world's economic development and the global fusing of cultural creativity industries. Chinese Nations have lasting history and profound traditional culture, which reflects various aspects of historical vicissitudes, consolidation of nations, social life, and cultural creations. Parts of the tangible and intangible traditional culture and historical relics formed in the long history of China have been protected well. Some parts are recreated and innovated into new products. However, a great number of such traditional culture and historical relics, such as ancient relics, folk dances and songs, performances, local dramas, original culture, traditional festivals, inherited recipes of biological and genetic types, and Chinese traditional medicine, are being lost, stolen, and overthrown. They wait for urgent protection of intellectual property rights (Li and Bai, 2008). In the age of cultural economy, the key is to construct and operate a sound system of intellectual property rights, which shall be used to limit and protect national traditional culture and historical relics that are being publicized and copied continuously around the world. One of the world's important issues is to expand the existing legal system, innovate laws and rules, and build an intellectual property rights system beneficial for the protection of traditional culture (Wu, 2007).

3. THE PUBLIC'S CULTURAL CONSUMPTION

3.1 Cultural Consumption Meets Spiritual Demand

As an art of grace, culture had ever been far apart from commerce, without intersection with consumption. Today, people are pride of their spending large money and excessive consumption and their enjoyment of the vainglory from luxury goods. However, they are not filled and satisfied in heart. The immense spiritual loss makes people find it difficult to fill the emptiness in their mind with whatever fortune and luxury. People are not happy even they spend money as soil and lead an extremely luxury life. People lost their interest in commodities when they become rich in material. The greatest salesman cannot retain their steps moving away. The public is tired of the commodities piled on shelves, and deafened to the noisy hawks. People walk around commercial streets, supermarkets, and first-class malls, meeting strangers, possibly with no goods in hand, for they are to enjoy the leisure time, the free life style without a purpose.

When the culture is viewed from the commercial angle, the solemn culture opens its closed door. The culture came to own the property of public entertainment, leisure, feeling, interest, interaction, and experience. Thus the public culture is formed. Among the most popular consumption ways offered to the public, as music of banrock, coffee of Starbucks, meeting at Be For Time, chat room of QQ, and club of killing game, culture is the only authentic product for consumption, and other services such as food supply and Internet access are just auxiliary items. People in the age of cultural economy cares not the product itself, but they care what the product means, where it is bought, how it is bought, and with whom it is bought.

Cultural creativity came to the fertile soil of commercial consumption. Culture is no more the proprietary thing of literary circles and artists. Cultural inspirations came to the minds of businessmen, salesmen, and entrepreneurs and formed countless business ideas. The before separated culture and commerce are now consolidating and forming new public culture and public consumption style gradually. Typical features of the environment of cultural economy are that the ideas of intelligence and wisdom are favored and adored by the market far more than any tangible thing; the designs, trademarks and patents with intellectual property rights, and other forms of products, inventions, and discoveries are all marked with "first-class"; their inventors and owners are the most favored people of this age. In such a situation the public is willing to pay more for the commodities of cultural creativity. In addition to their will to pay more money, consumers may want to spend more time and energy to keep interest after consumption; they even devote their emotions and make interactive communications.

3.2 Forms of Cultural Consumption

Group Cultural Consumption

Cultural consumption groups consumers of the same taste or interest and they consume their communication and discussion process, and get their memory, feeling, experience, and thinking.

Interactive Cultural Consumption

The consumption process is the process of participation. Culture has a mind, and people who consume culture have identification and recognition in their emotions and thinking. By cultural consumption they share and communicate with others, get satisfied, realize their personality, and find themselves successful and belonging to a group.

Experience Cultural Consumption

Cultural consumption can induce emotions in heart. People expect different things such as pleasure, compassion, excitement, sadness, courage, and self-test. They consume culture to experience such things and get satisfied in heart.

Entertainment Cultural Consumption

Digital art and multi-media technology richened the entertainment elements of culture. The cultural entertainment products as Karaoke, theme amusement parks, cartoon games, movies and performances move the audience to laugh and release their anxiety in work, and endow them with an optimistic and graceful attitude towards life.

Leisure Cultural Consumption

Leisure is a kind of state. Modern people suffering high spiritual pressure want the ambience to adjust their psychology and be relaxed, for example informal clothes, pleasant music, flowing water in countryside, and bridges and villages.

4. THE PROCESS OF CULTURAL COMMERCIALIZATION IN CHINA

Chinese tradition must be modernized, and the alien ideas shall be localized in China. Finally Chinese traditional culture and western civilization shall be consolidated in a creative way (Fu, 2007). In China, the concept and scope of "creative industries" has replaced "cultural industries" gradually. This replacement is based on the interpretation that creativity is private property. This replacement represents an important progress in politics (Bilton, 2010). Zhao (2009) claims that in China, cultural and creative industry is defined to be a industry cluster with intrinsic connections between firms, providing the public with a cultural experience, taking creation, creativity, innovation as the fundamental means, taking cultural content and creative results as core values, taking intellectual property achievement or consumption as transaction characteristics. The process of cultural commercialization in China is basically the same as the process of cultural industrialization in China (Zhao, 2010). The stages of exploration, expansion, and furthering of cultural commercialization are rightly the stages of start, forming, and prosperity of cultural industries in China, as shown in Figure 1.

For a long period after the foundation of the People's Republic of China in 1949, in a planned system the government controlled culture. From the Third Plenary Session of Eleventh Central Committee of CPC, basically the development process of cultural industries of China can be divided into the following three stages as following:

Figure 1. Stages of the process of cultural commercialization in China

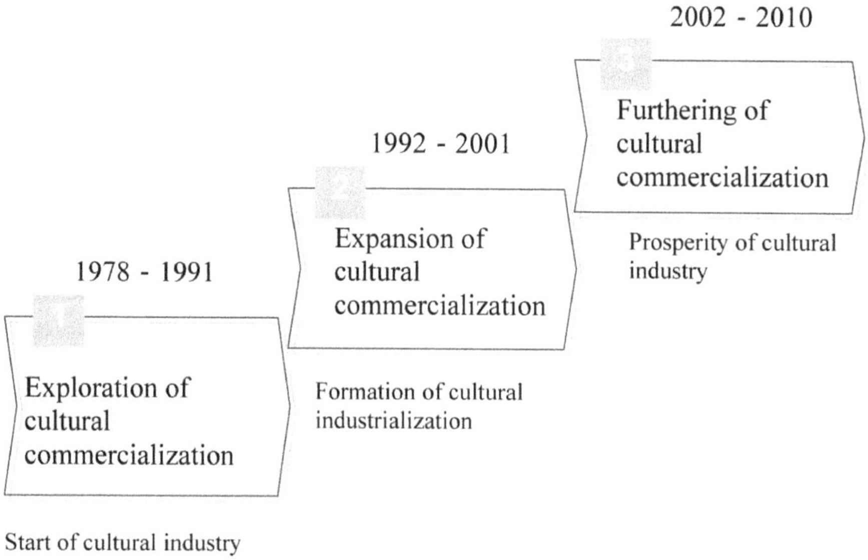

First Stage (1978-1991): The Stage for Initial Exploring and Building of Cultural Industries

After the Third Plenary Session of Eleventh Central Committee of CPC in 1978, China implemented the policy of reforming and opening to the world; thus Chinese people began to enjoy a life richer in materials and, influenced by foreign popular culture and public culture, Chinese entertainment industry began to recover and prosper increasingly.

In 1980 music teahouse appeared in Guangzhou. The entertainment industry began to resume. Later the profit-seeking entertainment activities such as video playing, dance performing, and song singing are increased gradually. In 1984 the Mainland China's first consultation company was founded in Shanghai. This is a market-oriented company that mainly sells spiritual products with knowledge and information as its features. At the beginning of 1980s, China's cultural industries started quietly.

During this stage China's advertising industry appeared and was booming. Soon people recognized the function of advertising as an agency in publicizing a product. More and more time and space of various media forms such as radio, newspaper, and television are devoted to advertising, which has quality and level improved step by step. The sport industry is smokeless industry, and it has a lot of associated industries, such as tourism, food industry, and gymnasium building etc. Meanwhile the development of sport industry can enhance people's body strength and spirit them up, and show a country's comprehensive strength in peacetime. So it became one of the important industries developed by many countries. China seized the chance that at Los Angeles Olympics 1984 Chinese athlete won the first gold prize, promoted the development of sport industry vigorously. In 1990 China hosted the eleventh Asia Games successfully. From this success China began the industrialization and market-oriented operation of sport. For example, the activity of Asia Games donation lottery drawing in fact pioneered the later issue of sport lottery. However, in conclusion at this stage the cultural industries of China are of exploration type, and it only occupied a small place with small influence in the whole social and economic system of China (Liu, 2007).

Second Stage (1992 - 2001): The Cultural Industries Began to Play an Important Role in National Economy

In 1992 the 14th National Congress of CPC clearly set the plan to build a socialist market economic system based on the facts of China. The building of market-oriented economic system constructed the foundation for the healthy development of cultural commercialization. In the same year the Central Committee of CPC and the State Council issued the *Decision on Accelerating Development of the Tertiary Industry*, which calls cultural industries as a member of tertiary industry formally. The cultural sectors are transferred from sectors wanting financial payout to sectors of production type, and thus to be prepared in policy and system for the development of cultural industries. This time the development of cultural industries entered a stage to have formal rules set up and industrial structure constructed. In 2000 the Central Committee of CPC in the proposal for the "Tenth Five Year" development planning mentioned the development of cultural industries for the first time in formal document of the state's central organ of power.

At this stage, newspapers exercised the power of cultural industries for the first time. As people have richer part-time activities and generally improved education level, newspapers became the important object of cultural consumption. Many newspapers increased their pages and produced special issues for culture and entertainment. Many newspaper offices created their own morning issue, evening issue, additional issue, and enlarged weekend issue, week issue, and auxiliary issue to increase circulation and attract readers, occupying and expanding market share. Meanwhile many newspapers grouped their target readers and specialized the contents they published to retain target readers in order to win them a competent position at the market of intensive competition. The expanded circulation of newspapers allured many advertising owners/agencies to put advertisements, which prospered newspaper industry and developed advertising industry. The movie industry of this period also made beneficial attempts in industrialization, and progressed the march to international market. The financing mechanism of movie production changed. Some movie companies and cultural companies went to the market consciously to absorb state-owned or private capital by stock system to invest in movie production and carry out commercialized operation. The movies produced during this period to celebrate the New Year harvested high turnouts, achieving success in commercialization and industrialization of movies produced in China. The production of TV plays also made a great progress. Many TV plays were produced and broadcasted separately in a new mechanism of market-oriented operation. Television stations purchased TV plays expected of rosy prospect to win high advertising income by high audience rate. Thus the television industry formed a healthy circulation. During this period, the tourism also was developed healthily. The government adopted various measures to promote the development of tourism. Three long holidays, "May Day", "National Day", and spring festival, added three golden highlights to the development of tourism and prospered the domestic travel routes and destinations greatly. The tourism industry enlarged service scope, improved service facilities of transport, food, and hotel, furnished quality service, and amended soft and hard environments to build true appeal for drawing domestic and foreign tourists, displaying image, and developing economy. During this period, new form of cultural industries– network industry appeared for the rapid development of information technology and the appearance of Internet. As a virtual society, the network provided new communication approach. The network economy features intensive intellect and high-tech, and is an industry of intellect and information-intensive type, also the key economic growth point in the age of knowledge economy and information technology. It is developing rapidly (Zhang, 2007).

Third Stage (From 2002 to Now): Cultural Industries Become a Pillar Support of National Economy

In 2002, the report of the 16th National Congress of CPC for the first time in its formal document stated explicitly the plan for vigorous development of cultural industries, and the definite differentiation of support to cultural undertaking and operation of cultural industries. Following the constantly rapid growth of China's national economy, the accelerated steps in overall construction of well-off society, and the continuously furthered reform in cultural system, the rising cultural industries are destined to grow fast and become the new highlight of China's economic development.

In 2007 Hu Jintao in the report of the 17th National Congress of CPC expressed the decision to "promote the great development and prosperity of socialist culture".

In Jul. 2009 an executive meeting of State Council examined and passed the *Plan to Adjust and Reinvigorate Cultural Industries* to accelerate and promote the development of cultural industries in China. The issue of the *Plan to Adjust and Reinvigorate Cultural Industries* means that the government has exalted the development of cultural industries to the level of national strategy.

At this stage the process of cultural commercialization is further developed.

5. THE PUBLIC'S CULTURAL CONSUMPTION IN CHINA

Viewed from consumption, cultural consumption has increased to over 30% in household consumption in America and some countries of Western Europe. But in China only 14.8% of household consumption is spent on cultural consumption. The public of China is now at the transition from material consumption to cultural consumption. Results from investigations in major cities show that, the ratio of cultural items and facilities in belongings of Chinese families is approaching that of developed countries rapidly. The public came to have more and more passion on the new items of cultural consumption as movie, entertainment, tourism, and network. However, viewed integrally the public's cultural consumption in China has two characteristics.

Demand of Cultural Consumption

The rapidly growing cultural consumption in China cannot keep up with the speed of economic growth. Using internationally universal criteria, a country with GDP per capita over 3,000 dollars will enter a period with rapid growth of cultural demand. In 2009 China's GDP per capita was 3,700 dollars. Authoritative researches show that when GDP per capita reaches 1,000 dollars and Engel's coefficient is 44%, the urban and rural cultural consumption shall occupy 18% of personal consumption. Estimated by this, the total amount of consumption demand of China shall be 1.09-trillion-Yuan. But the actual total demand of cultural products in China is less than 500-billion-Yuan. Additionally there are many blank points at the market of cultural consumption in China. The satisfaction degree is low. Integrally the total amount of cultural consumption of Chinese residents is too low. The satisfaction degree of residents' cultural demand is less than 1/4. Besides, the added-value of China's cultural industries is 700-billion to 800-billion-Yuan, indicating a huge potential of consumption.

Structure of Cultural Consumption

Structure of cultural consumption is a problem deserving care and research in China.

Firstly, the payout of cultural consumption is low. On the one hand, the public has a low satisfaction degree of cultural demand. People hope the market of cultural consumption will be richer and colorful. On the other hand, even in urban areas, people paid a low ratio of money for cultural consumption in the total payout of the family, not to mention rural areas. There are three reasons. Firstly, as the cities have increasingly quickened process of reforming and opening, the life quality is increased, life tension intensified, life cost higher and higher, and house, car, education, medical treatment, and insurance become huge pressure to life, the cultural consumption which is not necessary for life has to be diminished or neglected. Secondly, people may use some low-cost replacements for cultural products and items to satisfy and fulfill their demand to cultural consumption. The *Blue Book of China's Cultural Industries 2005* surveyed more than 20-thousand people in 21 major cities of China and found that the main leisure activities of Chinese urban residents are seven low-cost activities easy to take, namely: walking in a park, reading pastime books and movies, playing shuttlecock, climbing, mah-jong, planting flowers in pots, and playing machine games. Thirdly, the market of cultural consumption is still not a sound and consummate one. It has large gap to the public's requirements at quality and taste of products and projects, atmosphere of cultural environment, and level of service and management. This acts as a barrier to cultural consumption.

Secondly, the contents of cultural consumption are changing now. The traditional structure of China's cultural consumption with dominant knowledge education is now replaced by the co-existence of knowledge and intellectual consumption and amusement and entertainment consumption. Evidences show that as influenced by cultural factors of Hong Kong, Taiwan, America and Europe, and Japan, together with the rising demand of on entertainment consumption and enhanced research and development ability in Mainland

China, these three years the entertainment-type cultural consumption is growing vigorously, and the cultural consumption with dominant entertainment and leisure enjoyment become a favorite of young people. The consumption for the traditional cultural consumption items as subscribing newspapers, purchasing books, visiting museums, and watching operas is decreasing. On the one hand this shows that people are spending the consumption energy accumulated in the entertainment-type cultural consumption market that for a long period "no supply is available for demand". On the other hand, this reflects the contemporary public's recognition to the change of culture from highbrow to popular. It's right to say that according to the public's life, consumption ability, consumption motive, and appreciation habit in today's China, the leisure and entertainment consumption will for a long period be a dominant type of cultural consumption.

6. CONCLUSION

The world enters the age of cultural economy in the 21st century. Researches on cultural commercialization and the public's cultural consumption are significant. This paper demonstrates the diversification, virtual network nature, tradition, and history of the contemporary cultural commercialization. Cultural commercialization brings no cultural homogeneity. On the contrary, when to receive immense information and media publications, people will make diversified choices. The life of virtual network has become a part of the cultural life of some people, especially young people. The intellectual property rights are protection tool in the cultural and creative industries. Around the world countries have reinforced the protection on cultural niches and the respect and inheritance to unique cultures of different nations. The contemporary cultural commercialization has the characteristics of tradition and history. The process of cultural commercialization in China is basically the same as the process of cultural industrialization in China. Stages of exploration, expansion, and furthering of cultural commercialization are rightly the stages of start, forming, and prosperity of China's cultural industries.

Cultural consumption can satisfy people's spiritual demand. There are diversified forms of cultural consumption, including group cultural consumption, interactive cultural consumption, experience cultural consumption, entertainment cultural consumption, and leisure cultural consumption etc. Viewed from the angle of consumption, the public of China is changing from material consumption to cultural consumption. Integrally the total amount of cultural consumption of Chinese residents is too low, and the satisfaction degree of Chinese residents' cultural demand is less than 1/4. In the structure of cultural consumption, the payout of cultural consumption occupies a low ratio. Contents of cultural consumption are changing. The traditional cultural consumption structure with dominant knowledge education is replaced by the co-existence of knowledge and intellectual consumption and entertainment and leisure consumption.

Countries with developed cultural commercialization have a higher satisfaction degree of the public's cultural consumption demand. Currently China's cultural market is waiting for further opening, and it has a low maturity degree, with the satisfaction degree of the public's demand lower than that of developed countries. Therefore, cultural commercialization is the trend of development. There are many blank points in the public's consumption demand of China, and the potential of cultural commercialization and market is huge.

ACKNOWLEDGMENT

Thanks for the monetary support from the Beijing municipal excellent subject *Marketing Practice* and the subject *Demonstrative High Vocational*

Specialty of International Business of Beijing Union University.

REFERENCES

Bilton, C. (2010). *Management and Creativity: From Creative Industries to Creative Management*. Beijing, China: New Word Press.

Fu, Z. (2007). *Cultural Imagination and Humanist Criticism*. Beijing, China: Communication University of China Press.

Hartley, J. (2007). *Creative Industries*. Beijing, China: Tsinghua University Press.

Jiang, P. (2007). *Management on Intellectual Property Rights*. Beijing, China: Intellectual Property Press.

Li, Y. H. (2008). A Perspective Analysis into New Cultural Phenomena of Cultural and Creative Age. *China-Today Forum*, *6*, 53.

Li, Y. H., & Bai, Q. X. (2008). *Tutorial of Excellent Cultural Creativity Cases*. Beijing, China: China Economic Press.

Li, Y. H., & Lu, H. (2007). The Nature of Cultural Economy and Value of Intellectual Property Rights. *China-Today Forum*, *5*, 42.

Liu, M. Y. (2007). *Probing into Theory and Practice for Development of Cultural Creativity Industry of Beijing*. Beijing, China: China Economic Press.

Santos, E. D. (2008). *The Creative Economy Report 2008*. Beijing, China: People's Education Press.

Wu, D. H. (2007). *Study into International Protection System of Intellectual Property Rights*. Beijing, China: Intellectual Property Press.

Xu, B. T., Bi, L. Y., & Qi, M. (2007). *Protection on Intellectual Property Rights of Modern Enterprises*. Guangdong, China: Zhongshan University Press.

Zhang, J. C. (2007). *Report on the Development of Creative Industries in China*. Beijing, China: China Economic Press.

Zhang, X. M., Yin, C. L., & Li, P. (2007). *Report on the Development of International Cultural Industries*. Beijing, China: Social Science Academic Press.

Zhao, J. (2009). A survey on Beijing's cultural creative industries. *International Journal of Chinese Culture and Management*, *2*(2), 146–166. doi:10.1504/IJCCM.2009.029251

Zhao, J. (2010). *Management of Cultural Industries*. Beijing, China: Tsinghua University Press.

Zou, G. W., & Xu, Q. W. (2006). *Globalization and Development of China's Cultural Industries*. Beijing, China: Central Compilation and Translation Press.

This work was previously published in International Journal of Asian Business and Information Management, Volume 1, Issue 3, edited by Dasho Karma Ura and Patricia Ordóñez de Pablos, pp. 23-34, copyright 2010 by IGI Publishing (an imprint of IGI Global).

Chapter 5
The Interactive Relationship between Corporate Cultures, Leadership Style and Knowledge Management

Valerie (C. Y.) Zhu
Xi'an Jiaotong University, China

Linyan Sun
Xi'an Jiaotong University, China

ABSTRACT

In the knowledge economy era, increasing attention has been on the myths and secrets of knowledge management (KM). Being the core of the intangible assets of organizations, proper and effective KM is an inimitable and sustainable competitive advantage in today's competitive world. Taking this into consideration, this paper probes the dynamic relationship and impact mechanism between corporate culture, leadership styles and knowledge management in organizations based on literature review in this field. This research shows that corporate culture can serve as a propeller or inhibitor in KM. Knowledge-supportive and knowledge-sharing corporate culture has a positive influence on KM, while hierarchical and paternalistic corporate culture has a negative influence. Leadership style has a direct and indirect impact on KM through the formation and components of corporate culture. Leaders impact the organization's KM in various ways.

INTRODUCTION

With the embracement of the knowledge economy epoch, the development of science and technology and the advent of economic globalization, knowledge has become the most important strategic resources in today's fierce competition. Information management is no longer sufficient to meet the demand of knowledge for rapid growth rate, knowledge management (abbreviated as KM) as a new management mode has emerged to compensate for the insufficiency of information management. In reality, as early as 1890, Alfred Marshall had already exposed the importance of

DOI: 10.4018/978-1-4666-0276-2.ch005

knowledge within the organizations. However, his concept was so novel as not to be accepted at that time until 1990s (Quintas, 2002). Substantial research has revealed that the outcome of knowledge management will play a key role in the development of the organization performance (Bollinger & Smith, 2001; McDermott & O'Dell, 2001; Holowetzki, 2002; Chin-Loy, 2003; Chin; 2001; Crawford, 2004, Mustafa, 2002; Rowley, 1999; Gloet & Berrell, 2003). With the increasing awareness of the importance of knowledge management, research on KM has become a hot topic in academic fields. Studies on the relationship among the factors which influence the results of KM have become an important branch of study. DeTienne, Dyer, Hoopes and Harris (2004) have proposed the four most important factors which have positive influences on the effectiveness of knowledge management. To be specific, these four factors are: corporate culture, chief knowledge officer (hereinafter abbreviated as CKO), leadership style and education level. The proposed concept is rather similar to that of Davenport and Prusak who are two knowledge management gurus. In general, these scholars have laid a great emphasis on the importance of knowledge management and they have unanimously agreed that organizations with knowledge -supportive culture and sincere trust among organizational members will have a beneficial influence on knowledge management. Additionally, appropriate leadership styles will also have a favourable impact on organization's knowledge management. Celep and Cetin (2005) have conducted the research and follow-up research on a number of schools in Turkey. Their studies have listed seven most important factors which influence the outcome of knowledge management, among which leadership styles and corporate culture have been ranked as No. 1 and No. 2 respectively. Similarly, Mason and Pauleen (2003) have studied 71 middle managers in New Zealand and found that internal factors such as corporate culture, leadership styles and education level were the three most important factors which hindered knowledge management in their organizations. More specifically, the degree to hinder the organisational knowledge management originated from corporate culture, leadership styles and education level are 45%, 22% and 16% respectively. Therefore, it can be concluded that corporate culture and leadership styles are the two most important key factors to promote or hinder knowledge management in the organizations (Guptara, 1999; Holland, 1999; Marr, 2003; McDermott & O'Dell, 2001; Holowetzki, 2002).

Based on the literature review of concerning articles published in the research field of knowledge management, the authors of this paper have streamlined the organism and mechanism of the relationship between knowledge management and corporate cultures, as well as leadership style.

Furthermore, the direct and indirect impact of leadership styles on knowledge management (KM) through the formation and content of corporate culture have been discussed.

1. THE IMPACT OF CORPORATE CULTURE ON KNOWLEDGE MANAGEMENT

In his book "Organizational Culture and Leadership", Schein has defined the term "corporate culture" as this: corporate culture is the complete whole set of shared values systems, business ethics, behaviour norms and code of practices of all the members in a certain community through societal experiences under certain social and economical conditions. In the research field of organizational behaviours, scholars have adopted a very similar concept about corporate culture just as that proposed by Schein. These scholars have regarded corporate culture as a set of systems which includes management concepts, core values system, codes of practice and shared norms and customs. Management concept as a deeper level of cultural components instructs the value orientation and work behaviours of its group members; while core values are seen as much deeper cultural composite which have more direct influences on the various work behaviours of the members within the organization. However, code of practice refers to the concrete expressions of organizational members. The dynamic and interactive relationship among management concepts, core values system and codes of practice can be illustrated in Figure 1: in essence, core value systems covers the core values system and management concepts, which is a clear indication that personal core values system will not only influence the formation of his management concepts but also restrain his everyday real life behaviours.

To a great extent, corporate culture has demonstrated the common mindset and behavioral styles of most of the members in the organization. The higher the degree of such sharing among organizational members, the more advanced the corporate culture. Scholars have categorized corporate culture into three types: 1. hierarchical culture; 2. innovative culture; 3. supportive culture. Hierarchical culture as its name suggests is an upside down pyramid-shaped organizational culture, which is not favorable for the flow of

Figure 1. Dynamic relationship among core value systems, management concepts and codes of practice

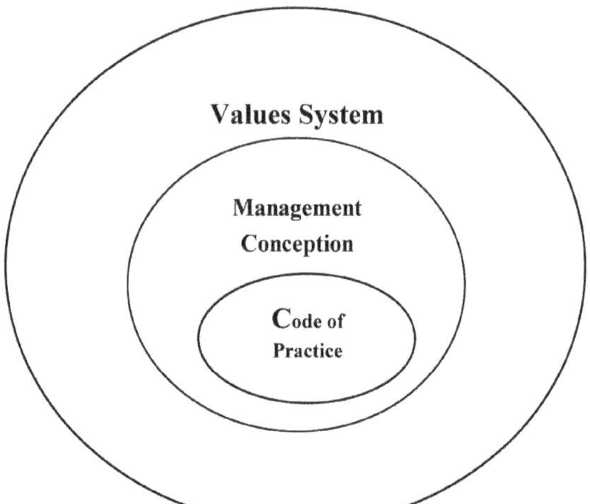

information and sharing of information among organizational members; whereas the later two types of corporate cultures have positive impact on the information flow and knowledge sharing. In the domain of the research of corporate cultures, scholars normally investigate corporate cultures from the following nine aspects, or rather, beliefs, values, recognition, attitudes, atmosphere, core values, norms and ideology (Sathem, 1983). More commonly-accepted views have seen corporate cultures as values beliefs, management philosophy, entrepreneurship, corporate norms and customs and corporate image. Researchers have carried out numerous studies on corporate cultures and their views on the definition of corporate culture. The influences of corporate cultures on the overall organizational performance can be summarized in Table 1.

In English, the three terms "company culture", "corporate culture" and "enterprise culture" are used interchangeably which can be observed from concerning literature. From the table, it is obvious that corporate culture has important influences on knowledge management of the organization, regardless whether influences are positive or negative. DeTienne, Dyer, Hoopes and Harris (2004) had pointed out that corporate culture had fatal influence on knowledge management of the organizations. Meanwhile, they have iterated that among members of the organization and among the various departments of the organization, there should formulate a suitable atmosphere to enable the free flow of information and knowledge. Baker and Baker (2001) also pointed out that though a series of technologies can support knowledge management activities, if an organization wants to achieve success in knowledge management, organizational behaviours which hinder the processes of knowledge management must be eliminated. The elimination of the ill behaviours will provide opportunities for other members of the organization to learn from their experiences. This simply means that corporate culture is the core for knowledge management.

Table 1. Definitions of Corporate cultures and their influences on organizational performance

Scholars	Corporate culture	Influence domains
Arogyswarny & Byles (1987)	the degree of recognition to organizational values beliefs	has great influence on the organizational behaviours and operations
Foil (1989)	corporate culture is the key source of organizational competition	influence the exertion of member performance and thus influence the organizational performance
Barney (1991)	corporate culture is an important intangible asset of the organization	bring about lasting competitive advantages to the organization
Goll & Sambharya (1990)	corporate culture is a crucial factor to influence organizational performance	have positive influence on organizational performance
Chatman & Jehn (1994)	corporate culture is the widely shared and recognized values system	influence the activities of in-groups and out-groups
William Ouchi (1984)	corporate culture is the tradition and style of the organization	influence member activities, viewpoints and behaviours
Charles & Gareth (2001)	corporate culture is the common values and code of practices of organizational members.	influence the communication patterns between organizational members and stakeholders
Jacalyn & James (1997)	corporate culture has its uniqueness	influence organization from all aspects
Mitchell & Tate (2002)	corporate culture can not be easily copied or imitated	long sustainability, strong stability and hard to change
Hope & Hendry (1995)	corporate culture has close tie with individuals	as an individual, he or she must adapt to the corporate culture, otherwise he or she will be kicked out of the organization
Rousseau (1995)	corporate culture is a kind of psychological incongruence between individuals and their organization	Influence employee morale and loyalty as well as their sense of belongingness
Gold, Malhotra & Segars (2001)	corporate culture synergizes with knowledge management	influence organizational performance from all aspects
Kayworth & Leidner (2003)	corporate culture synergizes with knowledge management	influence organizational performance from all aspects
Hendrinks (2004)	corporate culture impacts the whole processes of KM	influence KM from knowledge creation, knowledge retention to knowledge sharing
De Long & Fahey (2000)	corporate culture influences KM in four aspects	1. Assess the value and knowledge and the importance of knowledge; 2. Leverage the relationships horizontally among various types of knowledge; 3. Set up the interactive social setting; 4. mould new knowledge and improve previous knowledge

Source: compiled by the authors based on the literature reviews
On corporate cultures and knowledge management

As regards the relationship between corporate culture and knowledge sharing, if an organization has the competences to establish informal relations in the community, to promulgate the sense of equality among organizational members and shrink the organizational echelons, nurture an air of trust among members, cultivate innovative cultures, the organization has already established a favourable atmosphere for members to share their information with others and absorb new information from others. Sathe (1983), Schein (1983), Davenport and Prusak (2000), in their work have advocated that corporate culture was a kind of shared pattern of behaviours, shared values and beliefs. Therefore, common and unanimously agreed norms and customs should be established to enable organizational members willing to share information and knowledge. Traditional ideas sang high praise for the role of "broker" within the organization and under such corporate culture; members were encouraged to be ambitious and aggressive, focusing on the "self." However, in information age, the traditional corporate cultures "Me-centred" should be transformed into "We-centred." Under such a socialized culture, members of the organization have to establish a kind of communicative attitude to intercourse with others, to consider the benefit of others and to form a relationship of cooperation rather than merely competition. The former is to see other members as both rivalries and partners, whereas the latter sees other members only as rivalries. According to scholars like Preffer (1999) and others, corporate cultures which encourage members to air their views freely and to adopt the concept of lifelong learning were beneficiary to knowledge sharing. All in all, corporate culture which suits knowledge management accommodates such characteristic: being open-minded, innovative, of equal mindset, individual consideration and appeals to the emotions of the members, as well as co-exists with the ecological environment in which members are living.

1.1 The Influence of KM on Corporate Culture

The general procedure of KM can be categorized into three steps: knowledge acquisition, knowledge creation and knowledge dissemination. Additionally, knowledge can be divided into two general types according to whether it is easy to be acquired and absorbed: explicit knowledge and implicit knowledge. A KM guru Nonaka (1995) has described the transformational relationship between explicit knowledge and implicit knowledge as the knowledge creation process, which again has been termed as the spiral elevation of knowledge creation, i.e. the SCEI model (as can be seen in Figure 2).

This spiral process consists of socialization, combination, externalization and internationalization. De Long and Fahey (2000) in their research have pointed out the four approaches which have been influenced by corporate culture as knowledge creation, knowledge sharing and knowledge utilization. To be specific, they have presented the following hypotheses: 1. culture and sub-cultures influence characteristics of knowledge and whether it is worthwhile to be managed; 2. Culture restrains the relationship between individuals and organizations--- culture determines who will control a specific kind of knowledge, especially technical know-how and also who shares the knowledge and who stores it; 3. Culture determines the opportunities for social intercourse and how to use knowledge under a specific kind of environment; 4. Culture also determines the creation of new knowledge, the legalization of knowledge and the distribution of knowledge in the process of KM.

1. THE INFLUENCE OF CORPORATE CULTURE ON KNOWLEDGE ABSORPTION

As two of the most renowned KM gurus, as early as 1998, Davenport and Prusak had proposed the view that values and beliefs were inseparable parts of knowledge, which reflected the viewpoints of intellectuals, their ability to absorb knowledge and their inductive ability after the observation to knowledge. In a more extended way, corporate culture which accommodates the characteristics of being open, innovative, equal, considerate, the altruistic, as well as co-existent with nature will enable organizational members to share and learn knowledge with others broad-mindedly and honestly. In this way, the externalization of internal knowledge and the internalization of external knowledge will increase significantly the ability in knowledge absorption. However, the identification of knowledge, as well as the inductive ability and absorption ability will on the other hand prevent and even eliminate "the Knowledge is power

Figure 2. Spiral elevation of knowledge creation (adapted from Nonaka, 1995)

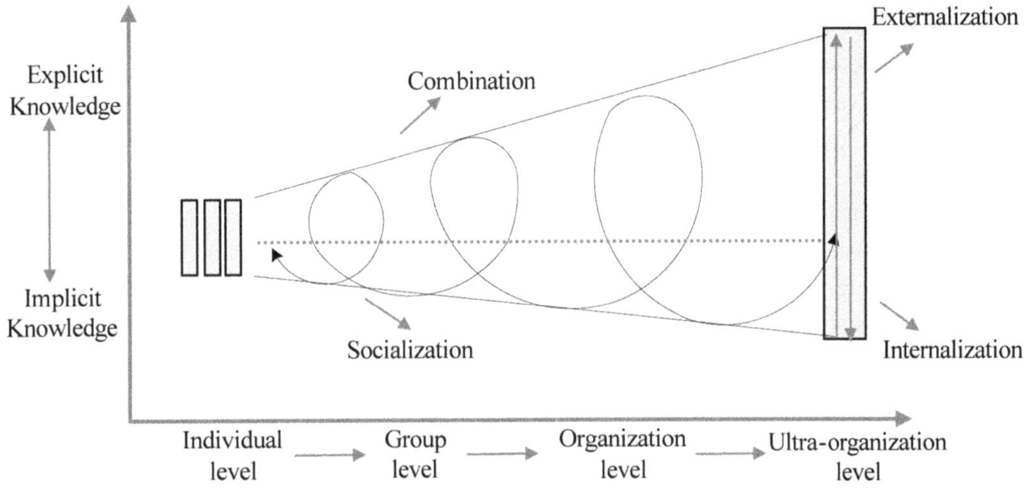

Syndrome" and "not invented here Syndrome". The former syndrome results in the unwillingness to share knowledge with others, while the latter results in the reluctance.

2. THE INFLUENCE OF CORPORATE CULTURE ON KNOWLEDGE CREATION

In the course of knowledge formation and knowledge creation, corporate culture may influence the organizational members on their ability to judge the value of knowledge. Based on the identification of the values, core benefits of the organization, organizational members usually compare these with their own core values and thus influence the formation and generation of corporate culture. To be more concrete, if the organizational core values are in conformity with that of the organizational members, the organizational members will have a strong feeling of belongingness to the organization and regard organization as "ours" and thus, effect proactive and positive influence on the formation and creation of knowledge. On the other hand, if the organizational core values are far away from that of the organizational members, organizational members will regard themselves as aliens to the organization, and as a result, have negative impact on the formation and generation of knowledge. In the course of knowledge formation and generation, the influences of corporate culture on knowledge management can be indicated from the following three aspects: First of all, corporate culture provides a suitable orientation to the formation and generation of knowledge. In essence, it composes of the commitment to learning, open atmosphere and organizational loyalty. Mistakes which will be unavoidable in the process of learning but such mistakes will not be punished, etc. (Friedman, Lipshitz & Overmeer, 2001). Secondly, inappropriate corporate culture may become a

Figure 3. Common syndromes in knowledge management

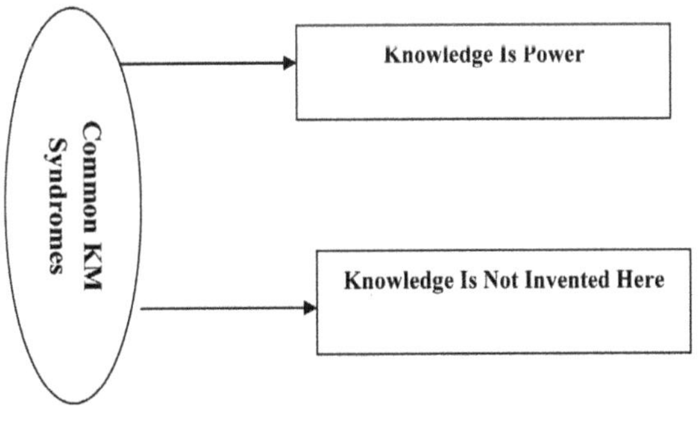

hurdle to organizational learning and thus influence the generation of new knowledge. If an organization relies heavily on the previous successful experiences, it will not be able to encourage innovation and hence refuse to adopt a positive attitude toward. Therefore, change programs, new knowledge will not be generated and identified (Antal, Lenhardt, & Rosenbrock, 2001). Thirdly, corporate culture also influences the channels for the formation and generation of new knowledge. In such corporate culture as advocating cooperation, trust and sharing of information and knowledge, organizational members will be willing to learn from one another. In such corporate culture with open and light-hearted atmosphere, as well as amiable interpersonal relationship, the mind of organizational members will be triggered and ignited from which many new ideas will emerge and as a result, new knowledge and new thoughts can be created.

3. THE INFLUENCES OF CORPORATE CULTURE ON THE KNOWLEDGE DISSEMINATION AND KNOWLEDGE SHARING

How organizational members organize and share knowledge depends on their values system, and the values system that has mainly reflected in the cultural factors during the process of knowledge management. In the works of Halowetzki (2002), he pointed out that cultural factors included items, such as, information system, organizational structure, remuneration systems, process management, human factor and leadership styles (can be seen in Figure 4).

First of all, information systems have been seen as the most important driving force in knowledge management, as it is closely related to the utilization of technologies; secondly, as for the organizational structure, learning organization enables knowledge to be more mobile and penetrable; thirdly, remuneration system consists of compensation system and performance evaluation system which also fall into the domain of cultural factors, from which corporate culture can be mirrored (Holland, 1999; Gloet & Berrel, 2003; Bollinger & Smith, 2001); fourthly, process management takes the lead to direct how and where knowledge should be shared and transferred in the organization. Organizations which focus on process management contain such characteristics: to focus on the continuous processes but not merely short-term & task-oriented objectives and thus enjoy greater autonomy. Process-oriented organizations normally adopt a flat-shaped organizational structure which is beneficiary for organizational members to lay more emphases on teamwork and inter- and intra-group collaborations. Organizations which stress the need for team spirit and team cooperation are capable of establishing a friendly platform for knowledge management and such a platform is helpful for knowledge mobility and sharing.

Only when organizations accommodate trust and collaborative culture, can knowledge be disseminated efficiently and effectively and ultimately resulting in better knowledge management.

4. THE INFLUENCE OF DIFFERENT TYPES OF CORPORATE CULTURES ON KNOWLEDGE MANAGEMENT

Chin-loy (2003), in his research has stressed that corporate culture can impact decision-making,

Figure 4. Components of organizational cultural factors

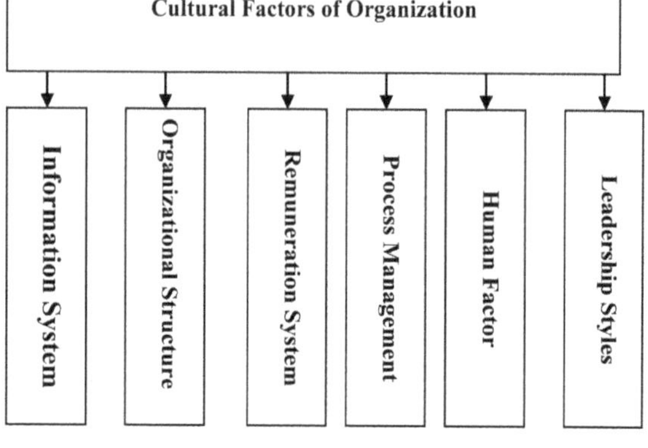

management styles and the relationship among organizational members, as well as influence the behaviour patterns of the organizational members. Therefore, corporate culture which compliments knowledge management will drive knowledge management further. DeTienne, Dyer, Hoopes, Harris and others (2004) have taken the view that if organizations want to guarantee knowledge management, they should adopt such kind of corporate culture: Members are willing to cooperate and collaborate; are willing to establish trust and participate into other members' activities. These scholars have paid much attention to the "soft factors" of the organization, but they have overlooked the "hard factors of the organization", such as organizational structure and organizational system. Based on the "soft and hard" factors of the organizations, corporate cultures can be divided into broad categories: hierarchical corporate culture and knowledge-sharing corporate culture.

4.1 The Influence of Hierarchical Corporate Culture on Organizational KM

Hierarchical corporate culture is a kind of stubborn and sluggish corporate culture. It emphasizes the dogmatic, routine organization rules and regulations and this kind of corporate culture is just the reflection of hierarchical organizational structure on corporate culture. Such a kind of corporate culture is detrimental to the mobilization and sharing of knowledge; it hinders the organizational members to be innovative in change programs. As regards organizational structure, such formal and hierarchical corporate culture has been seen as one of the obstacles for effective knowledge management. According to the research carried out by Guptara (1999), if an organization adopts hierarchical corporate culture, it will bring about the bad effect in knowledge management due to the void of interpersonal communication and very tense interpersonal relationships. This could be disastrous for knowledge management in the organization.

4.2. Influence of Knowledge-Sharing (Knowledge-Supportive) Corporate Culture on KM

The so-called knowledge-sharing corporate culture refers to such a kind of culture: organizational members see knowledge sharing as a daily activity (McDermott & O'Dell, 2001). Features of knowledge-supportive corporate culture can be summarized as: organizational members have been empowered to the largest extent and they are willing to disseminate information and knowledge (Ribiere & Sitar, 2003). In essence, knowledge-sharing corporate culture and knowledge-supportive corporate culture are of the same. A large number of researchers and scholars have taken the viewpoint that knowledge-sharing corporate culture and knowledge-supportive corporate culture can assure the success of organizational knowledge management (Ribiere & Sitar, 2003; McDermott & O'Dell, 2001; Bollinger & Smith, 2001). Knowledge-friendly corporate culture has the following features: organizational members adopt a positive attitude towards knowledge; they are curious about new knowledge and are anxious to explore the secrets of knowledge. As for superiors in the organization, they are eager to encourage their subordinates to create and apply knowledge; subordinates will not feel depressed while sharing knowledge with others. In such an organization, KM projects conform to the existing corporate culture.

All in all, the influences of corporate culture on the whole processes of knowledge management can be summarized in the following bullet points:

1). Appropriate corporate culture provides the right condition for the generation of technical knowledge, inclusive of commitment to learning, open and autonomous atmosphere and strong loyalty to the organization;

2). Inappropriate corporate culture can work as an obstacle for organizational learning and thus hinders the generation and creation of new knowledge;
3). Corporate culture also influences the channel from which knowledge is formed;
4). According to Glisby and Holden (2003), corporate culture has very significant effects on all four learning processes described in Nonaka's SECI model (Socialization, Externalization, Combination and internalization) (see Figure 5).
6.) Influences of leadership styles on knowledge management

In their research, Bollinger and Smith (2001) have stressed that knowledge should be seen as a kind of strategic resources for the organization and such strategic resources are closely related with organizational objectives. This assertion has been based on the proposition that knowledge comes from people, therefore the human factor is the most important key element of culture. Some scholars have argued that effective knowledge management should see people as the core (Salleh Yahya & Goh, 2002; Gloet & Berrel, 2003), as ambitious and industrious persons in the organization, leaders should be the most important element of culture. Leadership styles determine the organization's vision, mission, objectives, as well as the ethical orientation in the knowledge management system (DeTienne, Dyer, Hoopes, & Harris, 2004; Holland, 1999). On the other hand, chief knowledge officers (hereinafter shortened as CKOs) have played a detrimental role in clearing the hurdles for knowledge management (DeTienne, Dyer, Hoopes, & Harris, 2004), but if these CKOs want to exert their efforts in the processes of knowledge management but without the support of organizational leaders, they will not

Figure 5. Dynamic & interactive function among SECI model (adapted from Nonaka & Toyama, 2003)

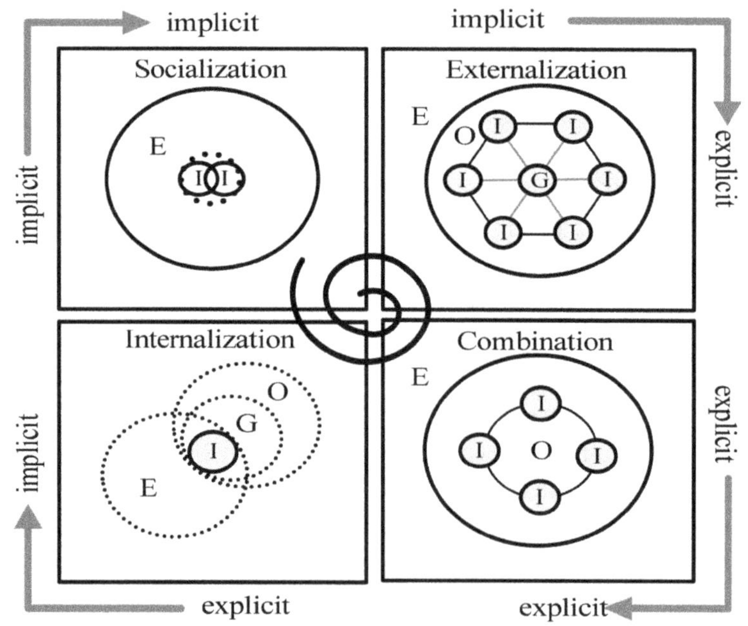

Legend in this figure: I: Individual; G: Group; O: Organization; E: Environment

procure the strategic resources which are essential for knowledge management (the decision-makers for the allocation of these strategic resources are no others but leaders). Therefore, without the consent and support of organizational leaders, no matter how hard they try, these CKOs will never achieve success in knowledge management activities (Nonaka, Toyama & Konna, 2002).

Therefore, leadership styles not only lay direct influences on the complete processes of knowledge acquisition, knowledge creation and knowledge sharing, but also play an essential role in the formation and change of corporate culture and thus have indirect impacts on knowledge management.

2.1 The Direct Influences of Leadership Styles on KM

Leaders are the founders and creators of organizational atmosphere. The corporate vision that leaders have established will influence every aspect in the course of knowledge creation which is a direct reflection of the organizational values system and corporate culture. In other words, it also makes the right identification to knowledge created. Based on the above argument, it is obvious that leadership styles have direct influences on organizational knowledge management. Viitala (2004) has postulated that knowledge leadership is the most effective leadership style in the course of knowledge management. Knowledge leadership is also called participative leadership. Leaders with such leadership style have the ability to create a good atmosphere to facilitate organizational learning. In the research of Riebera and Sitar (2003), it has been exposed that leaders are of the greatest importance in creating a knowledge-supportive corporate culture. Politis (2001) has studied the various kinds of causations between leadership styles and knowledge management and found that leadership styles which advocate and encourage employees to participate in the decision-making processes have positive effect on knowledge management. If leaders can establish trustworthy relationship with their employees, respect the sentiments of their employees and are willing to accept employee suggestions, this will have positive influence on knowledge absorption and knowledge acquisition. At the mean time, leadership styles lay a solid foundation for the processes of knowledge creation. Nonaka, Toyama & Konno (2002) have taken the viewpoint that leadership styles influence the processes of knowledge creation through basic activities, such as providing knowledge vision, developing and promoting knowledge assets, as well as creating knowledge Ba at which knowledge creation will follow all the way along the spiral so as to conduct a better knowledge management. Ba provides the space for knowledge transfer at specific stages and provides space for knowledge sharing. To a great extent, ba provides a platform for individuals, teams and organizations to create knowledge and share knowledge. In the dynamic process of knowledge management along the spiral, in Nonaka's SECI model, socialization, externalization, combination and internalization correspond to Originating Ba, Interacting Ba, Cyber Ba and Exercising Ba respectively. Every Ba speeds up the knowledge creation activities in the way of supporting a specific knowledge sharing process and thus increases the comprehensive performance of the organization (see Figure 6 SECI Ba).

2.2 The Indirect Influence of Leadership Style on KM — Influence the Formation of Corporate Culture

As early as 1992, a world-renowned research called Schein had pointed out that no matter which stage the organization lies in/ posits? (according to organizational life cycle, any organization can be divided into stages as formation, growth period, maturity period, decline period), leaders play a vitally important role in the creation of corporate culture and in the processes of knowledge management. In other words, leadership styles have a direct impact on the formation and development

Figure 6. Concept of Ba in KM (adapted from Nonaka & Toyama, 2003)

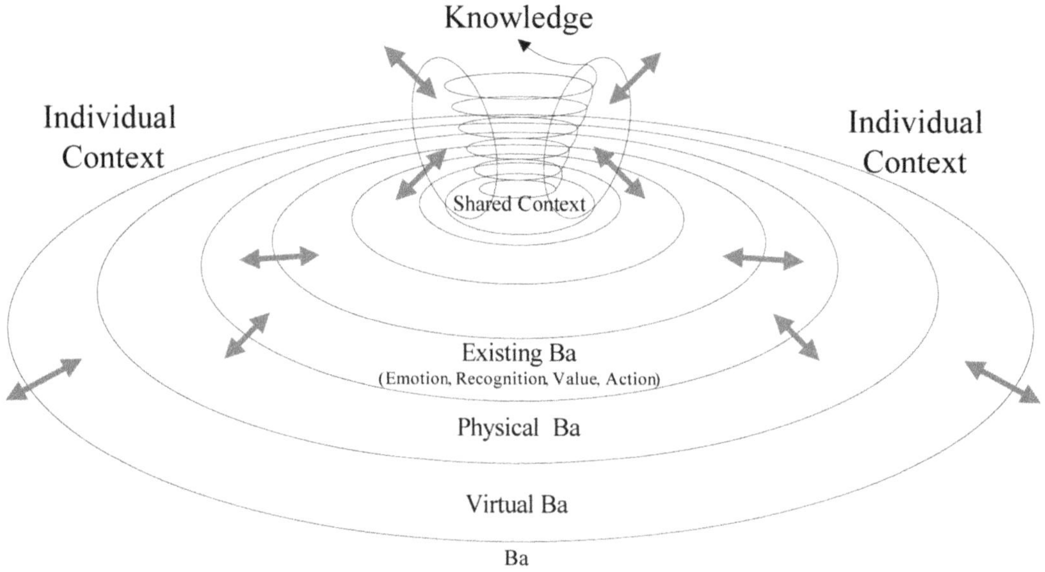

of corporate culture. Through the formation and change activities of corporate culture, leadership styles lodge indirect influence on knowledge management.

In the formation period of corporate culture, leaders or the founders of the organization influence the external adaptability and internal integration of the organizational members at the early stage. At that time, leaders or founders of the organization are usually entrepreneurs, who have strong confidence and great perseverance and also have high expectations on the futures of the organization. When their expectations come true, the entrepreneurship of these initial leaders of the organization will gain recognition and congruence among other members of the organization and such entrepreneurship will become part of the corporate culture. Founders of the organizations will become the idols of their employees and take for granted that what the leaders have said is all correct. At this stage, leaders or founders of the organization will normally select those though outside of the organization but share similar or same visions with them to join the organization to enhance such corporate culture. Those who hold different or quite opposite views will choose to leave the organization which will push the organization into the sate of "groupthink" and as a result, corporate culture will become stubborn and dogmatic due to lack of "new bloods". Based on the common observation of such phenomenon, Schein (1992) has proposed two mechanisms for leaders and founders of the organization to choose from. These two kinds of mechanisms can help leaders and founders of the organization to embed their mind into corporate culture and such mechanisms have been called "the embedded mechanism" (as can be seen in Table 2).

In an organization, the primary mechanism of the corporate culture has sometimes been named as organizational atmosphere, which is normally the implication and reflection of the leader's mind, especially at the initial stage when the organization is just founded. The measures in the secondary mechanism are usually adopted by leaders to strengthen the primary mechanism. The principle to implement secondary mechanism is that approaches in the secondary mechanism must go along with that in the primary mechanism. In addition, leaders must play the role of model and

Table 2. Embedded mechanism of corporate culture

Primary mechanism	*Secondary mechanism*
a. Normally, where do leaders lay their attention? How to control it? b. How leaders deal with occasional events and organizational crises? c. how do leaders allocate scarce resources? d. leaders may consider to be a model, a coach and mentor e. how do leaders deal with remuneration issues and promotion in the organization? f. how do leaders recruit, select and promote their employees? How do they communicate with the subordinates?	a. organizational design and structure b. organizational system and procedures c. organizational modes and patterns d. specific space and design of the building e. related persons, legendary stories and mysteries f. formal values system, ethics and other cultural aspects

Source: Schein (1992)

example for subordinates to follow. However, at the stages of primary developing stage, maturity stage and decline stage, the influence of leaders will be quite different as against that in the formation stage. At such times, the original corporate culture is already out of date and organizations must conduct change programs accordingly. At such moments, transformational leaders will adapt to the new situation easily (Politis, 2001; Schein, 1992). No doubt, in order to establish a knowledge-friendly corporate culture, it is essential for leaders to grant their support and hold a strong will in knowledge management. Organizational resources, organizational system, as well as organizational structures will reflect the leaders' cultural mentality through the adoption of the primary mechanism. Meanwhile, leaders must be fully aware of how to assess the organizational development and how to deploy the cultural change mechanisms at different stages of the organizational development.

3. CONCLUSION

Corporate culture and leadership styles are the two most important factors which influence knowledge management. In one way, corporate culture has a significant impact on the whole process of KM from knowledge acquisition, knowledge creation, down through knowledge dissemination. In knowledge-supportive culture and knowledge sharing culture, organizational members can establish trust among them, and thus allows the free flow of information and knowledge. As a result, it will promote knowledge sharing and knowledge management in the organization. KM activities within the organization should match cultural mechanisms of the organization and the whole process of organizational activities. As for cultural mechanisms, it can include ingredients such as organizational structure, remuneration systems, compensation system, as well as employee promotion mechanisms, etc. Additionally, leaders play a key role in the course of KM within organizations. He or she stimulate and develop corporate culture, establish corporate visions and missions, as well as allocate the relevant resources which are required for KM in the organization. Leadership styles have close causation with the formation and connotation of corporate culture. It can be said that corporate culture is normally the value orientation and the reflection of personal culture of the leader himself or herself. This is especially so at the very beginning of organizational development period, the leader enhances the secondary mechanism of corporate culture through the insertion of primary and secondary mechanism of corporate culture, and furthermore, has an indirect impact on KM in the organization. As argued above, it can be deduced that corporate culture and leadership styles both have influences on KM in the organization from different degrees. As for the internal factor of the organization, if an organization wants to achieve effective and successful knowledge management, it must simultaneously satisfy two essential conditions: have a leader who encourages and support KM + have a corporate culture which has been forged by the leader.

ACKNOWLEDGMENT

The authors of this paper extend their sincere thanks for the kind and generous help of Professor W. B. Lee and Mr. Gareth Blackett in correcting and proofreading this paper. This research has been sponsored by the National Natural Science Foundation of China (grant No. NNSFC No. 70433003) and National Social Science Foundation of China (grant No. NSSFC 08XJY016).

REFERENCES

Argote, L., & Ingram, P. (2000). Knowledge transfer: A basis for competitive advantage in firms. *Journal of Organizational Behaviour and Human Decision Processes*, *82*(1), 150–169. doi:10.1006/obhd.2000.2893

Baker, J. H., & Baker, G. A. (2001). *Leadership, Culture and Knowledge Management.* Retrieved from http://leadendtoday.org.

Bollinger, A. S., & Smith, R. D. (2001). Managing Organizational Knowledge as a Strategic Asset. *Journal of Knowledge Management*, *5*(1), 8–18. doi:10.1108/13673270110384365

Celep, C., & Cetin, B. (2005). Teachers' Perception about the Behaviours of School Leaders with regard to Knowledge Management. *International Journal of Educational Management*, *19*(2), 102–117. doi:10.1108/09513540510582408

Chin, K. F. (2001). *Developing a Knowledge Management Strategy*. Price Water House Coopers.

Chin-loy, C. (2003). *Assessing the Influence of Organizational Culture on Knowledge Management Success.* Unpublished UMI dissertation, Wayne Huizeng School of Business and Entrepreneurship. Nova Southeastern University.

Crawford, C. B. (2004). *Transformational Leadership, Innovation and Knowledge Management: Empirical Findings and Emergent Conclusions*. Hays, KS: Fort Hays State University.

Danieir, T. (1998). Networking your knowledge. *Management Review.*

Davenport, T. H. (1997). Ten principles of Knowledge management and four case studies. *Knowledge and Process Management*, *4*, 187–208. doi:10.1002/(SICI)1099-1441(199709)4:3<187::AID-KPM99>3.0.CO;2-A

Davenport, T. H., & Prusak, L. (1998). *Working Knowledge: How Organizations Manage What They Know*. Boston: Harvard Business School Press.

De Long, D. W., & Fahey, L. (2000). Diagnosing Cultural Barriers to Knowledge Management. *The Academy of Management Executive*, *14*(4), 113–127.

DeTienne, K. B., Dyer, G., Hoopes, C., & Harris, S. (2004). Toward a Model of Effective Knowledge Management and Directions for Future Research: Culture, Leadership, and CKOs. *Journal of Leadership & Organizational Studies*, *10*(4), 26–43. doi:10.1177/107179190401000403

Gloet, M., & Berrel, M. (2003). The Dual Paradigm Nature of Knowledge Management: Implications for Achieving Quality Outcomes in Human Resource Management. *Journal of Knowledge Management*, *7*(1), 78–89. doi:10.1108/13673270310463635

Guptara, P. (1999). Why Knowledge Management Fails: How to Avoid Common Pitfalls. *Knowledge Management Review*, *9*, 26–29.

Holland, D. (1999). Ten Ways to Embed KM into Organizational Culture. *Knowledge Management Review, 7.*

Holowetzki, A. (2002). *The Relationship between Knowledge Management and Organizational Culture: An Examination of Cultural Factors that Supports the Floe and Management of Knowledge within an Organization.* Eugene, OR: University of Oregon Applied Information Management Program.

Ilkka, T. (1999). Data is more than knowledge: implications of reserved knowledge hierarchy for knowledge management and organizational theory. *Journal of Management Information Systems, 16*(3), 103–117.

Knapp, E., & Yu, D. (1999). Understanding Organizational Culture: How Culture Helps or Hinders the Flow of Knowledge. *Knowledge Management Review, 7*, 16–21.

Lahtl, R. K., & Beyerlein, M. M. (2000). Knowledge transfer and management consulting: A look at the firm. *Business Horizons, 43*(1), 65–74. doi:10.1016/S0007-6813(00)87389-9

Luo, M. (2005). On the Paradigm of Knowledge Management. *Journal of Fiscal Science, 3.*

Mansur, M. (2003). *Entrepreneurship and Innovation in the Knowledge-based Economy: Challenges and Strategies.* Paper presented at the APO Symposium on Entrepreneurship in Knowledge-based Industry, Tokyo.

Marr, B. (2003). Consider the Culture when Benchmarking KM Processes. *KM Review, 5*(5), 6–7.

Mason, D., & Pauleen, D. J. (2003). Perceptions of Knowledge Management: A Qualitative Analysis. *Journal of Knowledge Management, 7*(4), 38–48. doi:10.1108/13673270310492930

McDermott, R., & O'Dell, C. (2001). Overcoming Cultural Barriers to Sharing Knowledge. *Journal of Knowledge Management, 5*(1), 76–85. doi:10.1108/13673270110384428

Nonaka, I., & Noboru, K. (1998). The concept of "Ba": Building a foundation for knowledge creation. *California Management Review, 40*(3), 40–54.

Nonaka, I., & Takeuchi, H. (1995). *The knowledge – creating company: How Japanese companies create the dynamics of innovation.* Oxford, UK: Oxford University Press.

Nonaka, I., Toyama, R., & Konno, N. (2002). SECI, Ba and Leadership: a Unified Model of Dynami Knowledge Creation. In Little, S., Quitas, P., & Ray, T. (Eds.), *Managing Knowledge: An Essential Reader*. London: Sage.

Polanyi, M. (1958). *Personal knowledge: toward to post-critical philosophy* (1st ed.). Chicago: University of Chicago Press.

Polanyi, M. (1966). *The tacit dimension.* London: Routledge.

Politis, J. D. (2001). The Relationships of Various Leadership Styles to Knowledge Management. *Leadership and Organization Development Journal, 22*(8), 354–364. doi:10.1108/01437730110410071

Preffer, J., & Robert, I. S. (1999). Knowing "what" to do is not enough: turning knowledge into action. *California Management Review, 42*(1), 83–108.

Quintas, P. (2002). Managing Knowledge in a New Century. In Little, S., Quitas, P., & Ray, T. (Eds.), *Managing Knowledge: An Essential Reader*. London: Sage.

Ribiere, V.M., & Sitar, A.S. (2003). Critical Role of Leadership in Nurturing a Knowledge-supporting Culture. *Knowledge Management Research & Practice*, 39-48.

Robbins, R.F. (2003). Harnessing Group Memory to Build a Knowledge-Sharing Culture. *Of Counsel, 22*(6), 7-11.

Rowley, J. (1999). What is Knowledge Management? *Library Management, 20*(8), 416–420. doi:10.1108/01435129910291175

Salleh, Y., & Goh, W. K. (2002). Managing Human Resources towards Achieving Knowledge Management. *Journal of Knowledge Management*, *6*(5), 457–468. doi:10.1108/13673270210450414

Schein, E. H. (1983). The role of founder in creating organizational culture. *Journal of Organizational Dynamics*, *12*(1), 13–28. doi:10.1016/0090-2616(83)90023-2

Schein, E. H. (1992). *Organizational Culture and Leadership* (2nd ed.). San Francisco: Jossey-Bass.

Senge, P. (1997). *The fifth Discipline, the art and practice of the learning organization*. New York: Currency Doubleday.

Stauffer, D. (1999). Why people hoard knowledge. *Across the Board*, *36*(8), 16–21.

Szulanski, G. (1996). Exploring internal stickiness: impediments to the transfer of best practice with the firm. *Strategic Management Journal*, 17.

Szutanski, G. (2000). The process of knowledge transfer: A diachronic analysis of stickiness. *Organizational Behavior and Human Decision Processes*, *82*(1), 9–27. doi:10.1006/obhd.2000.2884

Viitala, R. (2004). Towards Knowledge Leadership. *Leadership and Organization Development Journal*, *25*(26), 528–544. doi:10.1108/01437730410556761

Wang, Y., & Qin, J. (2002). Empowerment and Knowledge-Sharing. *Economic Forum, 15*.

Wilg, K. M. (1994). *Knowledge Management: The central management focus for intelligent-acting organizations*. Arlington, TX: Schema Press.

Xiang, X. P., & Wang, L. J. (2001). Core Competitiveness of Enterprises in the Perspective of Knowledge. *Journal of Economy and Management of Foreign Countries, 3*.

Xu, S. (1996). Management and Leadership Style in Knowledge Society. *World Manager's Digest*, *12*, 44–54.

This work was previously published in International Journal of Asian Business and Information Management, Volume 1, Issue 3, edited by Dasho Karma Ura and Patricia Ordóñez de Pablos, pp. 54-68, copyright 2010 by IGI Publishing (an imprint of IGI Global).

Chapter 6
Interactive Strategies for Cultural Heritage Tourism and Macau City Development:
The Tourists' Perspectives

Aliana M. W. Leong
Macau University of Science and Technology, China

Jacky X. Li
Macau University of Science and Technology, China

ABSTRACT

This paper examines the cultural heritage resources of Macau in tandem with the city's development. The investigation encompasses the actual condition of Macau's cultural heritage tourism and the challenges and opportunities for its future development. Based on the experiences of other countries, the authors set strategic goals, provide an overall path and developmental strategies, and create a scientific structure with a framework built on theory and practice. The necessity and viability of cultural heritage tourism in cultivating and enhancing the city's competitiveness as a tourism destination is argued. Based on a survey of visitors to Macau, the researchers gain an understanding of the tourists' behavior, awareness and evaluation of scenic spots that are part of the cultural heritage tour.

INTRODUCTION

This study is designed as part of the whole research titled, "A Primary Study on the Interactive Strategies for Cultural Heritage Tourism and City Development: A Case Study of Macau". A part of the study seeks to understand the perceptions of Macau local community residents, their evaluation and demands that influence of the development of Macau's cultural heritage tourism. It aims to identify the direction and ways for the betterment of the cultural heritage development in Macau from the local community residents' perspectives in order to promote the development of Macau

DOI: 10.4018/978-1-4666-0276-2.ch006

with the hope of laying down a concrete foundation on the interactive development between cultural heritage tourism and city development.

LITERATURE REVIEW

Cultural Heritage Tourism

According to Raivo (2002), heritage is the use of the past for the present through the 'production, consumption and regulation of the cultural, political and economic meanings of the past'. Lowenthal (1994, p. 43) indicates that 'heritage distils the past into icons of identity, bonding us with precursors and progenitors, with our own earlier selves, and with promised successors'. In this case, heritage is a set of ideas, symbols and events that establishes and strengthens social unity and identity, real or imagined, of a group of individuals.

It is crucial to be aware that heritage is not history. 'Heritage has greater symbolic meaning than the object, time or place that is the historical reference' (Edson, 2004, p.338). The term 'heritage' is naturally controversial since it implies at its best the offer of a new kind of interest in and understanding of the past and at its worst, and subjective and selective false history that underestimates the historical variety of social experiences of class, gender, and ethnicity (Johnson, 1996; Raivo, 2002). It is hardly surprising that heritage tourism is equally vexed.

Tourism is the 'transformation of the object and place into attractions, their gradual movement from a setting to a representation of a setting' (Wedow, 1977, p.201). Heritage tourism involves the connection of tourists with a sometimes constructed, often mythical, past by promoting 'a vicarious experience that depends on using objects or locations as means of entering into or living in the past' (Edson, 2004, p.337; Voase, 1999).

Tourism, as an international activity, is a defining characteristic of contemporary societies (Jolliffe & Smith, 2001). The issue of understanding, exploring and conquering the mystery of the past and seeking answers to the questions posed by ancient monuments is something in human nature. For many people, the remains of the past provide a sense of security, belonging and continuity in an uncertain changing world (Darvill, 1987; Millar, 1989).

Historic resources - the surviving physical leftovers from past civilizations and past eras such as architecture, landscape, archaeological features and artifacts constitute some of the foundations for heritage tourism. Places associated with significant historical and religious or mythical events have always been points of interest (Burton, 1995). A number of these world heritage resources have been developed into attractions and opened to the general public. Worldwide attention has been given to the implication of human visitations on artifacts that have survived for hundreds of years; however, little attention has been given to the effects of such visitation on individuals and groups.

Visitors' books generally represent a major source of feedback for site organizers. They permit the analysis of views and attitudes, since their communicative function consists in the expression of comments, remarks and suggestions of people visiting hotels, museums, and heritage sites. As Robinson (2001) states, they constitute a source of rich qualitative data in that they are unsolicited and spontaneous personal narratives. However, to our knowledge, few studies have examined this particular genre in tourism.

Heritage tourism is a reflexive action that both reaffirms and constructs identity and allows the telling of a "national story" through museums and other heritage sites (Light, 2000, p. 158; Worden, 2003). And heritage is itself the result of a discourse over 'which representation (of a place) from a variety of interpretations of place, will dominate' and the sites themselves are this discourse materialized (Hollinshead, 1997; Kruse, 2005, p. 90; Worden, 2003).

Silberberg (1995) defines heritage tourism as "visits by persons from outside the host community motivated by the interest in historical, artistic, and scientific or lifestyle/heritage offerings of a community, region, group or institution"

Heritage tourism is not merely tourist activity in a space where historic artifacts are presented. Rather, heritage tourism should be understood based on 'the relationship between the individual and the heritage presented and, more specifically, on the tourists' perception of the site as part of their own heritage' (Poria et al., 2004, p. 20). Furthermore, the 'differences in perceptions of a site are reflected in differences in reasons for visiting a site. .. the link between the individual and the site is at the core of the understanding of heritage tourism as a social phenomenon' (Poria et al., 2004, p. 26). At heritage sites, visitors experience a 'merging of the real and imagined which makes the visit more meaningful' (Kruse, 2005, p. 89). As such, heritage sites are highly symbolic, for they connect visitors to the personal and collective memories that comprise their identities.

Heritage Resources are Attractive for a Number of Reasons

Firstly, tourists may simply be attracted by the age of the ancient relics of past civilization. Typical examples are travels to Egypt and China to see pyramids, tombs, sphinx, temples and the Great Wall of China.

A second important incentive is for tourists to witness past technological achievement and the physical skills of the people that constructed them without the aid of modern technology. This has been identified as one of the major factors that draw North Americans to visit Western Europe.

Thirdly, tourists often enjoy visiting historic sites purely for the appreciation of their beauty of art and architecture.

Fourthly, visitors may be attracted to buildings and places for their historic meaning other than their visual appeal. Thus, places associated with historical and mythical exploits and events can be of great interest for tourism. For instance, some African–Americans visit the Forts and Castles in Ghana because they regard them as their root - sites from where their ancestors were taken to the other parts of the world.

Finally, places may hold great religious meanings for their visitors. For example, Christians and Moslems travel to Jerusalem and Mecca respectively for their spiritual well-being and inspiration (Prentice, 1993; Swarbrooke, 1995).

Heritage attractions have become popular because of their ability to combine their traditional role of teaching visitors of the past whilst offering entertainment and fascinating experiences (Faulkner et al., 1999). Given the increasing interest, countries have evolved systems, legislation and new approaches to the planning and protection/conservation of buildings and features of historic importance. Some of these efforts are being encouraged and supported by UNESCO through the World Heritage Convention Program started in 1972 and designed to protect the most outstanding of the world's cultural and natural heritage.

Cultural Heritage Tourism (Tourists' Perspectives)

Visitors' outlook on scenic spots are researched for a variety of reasons and for those involved in management, an important aspect is to gain feedback that can be used to monitor and improve future visitors' experiences (Swarbrooke, 1995). Managers often attempt to gain an insight into visitors' views of their level of satisfaction (Ryan, 1995; Shackley, 1998), and to a great extent, this is concerned with the fulfillment of motivations, and visitors behave in particular ways in an attempt to ensure a satisfying visit (Ryan, 1995).

Motivations of those visiting heritage attractions are closely linked to expectations of the visit (Ryan, 1995; Timothy & Boyd, 2003), but these are very varied (Hayward & Brydon-Millar, 1984). As

Shackley (1999) indicates, many first time visitors are likely to have stereotype impressions of the attractions. The nature of interpretation, signage and guiding are key factors associated with visitors' expectations of a heritage site (Shackley, 1999).

Research suggests that visitors prefer a themed route (Shettel-Neuber & O'Reilly, 1981) starting near the entrance to the attraction. The length of this is important; approximately half to three quarters of a mile, taking 40–45 minutes being seen as optimum. There should be several marked stopping points (7–10 is preferable) with visitors returning, as close as possible, to the starting point (Shackley, 1999).

At most heritage attractions, conventional guiding is still by hand-held map and/or guide book. In addition, expectations are likely to focus on the possible number of visitors at the site and issues concerned with parking (Shackley, 1999). Hence, at major heritage sites, many visitors will probably anticipate that there will be crowding, congestion and parking problems.

Visitors tend to behave in particular ways at heritage attractions. At major attractions visitors are likely to be particularly respectful and may feel that they should look but not touch (Shackley, 1999). Nevertheless, different types of visitors behave in different ways at heritage sites. For example, young adults usually do not have the same behavioural norms as middle-aged visitors. Groups tend to act differently from families, couples or unaccompanied visitors (Dawes & Rowley, 1996). Repeat visitors also tend not to follow the pattern of first time visitors.

According to Falk and Dierking (1992), first time visitors' pattern of behaviour has four stages including orientation (taking 4–10 minutes), intensive looking (approximately 15–40 minutes), exhibit cruising (between 20–45 minutes) and leave taking (involving 3–10 minutes).

In relation to major attractions, queuing and bottlenecks are common anticipated problems. However, if these actually occur on site, they may affect visitors' enjoyment and assessment of satisfaction. Some popular attractions have resorted to high charges at particular times to reduce queuing or crowding, while others have introduced a self-guiding system, to increase throughout and maintain a regular flow of visitors (Shackley, 1999).

Poria et al. (2001) suggest that in the tourism literature, it is common to define and discuss subgroups solely in terms of the presence of tourists or their activity in certain spaces or physical environments. This approach has two main effects. First, we come across many situations where the same activity could be classified under various tourism subgroups. A visit to a winery, for example, could be classified as industrial tourism, cultural tourism, heritage tourism, culinary tourism, wine tourism, etc.

Secondly, a vast array of subgroups has emerged in tourism research. Poria et al. (2001), despite suggesting that it is easier to quantify tourism subgroups based on this approach (while noting that overlapping may occur), also point out that it has substantial disadvantages.

A review of the literature leads to the assumption that heritage tourism itself has been divided into further 'sub-subgroups'. Examples include the emergence of subgroups based on the attributes or location of the heritage in question. Thus a literature search reveals work on cultural heritage (Alzua et al., 1998; MacDonald & Jolliffe, 2003), built heritage (Black, 1990), urban heritage (Chang et al., 1996; Gibson & Hordman, 1998), natural heritage (Keller, 1999), industrial heritage (Edwards & Coit, 1996; Jansen-Verbeke, 1999; Kerstetter et al., 1998), maritime heritage (Smith, 2001), rural heritage (Prideaux & Kininmont, 1999), local heritage (Teo & Yeoh, 1997), culinary heritage (Reynolds, 1994; Romagnoli, 1995), dark tourism (Strange & Kempa, 2003), ethnic heritage (Moscardo & Pearce, 1999) and human heritage (Simic, 1994). It is argued here that this 'over-specification' may lead to confusion in the understanding of heritage tourism and to unnecessary over complication (if heritage tourism itself is to be regarded as a subgroup).

Different people experience heritage artifacts as well as heritage spaces differently due to their constructed symbolic meaning (e.g. Howard, 2003; Lowenthal, 1998). A question will be asked as to whether it is simply what the tourists look at, or the meaning they ascribe to it. With this in mind, three working approaches to 'heritage tourism' are first presented. The first stance is to regard heritage tourism as tourists located at heritage sites, with little emphasis on the actual tourists and their visitation patterns (e.g. Cheung, 1999; Cossons, 1989; Garrod & Fyall, 2000). This means that all tourists who enter a space categorized as 'heritage' are seen as heritage tourists or taken to be engaged in heritage-related activities, even if they are not aware that the place being visited is a heritage site or have not observed the artifacts presented, the same argument was made by McKercher and du Cros (2002) in relation to ecotourism.

The second approach goes one stage further and emphasizes the tourists rather than the site attributes (Swarbrooke, 1993). This approach focuses upon the tourist motivations to visit the site, primarily emphasizing a willingness to learn and be involved in an educational experience because of the historic attributes of the site. Other studies highlighted religious motivation (among others) for a visit to heritage sacred sites (Graham & Murray, 1997; Shackley, 2001). However, it is argued that this approach still does not get to the heart of the phenomenon, namely heritage. This approach could be useful for defining historic tourism, but does not give sufficient grounds to identify heritage tourism as an actual subgroup of tourism, distinct from historical tourism, cultural tourism or educational tourism.

An approach that examines visitation patterns in light of tourist perceptions of a site relative to their own heritage (before, during and after the visit) may be one way of meaningfully distinguishing heritage tourism. This approach is based on interaction between the tourist and the site attributes. This approach attempts to highlight the point that the meaning the tourists ascribe to what they see is of importance to a better understanding of heritage tourism. This is because tourism, in essence, is a social phenomenon taking place at a specific location, and to understand it both the site attributes and the visitors must be considered. Trying to understand tourism by examining site attributes alone is equivalent to examining a social phenomenon without discussing the individuals involved in the activity.

The following definition of heritage tourism attempts to integrate the three approaches described above (Poria, 2001; Poria et al., 2000, 2003): A subgroup of tourism, in which the main motivation (demonstrated by the tourists' visitation patterns) for visiting a place is based on the heritage characteristics of the place (demonstrated by the site's attributes) according to the tourists' perception of their own heritage (demonstrated by the perceived link between the individual and the site – the tourists' perception of the site).

The crucial issue that differentiates the above definition from others is that it is the perception of site attributes rather than site attributes per se that are important. Such perceptions are key, because of their influence on tourist visitation patterns and their role in determining tourist views of a site in terms of it being part of their own heritage.

To illustrate the differences between these three approaches, the complexity of heritage tourism is discussed. The first approach includes all tourists who visit a site with heritage attributes classifying it under 'heritage tourism'.

The second includes tourists who are aware of the heritage attributes of the site. In contrast, Poria et al. (2000, 2003, 2004), when considering the phenomenon of 'heritage tourism', include only tourists who are motivated by the heritage attributes of the site and who consider the site to be part of their own heritage.

It is argued that distinguishing between such groups of tourists may lead to a better understanding of tourism in general and heritage tourism in particular. This is due to the fact that tourist perceptions of a site relative to their own heritage are considered significantly to influence their visitation patterns.

The concept put forward is that the link between the individual (the tourist) and the site (the heritage presented) lies at the heart of understanding heritage tourism (as well as other tourism subgroups). If accepted, this concept, which has already been identified in theoretical studies, could lead to segmentation of tourists visiting heritage sites in order to better reflect tourist interests.

RESEARCH DESIGN

Questionnaire Design

The creation of the survey questionnaire is based on the interactive development strategy for cultural heritage tourism and Macau. The questionnaire is an academic survey of local tourists as well as visitors coming from Mainland China. It is comprised of questions geared towards understanding the perceptions, requirements, and customer behavior in order to provide effective suggestions on the interactive development of cultural heritage tourism in Macau.

The survey questionnaire has three parts with a total of 26 questions. Part one and part two are directed towards tourists' behavior, perceptions and evaluations of Macau cultural heritage tourism. Part three identified tourists' personal information. For details, please refer to appendix one: Questionnaire- Macau Cultural Heritage Tourism Market Research.

Questionnaire Distribution and Feedback

The main approach of this research is through the use of survey questionnaires. This questionnaire survey was conducted on the spot at several Macau cultural heritage sites with local tourists as well visitors from Mainland China, Hong Kong, Taiwan and overseas. The on-the-spot survey was conducted from January to February 2008. There were 514 questionnaires distributed and all questionnaires were returned, representing a 100% response rate. The 500 valid samplings had a response rate of 97.3%, giving validity and credibility to the research results.

Research Method

The research result was analysed through description statistics, factor analysis, variance analysis via EXCEL, and SPSS. It intended to understand the character of basic population statistics of survey samplings; the perception factor and the influence degree of community citizens to the main influence of cultural heritage tourism; dissection through comparing difference evaluation to cultural heritage recourses development by community citizens.

SURVEY ANALYSIS

Sampling Analysis

Of the gender distribution: there were 264 males (51.36%) and 236 females (45.91%), a ratio of 66:59; On marital status: 192 were married and 308 were unmarried or 37.35% and 59.92% respectively. On nationality: 213 were from mainland China (41.44%), 144 from Hong Kong (28.02%), 66 from Macau (12.84%), 31 from Taiwan (6.03%), and 46 were from overseas (8.95%). Of the overseas tourists: 8 were from the U.S.A., 1 was from Morocco, 2 were from Switzerland, 6 were from the U.K., 2 were from Thailand, 1 was from Turkey, 2 were from Brazil, 3 were from Australia, 3 were from the Philippines, 2 were from Ireland, 5 were from Malaysia, 5 were from Singapore, 2 were from Canada, 1 was from France, 1 was from Cambodia,

1 was from Portugal, and 1 did not answer this item which occurred as 17.39%, 2.17%, 4.35%, 13.04%, 4.35%, 2.17%, 4.35%, 6.52%, 6.52%, 4.35%, 10.87%, 10.87%, 4.35%, 2.17%, 2.17%, 2.17%, 2.17% respectively. On occupation: there were 19 civil servants, 20 private enterprise owners, 129 employees, 12 researchers, 21 teaching staff, 11 doctors, 99 freelancers, 162 students, 27 others, which occurred as 3.7%, 3.89%, 25.1%, 2.33%, 4.09%, 2.14%, 19.26%, 31.52%, 5.25% respectively; On monthly income, 294 had income below 10,000 HKD, 153 had between10, 000 and 30,000 HKD, 41 had between 30,000 and 60,000 HKD, and 12 had over 60,000 HKD, or 57.2%, 29.77%, 7.98% and 2.33% respectively. On education level: 13 completed primary education, 37 reached junior secondary, 158 reached senior secondary, 144 had a higher diploma, 121 were undergraduates, and 27 had postgraduate education or above, which occurred as 2.53%, 7.2%, 30.74%, 28.02%, 23.54%, 5.25% respectively.

The major characteristics analyzed can be summed up as the following:

- Males were present in the majority of samplings;
- Singles were present in the majority of samplings;
- Mainland Chinese were present in the majority of samplings;
- Ages between 15 and 44 were present in the majority of samplings;
- Employees and students were present in the majority of samplings;
- Monthly income below 10,000 HKD interviewees were present in the majority of samplings;
- Education levels of senior secondary, higher diploma and undergraduate were present in the majority of samplings.

Analysis of Customer Behavior towards Macau Cultural Heritage Tourism

Frequency of Travel

According to the survey results of the frequency of travel to Macau heritage tourism (see Figure 1), the maximum portion number was second or third time travelers, or 38.72%; second were first time travelers, with 35.99%; following with this was more than ten times, fourth to sixth times and seventh to tenth times, which occurred as 10.60%, 10.00% and 2.60% respectively.

According to the survey results of travelers, there were more second or third time visitors,

Figure 1. The frequency of tourists' visits to Macau cultural heritage tourism attractions (spots)

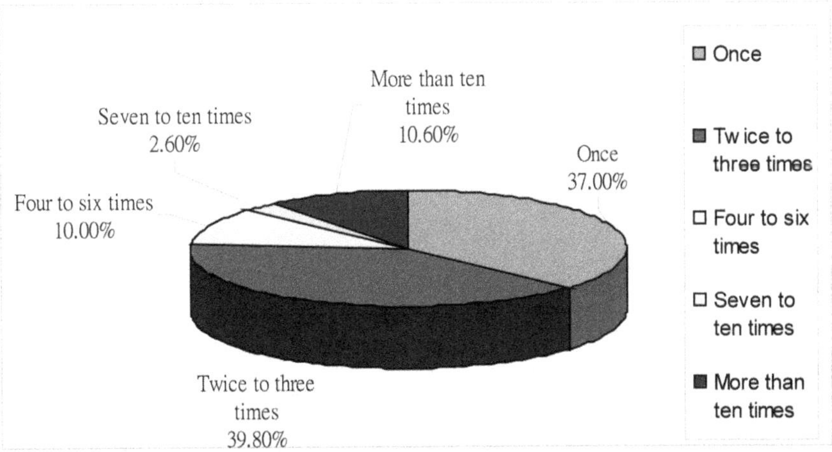

which occurred as 38.72%; second were first time travelers. This indicates that the revisit rate of Macau heritage tourism was relatively high with 10.31% of tourists visited ten or more than ten times. The high revisit rate of Macau heritage tourism indicates a close relationship among the increasing need of cultural tourism products, the rapid development of Macau tourism, and Macau's economic development.

Modes of Travel and Length of Stay Schedule

According to the gathered statistics on modes of travel (see Figure 2), 81.30 of tourists came as individual travelers; second were those who came in tours organized by travel agencies, with 13.58 came via company organized tours and others occurred with 2.95% and 2.17% respectively.

According to the statistical analysis on result pertaining to travel partners (see Figure 3), tourists who came with their family occurred the most with 39.80%; second were those who came with relatives and friends, with 36.60%; third were those who came with classmates/colleagues as 11.00%; and individuals, business partners and others as 8.00%, 2.40% and 2.20% respectively.

It can be indicated that majority of visitors to Macau cultural heritage tourism were individual travelers and those who came accompanied by their partners.

On the aspect of length of stay (see Figure 4), 51.00% tourists came on a one-day trip; 31.80% stayed for two days and one night; 11.40% stayed for three days and two nights; 5.80% stayed for more than four days. This showed that a short-line trip was the main mode of visits to Macau cultural heritage tourism.

Motivation of Travel

According to the survey (Figure 5), sightseeing was the main tourism objective which occurred as 37.20%; second was leisure, as 27.90%; following the above was cuisine and gourmet experience of Chinese and Western culture, to acquire knowledge, to educate children, to conduct scientific research, as a celebrity, due to nostalgia, or religious, which occurred as 16.00%, 6.60%, 3.70%, 2.20%, 1.60%, 1.50%, 1.10% respectively.

Synthesizing the results, multiple motivations of Macau cultural heritage tourism are mainly focused on two aspects: sightseeing and leisure. On one hand it indicated that tourists' main motive to visit Macau

Figure 2. percentage of tourist mode of travel

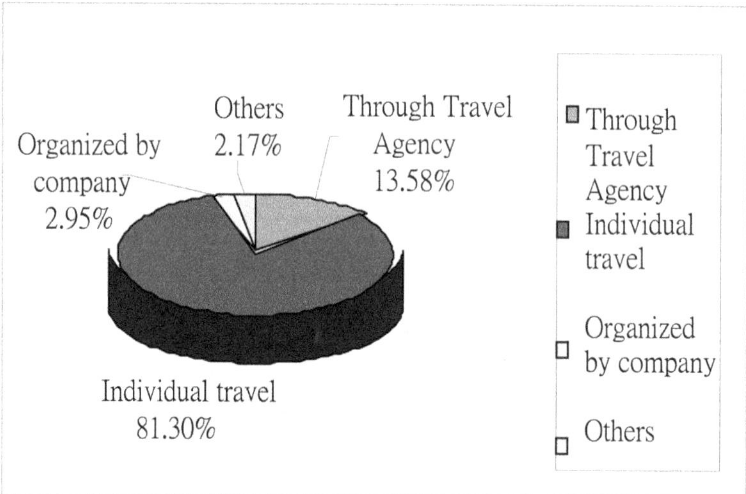

Figure 3. Percentage of tourists traveling with partners/companions

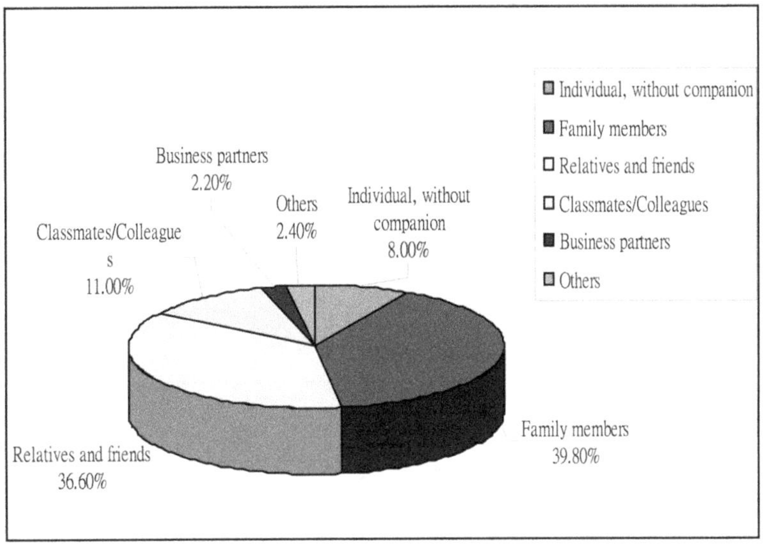

cultural heritage was sightseeing. On the other hand, it indicated that leisure was also favored, along with the strong attraction of Macau's local food. Therefore, to meet individual and multiple needs of tourists, it is suggested that sightseeing, leisure, experience and cultural travel products be created.

Information Collection

The information that tourists collected as indicated in Figure 6, ranged according to the following ratio: television or broadcasting, internet, guide books, magazine and newspaper, books and journals, recommendation of friends and relatives, advertising and promotional materials, recommendation of tour guides, tourism expositions, outdoor billboards and others occurred as 22.43%, 15.44%, 13.98%, 13.69%, 11.84%, 8.25%, 4.17%, 4.08%, 3.11%, 1.84%, 1.17% respectively. It can be seen that television or broadcasting, the internet, guide books, magazines and newspapers are the main information channels of the current Macau cultural heritage tourism. Thus, it is suggested that investment on the four information channels mentioned above and selected different

Figure 4. Distribution of length of stay

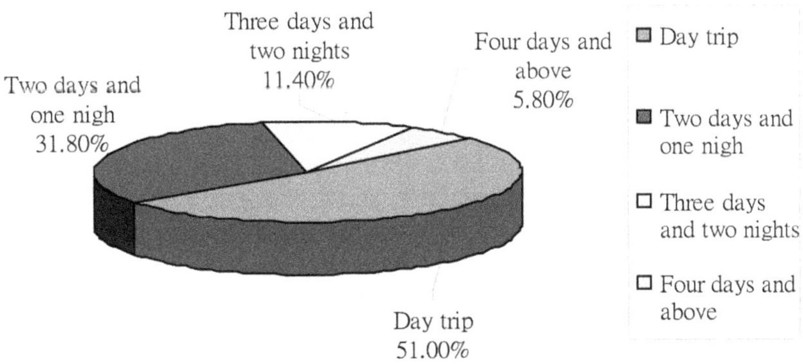

Figure 5. Motivation of travel

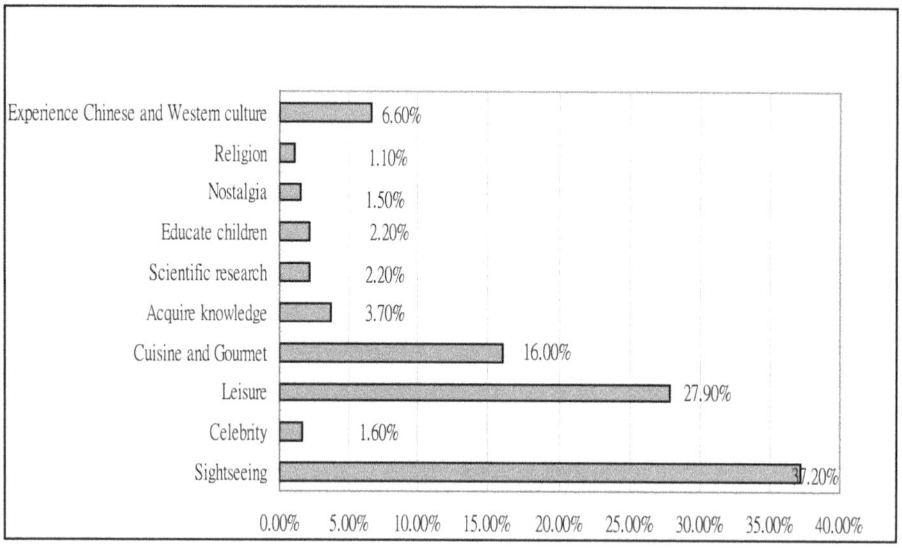

marketing manner to promote market popularity according to a variety of information channel be strengthened.

Consumption Structure

In accordance with the survey statistics (Figure 7), the average of daily spending of each tourist, was noted to reach the maximum of 500 HKD, which occurred as 44.36%; second was 500 HKD to 1500 HKD, which was 38.335; following the above were 1501-2500 HKD, 2501-4000 HKD, over 4000 HKD, which occurred as 9.73%, 3.31%, 1.56% respectively.

It is indicated that the total spending of current Macau cultural heritage tourists was somewhat low.

Figure 6. Information collection channels of tourists

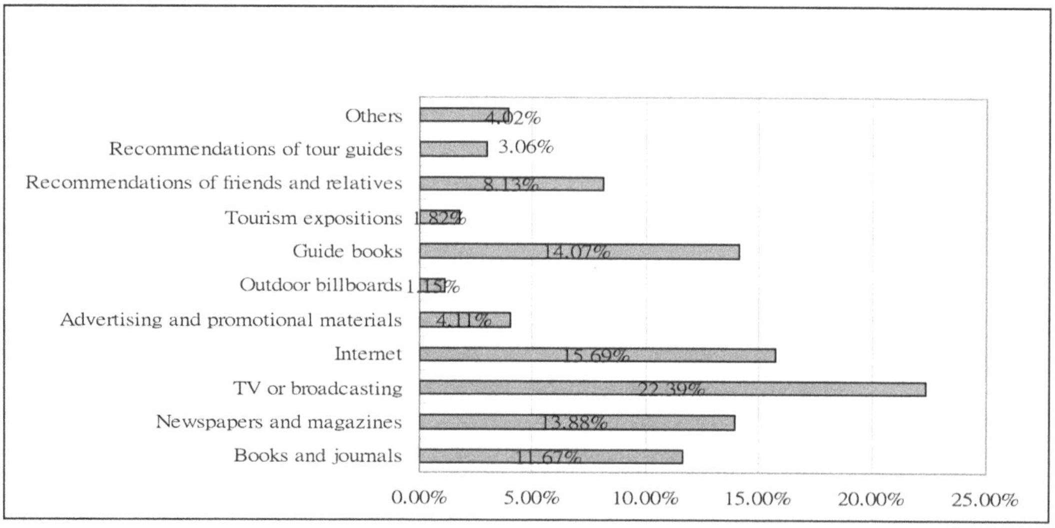

Figure 7. Total daily expenditure distribution of tourists

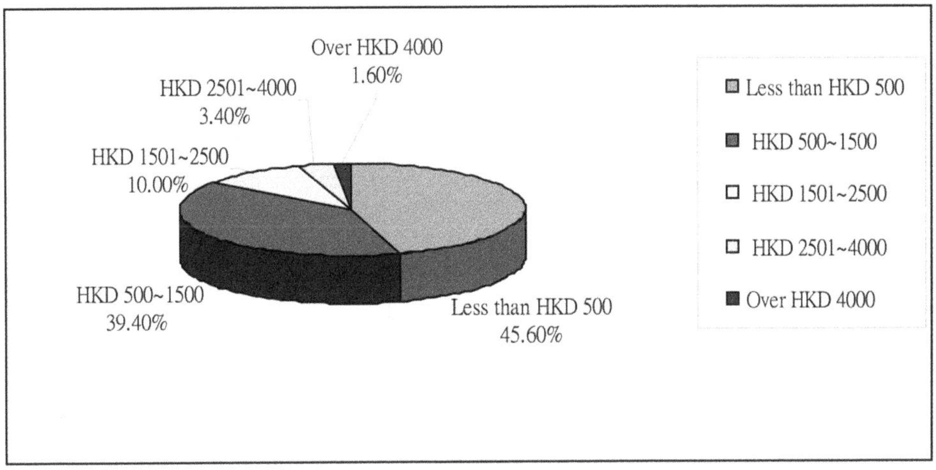

It has been mentioned above that most tourists only visited Macau for one day in accordance with the length of stay schedule analysis. On the development of Macau cultural tourism, it is necessary to generate more creative and attractive tourism products and services to encourage tourists to increase their length of stay.

In accordance with the survey statistics (Figure 8), the consumption structure of tourists spent on Macau cultural heritage tours: 27.24% tourists listed accommodation as most expensive; 21.01% tourists listed catering as number one; while 20.43% tourists listed shopping as the first, 17.90% tourists listed entertainment; 9.53% tourists listed transportation; and only 0.97% tourists listed entrance ticket as the first place. It indicated that on the tourists' consumption, accommodation, catering, shopping played the leading role whereas entertainment and entrance ticket were less. It is clear that on the future development of Macau cultural heritage tourism, strengthening the development of entertainment and tourism attractions is necessary. Relying on more original and characteristic tourism activities to attract tourists, to stimulate their interests to ensure the six main elements of tourism can be achieved for an all-round and concerted development.

Analysis of Perception and Evaluation of Tourists

Tourists' Perceptions of Macau Cultural Heritage Tourism

On the aspect of perception and evaluation of this survey, 25 typical cultural heritage scenic spots and listed by tourists, most chose "heard of the scenic spots" and "the scenic spots visited" to understand the perception level of Macau cultural heritage tours, see Table 1.

It can be seen from the table:

(1) Amongst all Macau cultural heritage spots tourists knew, the Ruins of St. Paul's was the most popular with 484 votes, corresponding to 13.35%; second were A-Ma Temple and Senado Square with 383 and 364 votes, corresponding to 10.57% and 3.64% respectively. Third were Mount Fortress, "Leal Senado" Building, St. Dominic's Church with 283, 271 and 210 votes, corresponding to 7.81%, 7.48% and 5.79% respectively; Votes between 100 to 200 were the Cathedral, the Lou Kau Mansion, the St. Joseph's Seminary and the Church (including Guia Chapel and Guia Lighthouse), the Holy House of Mercy, the Na Tcha Temple, along with the rest that received votes below 100.

Figure 8. Structure of expenditures

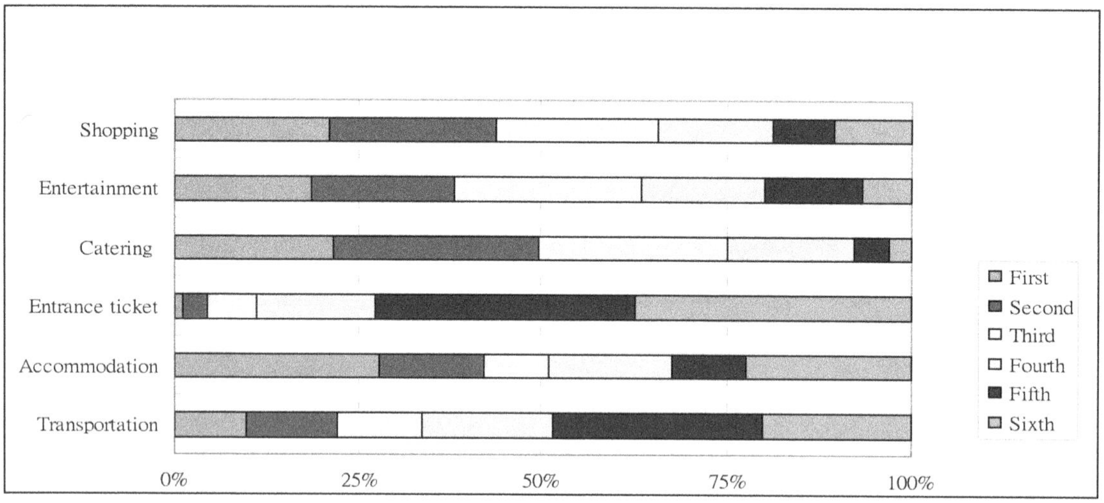

(2) Amongst the scenic spots pertaining to Macau's cultural heritage, the Ruins of St. Paul ranked first with 438 votes, corresponding to 15.78%; Second was the Senado Square with 325, corresponding to 11.71%; third were the A-Ma Temple, the "Leal Senado" Building, the Mount Fortress with 292, 236 and 225 votes, corresponding to 10.52%, 8.50% and 8.11% respectively. Votes between 100 - 200 were: the St. Dominic's Church, the Holy House of Mercy. The rest received below 100 votes.

Drawing a conclusion from the statistic above, there is a strong relevance between the scenic sports that tourists had heard about and the ones they visited. Basically, the more popular the spot, the more tourists went there. Thus, for the future development of Macau cultural heritage tourism, advertising and marketing will definitely increase popularity and attract more tourists.

Tourists' Evaluation on the Value of Macau Cultural Tourism Scenic Spots

According to the sampling statistics (Figure 9), historical value and tourism value should be stressed on the development of Macau cultural heritage products. They were accounted with 312 and 300 votes, corresponding to 26.28% and 25.27% respectively. Second was cultural heritage value, artistic value and archaeological value with 166, 162, 119 votes, corresponding to 13.98%, 13.65% and 10.03 respectively. Following the above were social influence, aesthetic, scientific and others with 53, 52, 20 and 3 votes, corresponding to 4.47%, 4.38%, 1.685 and 0.25% respectively.

According to the statistics above, tourists paid a great attention to the historical and tourism value of Macau cultural heritage spots, therefore, on the process of Macau cultural heritage development, a special development on the historical genre will be an advantage.

Factors Influencing Historical Blocks Tourism

According to the survey results, (see Figure 10), factors which qualified into historical blocks tourism according to importance are: 48.05% of tourists listed well-preserved historical heritage sites as being of primary importance; 5.64% tourists listed interactive entertainment events; 2.14% tourists

Table 1. Perceptions on Macau cultural heritage scenic spots

Heard of the scenic spots			**The scenic spot visited**		
Spots	**Number of *tourists***	**(%)**	**Spots**	**Number of *tourists***	**(%)**
Ruins of St. Paul's	**484**	13.35%	Ruins of St. Paul's	**438**	15.78%
Moorish Barracks	**78**	2.15%	Moorish Barracks	**62**	2.23%
St. Lawrence's Church	**63**	1.74%	St. Lawrence's Church	**44**	1.59%
Cathedral	**106**	2.92%	Cathedral	**79**	2.85%
The Protestant Cemetery	**65**	1.79%	The Protestant Cemetery	**32**	1.15%
Lou Kau Mansion	**132**	3.64%	Lou Kau Mansion	**97**	3.49%
St. Anthony's Church	**69**	1.90%	St. Anthony's Church	**38**	1.37%
Mandarin's House	**80**	2.21%	Mandarin's House	**57**	2.05%
St. Joseph's Seminary and Church (including Guia Chapel and Guia Lighthouse)	**159**	4.39%	St. Joseph's Seminary and Church (including Guia Cha-pel and Guia Lighthouse)	**89**	3.21%
Senado Square	**364**	10.04%	Senado Square	**325**	11.71%
Holy House of Mercy	**162**	4.47%	Holy House of Mercy	**122**	4.39%
Lilau Square	**73**	2.01%	Lilau Square	**45**	1.62%
Section of the Old City Walls	**74**	2.04%	Section of the Old City Walls	**44**	1.59%
Casa Garden	**45**	1.24%	Casa Garden	**33**	1.19%
Sir Robert Ho Tung Library	**61**	1.68%	Sir Robert Ho Tung Library	**47**	1.69%
St. Augustine's Square	**84**	2.32%	St. Augustine's Square	**54**	1.95%
St. Joseph's Seminary and Church	**58**	1.60%	St. Joseph's Seminary and Church	**36**	1.30%
"Leal Senado" Building	**271**	7.48%	"Leal Senado" Building	**236**	8.50%
St. Dominic's Church	**210**	5.79%	St. Dominic's Church	**164**	5.91%
St. Augustine's Church	**55**	1.52%	St. Augustine's Church	**37**	1.33%
Mount Fortress	**283**	7.81%	Mount Fortress	**225**	8.11%
Sam Kai Vui Kun Temple	**61**	1.68%	Sam Kai Vui Kun Temple	**45**	1.62%
Na Tcha Temple	**139**	3.83%	Na Tcha Temple	**97**	3.49%
St. Augustine's Theatre	**66**	1.82%	St. Augustine's Theatre	**38**	1.37%
A-Ma Temple	**383**	10.57%	A-Ma Temple	**292**	10.52%
	3625	100.005		**2776**	100.00%

listed performance events; 6.61% tourists listed historical cultural heritage museum; 6.81% tourists listed a comprehensive interpretation system as being of primary importance; 6.03% tourists listed design of the travel itinerary; 10.89% tourists listed service quality; 11.48% tourists listed the accompanying tourism facilities as being of primary importance. Therefore, when tourists chose historical blocks of Macau cultural heritage, they paid a great value on the completeness of historical heritage sites along with the accompanying tourism facilities. It is therefore, essential to protect the completeness of historical heritage sites as being of primary importance; second were the accompanying tourism facilities and along with that was enhancing service quality.

Figure 9. Tourists' evaluation on the value of Macau cultural tourism scenic sports

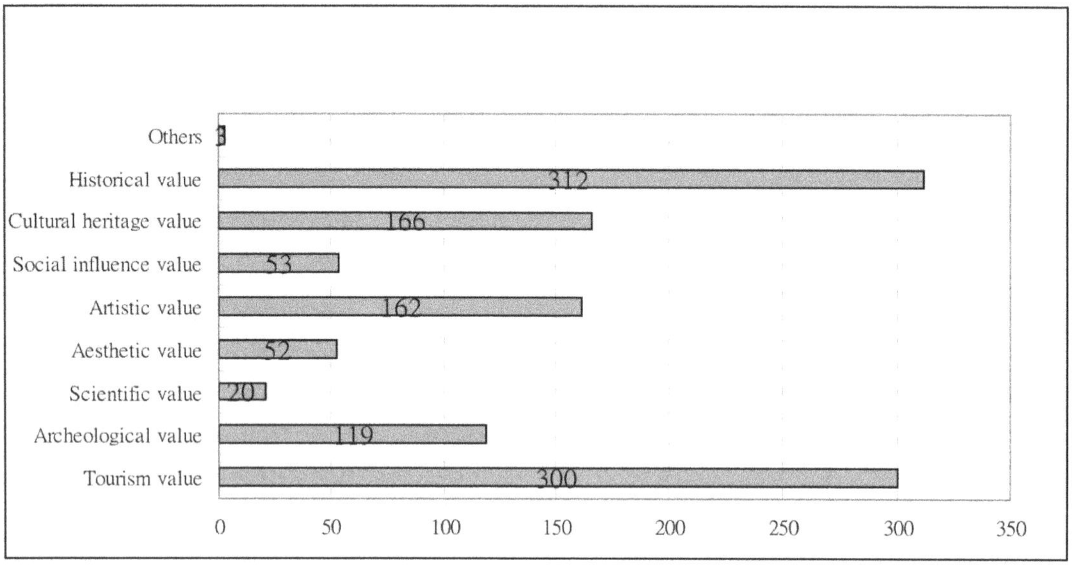

Overall Evaluation of Macau Cultural Tourism Scenic Spots by Tourists

According to the survey statistics (see Figure 11), amongst the overall evaluation of Macau cultural tourism scenic spots by tourists, 57.80% said they were quite good; 34.60% evaluated average; 6.00% evaluated very good; and 1.40% evaluated quite bad.

Majority of tourists had relatively good estimation on evaluation towards Macau cultural tourism scenic spots. It is recommended that this situation be maintained.

Tourists' Evaluation of Macau Cultural Tourism Products

According to the survey statistics (see Table 2 and the Figure 12), on the characteristics of scenic spots, 56.61 said they were very clear; on the extent of participation in the tourism event, 58.75% rated it average; on the originality or authenticity of the tourism event, 44.55% said they were quite clear; on the cultural value of the travel product, 48.44% rated it quite clear; on the rationality of the tourism itinerary, 53.89% rated it average.

According to the survey statistics and Figure 12, showing the evaluation of Macau cultural tourism products, majority of tourists considered the characteristics of scenic spots, originality or authenticity of the tourism product, cultural value of the travel product as having clear distinguishing features. This means that Macau cultural tourism products were favorably considered, but most tourists deemed the extent of participation in the tourism event and the rationality of the tourism itinerary as average. Thus, on the development of Macau cultural tourism products, an enhancement of tourism events and improvement on the design of the tourism itinerary are needed.

Tourists' Evaluation of Comprehensive Facility of Macau Cultural Tourism Historic Block

According to the survey statistics (see Table 3 and Figure 13), on accommodation, 41.25% tourists rated it quite good while 41.63% rated it average; on the aspect of catering, 52.14% said it was quite good; on the aspect of transportation, 43% rated it average; on the aspect of spots, 49.81% rated it quite

Figure 10. Factors influencing historical blocks tourism

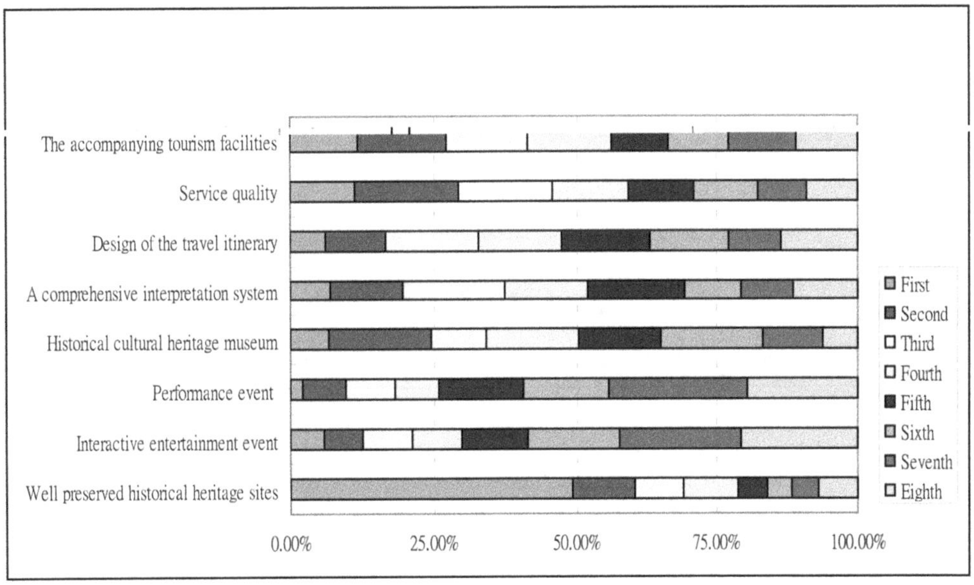

Figure 11. Overall evaluation of tourists

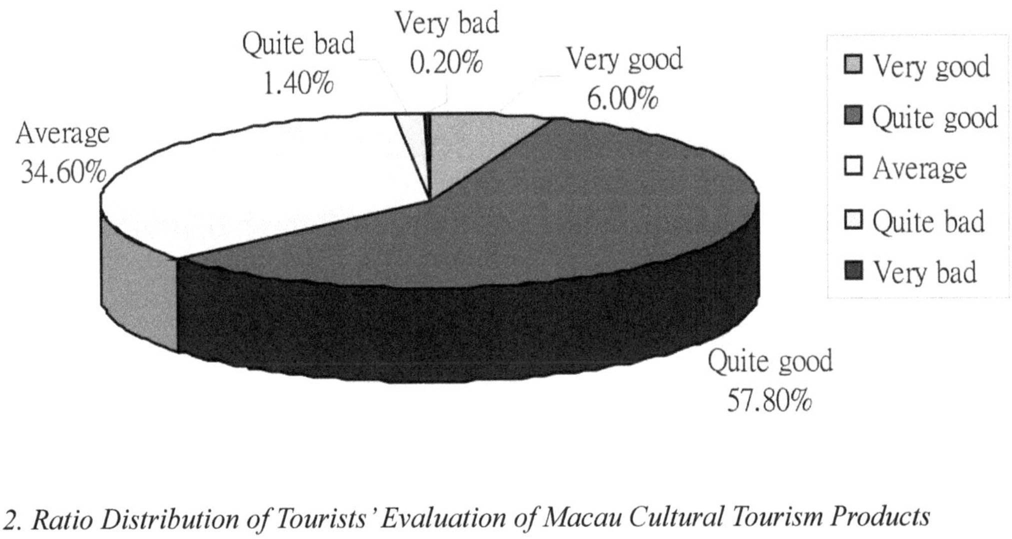

Table 2. Ratio Distribution of Tourists' Evaluation of Macau Cultural Tourism Products

Evaluation	Very clear	Quite clear	Average	Quite unclear	Very unclear
Characteristic of the spot	10.20%	58.20%	28.20%	3.20%	0.20%
The Extent of Participation in the tourism event	4.00%	29.00%	60.40%	6.40%	0.20%
Originality/Authenticity of the tourism product	8.20%	45.80%	43.80%	1.80%	0.40%
Cultural value of the travel product	10.80%	49.80%	36.20%	3.00%	0.20%
Rationality of the tourism itinerary	3.00%	32.80%	55.40%	8.20%	0.60%

Figure 12. Distribution of tourist products

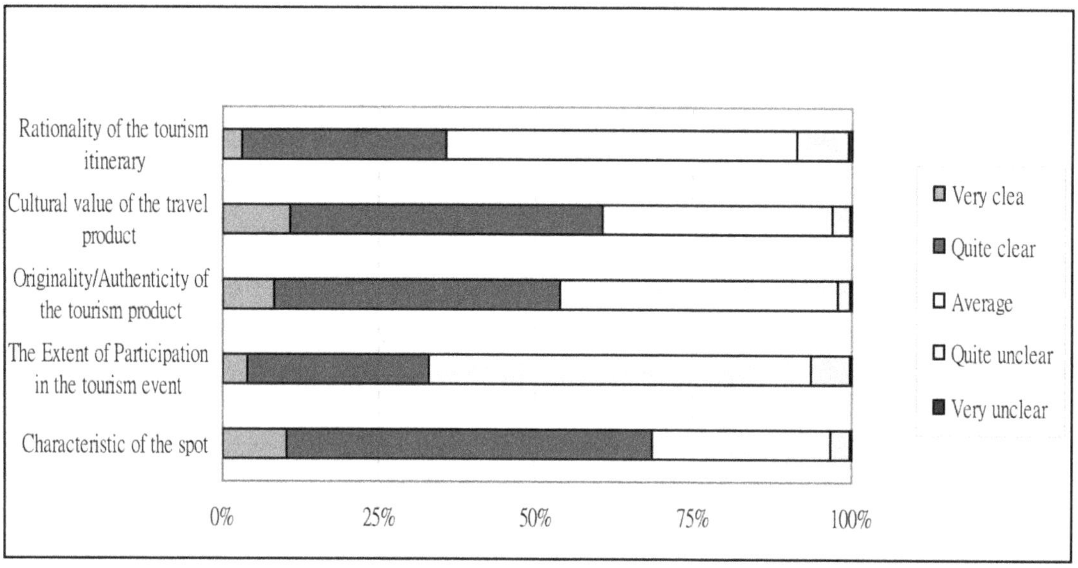

good; on the aspect of entertainment, 41.63% rated it quite good; on shopping, 46.5% said it was quite good; on tourist centre, 54.86% rated it average.

The evaluation of the comprehensive facilities of Macau cultural tourism historic block showed that more than 10% of tourists gave very good feedback of accommodation, catering, spots, entertaining and shopping, and only 25.29% selected tourist centre. Some 10% marked transportation and tourist centre as quite poor. The overall evaluation of the comprehensive facilities of Macau cultural tourism historic block was quite good, but with relatively poor rating on tourist centre and transportation. It is suggested that transportation and tourist centre be improved to promote the coordinated development of tourism facilities.

CONCLUSION

Tourists' Personal Information

On the age structure, young tourists comprised the majority (over 80%), thus, during the evaluation

Table 3. Ratio distributions of tourists' evaluation on comprehensive facilities of Macau cultural tourism historic block

Items	Very good	Quite good	Average	Quite poor	Very poor
Accommodation	13.40%	42.40%	42.80%	1.20%	0.20%
Catering	14.00%	53.60%	30.00%	2.00%	0.40%
Transportation	7.20%	36.60%	44.20%	9.60%	2.40%
Spots	8.40%	51.20%	38.00%	2.40%	0.00%
Entertainment	17.40%	42.80%	36.40%	2.80%	0.60%
Shopping	11.80%	47.80%	37.20%	3.00%	0.20%
Tourist Centre	5.80%	26.00%	56.40%	11.20%	0.60%

Figure 13. Evaluation of tourism facilities

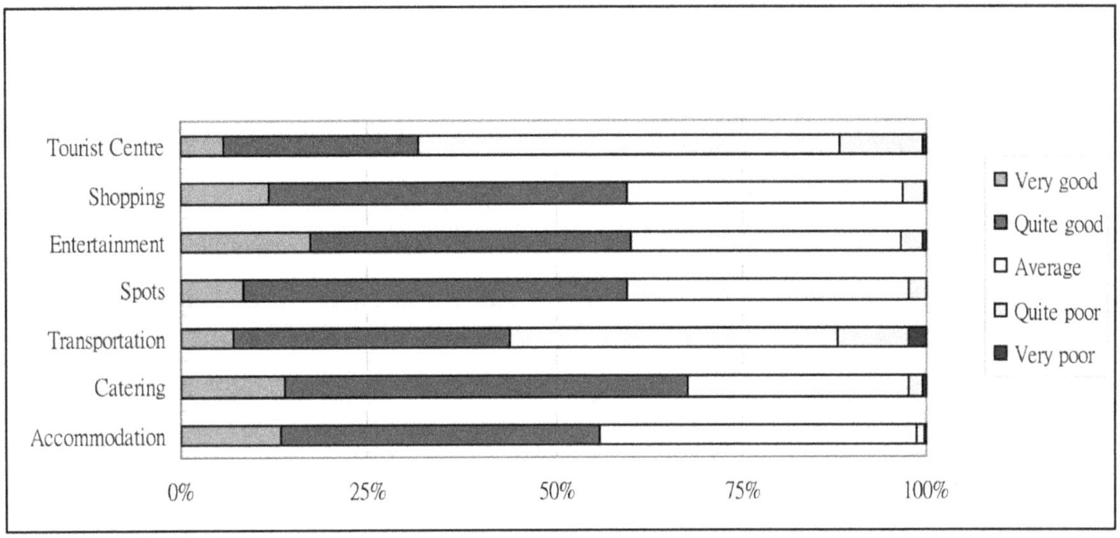

of tourist products and event development, it is suggested to take into consideration more features that cater to the preferences of a young market. On the tourists' nationality, Mainland China is the source that dominates the Macau tourism market. It is strongly suggested that features catering to mainland tourists be taken into consideration. On occupation and education, employees and students are the dominant ones. Thus the development of tourist product and event should emphasize their cultural value: not only develop practical products and events but also take into account those with cultural value.

Tourists' Consumption Behavior

Nearly 40% of tourists had visited Macau cultural heritage scenic spots for the second or third time. Obviously, the re-travel rate was relatively high. First time travelers also recorded a significant number. The way of travel was mainly individual travel (over 80%). The length of stay was relatively short, mainly day trips and two days plus one night (occurring over 80%) and the motivation was sightseeing and leisure. In the development of tourist products and events, it is suggested that while tourists' consumption is regarded, it should also include the two tourism motivations into the scope of consideration. On the information aspect, the source was mainly television and broadcasting, internet, guiding book and newspaper magazine. Thus, these channels are suggested to be the main channel of advertising Macau cultural tourism scenic spots. On the aspect of consumption structure, entertaining and entrance ticket occurred with less percentage. It is suggested to strengthen entertainment and development of scenic spots events in order to attract tourists and to stimulate their interests by offering more innovative and more distinguishing features and activities.

Tourists' Perception and Evaluation

Through the questionnaire survey and statistical analysis, a strong correlation between the scenic spots tourists had heard about and those they visited was identified. Therefore, advertising and marketing of Macau cultural tourism scenic spots is significant. On the tourists' evaluation of the Macau cultural tourism, majority of visitors believed in the historical and tourism value of Macau scenic spots, along with the completeness of Macau's historical heritage sites. In addition, tourism facilities and service quality are two factors that were paid attention to by tourists. On the aspect of tourism products, a majority of tourists considered that the participation of Macau cultural tourism events was not enough while rationality of the tourism itinerary was poor. On the accompanying tourism facilities, a majority of tourists considered the unavailability of a tourist centre and lack of transportation and hoped these aspects to be improved.

REFERENCES

Alzua, A., O'Leary, J. T., & Morrison, A. M. (1998). Cultural and heritage tourism: Identifying niches for international travelers. *Journal of Tourism Studies*, *9*(2), 2–13.

Black, N. L. (1990). A model and methodology to assess change to heritage buildings. *Journal of Tourism Studies*, *1*(1), 15–23.

Burton, R. (1995). *Travel Geography*. Pitman Publishing.

Chang, T. C., Milne, S., Fallan, D., & Pohlmann, C. (1996). Urban heritage tourism: The global-local nexus. *Annals of Tourism Research*, *23*(2), 284–305. doi:10.1016/0160-7383(95)00064-X

Cheung, S. C. H. (1999). The meanings of a heritage trail in Hong Kong. *Annals of Tourism Research*, *26*, 570–588. doi:10.1016/S0160-7383(99)00006-7

Cossons, M. (1989). Heritage tourism: Trends and tribulations. *Annals of Tourism Research*, *10*, 192–194.

Darvill, T. (1987). *Ancient Monuments in the Countryside*. London: English Heritage.

Dawes, J., & Rowley, J. (1996). The waiting experience: Towards service quality in the leisure industry. *International Journal of Contemporary Hospitality Management*, *8*(1), 16–21. doi:10.1108/09596119610108608

Edson, G. (2004). Heritage: Pride or passion, product or service? *International Journal of Heritage Studies*, *10*(4), 333–348. doi:10.1080/1352725042000257366

Edwards, J. A., & Coit, J. C. L. (1996). Mines and quarries: Industrial heritage tourism. *Annals of Tourism Research*, *23*, 341–363. doi:10.1016/0160-7383(95)00067-4

Falk, J., & Dierking, L. (1992). *The Museum Experience*. Washington, DC: Whalesback Publications.

Faulkner, B., Opperman, M., & Fredline, E. (1999). Destination Competitiveness: An Exploratory Examination of South Australia's Core Attractions. *Journal of Vacation Marketing*, *5*(2), 125–139. doi:10.1177/135676679900500202

Fridgen, J. D. (1992). *Dimensions of Tourism*. The Educational Institute.

Garrod, B., & Fyall, A. (2000). Managing heritage tourism. *Annals of Tourism Research*, *27*, 682–708. doi:10.1016/S0160-7383(99)00094-8

Gibson, C., & Hordman, D. (1998). Regenerating urban heritage for tourism. *Managing Leisure*, *3*(1), 37–54. doi:10.1080/13606719808559324

Graham, B., & Murray, M. (1997). The spiritual and the profane: The pilgrimage to Santigo de Compostela. *Tourism Management*, *9*, 44–62. doi:10.1016/0261-5177(88)90057-X

Hollinshead, K. (1997). Heritage tourism under post-modernity: Truth and the past. In Ryan, C. (Ed.), *The Tourist Experience: A New Introduction* (pp. 170–193). London: Cassell.

Howard, P. (2003). *Heritage: Management, Interpretation, Identity*. London: Continuum.

Jansen-Verbeke, M. (1999). Industrial heritage: A nexus for sustainable tourism development. *Tourism Geographies*, *1*(1), 70–85. doi:10.1080/14616689908721295

Johnson, N. C. (1996). Where geography and history meet: Heritage tourism and the big house in Ireland. *Annals of the Association of American Geographers. Association of American Geographers*, *86*(3), 551–566. doi:10.1111/j.1467-8306.1996.tb01766.x

Jolliffe, L., & Smith, R. (2001). Heritage, Tourism and Museums: the case of the North Atlantic Islands of Skye, Scotland and Prince Edward Island, Canada. *International Journal of Heritage Studies*, *7*(2), 149–172. doi:10.1080/13527250120060187

Keller, A. (1999). Jobs and the natural heritage in Scotland. *Countryside Recreation*, *7*(1), 6.

Kerstetter, D., Confer, J., & Bricker, K. (1998). Industrial heritage attractions: Types and tourists. *Journal of Travel & Tourism Marketing*, *7*(2), 91–104. doi:10.1300/J073v07n02_05

Kruse, R. J. (2005). The Beatles as place makers: Narrated landscapes in Liverpool, England. *Journal of Cultural Geography*, *22*(2), 87–114. doi:10.1080/08873630509478240

Light, D. (2000). Gazing on communism: Heritage tourism and post-communist identities in Germany, Hungary and Romania. *Tourism Geographies*, *2*(2), 157–176. doi:10.1080/14616680050027879

Lowenthal, D. (1994). Identity, heritage and history. In Gillis, J. R. (Ed.), *Commemorations: The Politics of National Identity* (pp. 41–57). Princeton, NJ: Princeton University Press.

Lowenthal, D. (1998). *The Heritage Crusades and the Spoils of History*. Cambridge, UK: Cambridge University Press. doi:10.1017/CBO9780511523809

MacDonald, R., & Jolliffe, L. (2003). Cultural rural tourism: Evidence from Canada. *Annals of Tourism Research*, *30*(2), 307–322. doi:10.1016/S0160-7383(02)00061-0

McKercher, B., & Du Cros, H. (2002). *Cultural Tourism: The Partnership between Tourism and Cultural Heritage Management*. New York: The Haworth Press.

Millar, S. (1989). Heritage Management for Heritage Tourism. *Tourism Management*, *10*(1), 9–14. doi:10.1016/0261-5177(89)90030-7

Moscardo, G., & Pearce, P. L. (1999). Understanding ethnic tourists. *Annals of Tourism Research*, *26*(2), 416–434. doi:10.1016/S0160-7383(98)00101-7

Poria, Y. (2001). Challenging the present approach to heritage tourism: Is tourism to heritage places heritage tourism? *Tourist Review*, *56*(1/2), 51–53. doi:10.1108/eb058358

Poria, Y. (2001). The show must NOT go on. *Tourism and Hospitality Research*, *3*(2), 115–119.

Poria, Y., Butler, R., & Airey, D. (2000). Clarifying heritage tourism: A distinction between heritage tourism and tourism in historic place. *Annals of Tourism Research*, *28*(4), 1047–1049. doi:10.1016/S0160-7383(00)00069-4

Poria, Y., Butler, R., & Airey, D. (2001). Tourism subgroups: Do they exist? *Tourism Today*, *1*(1), 14–22.

Poria, Y., Butler, R., & Airey, D. (2003). The core of heritage tourism: Distinguishing heritage tourists from tourists in heritage places. *Annals of Tourism Research*, *30*(1), 238–254. doi:10.1016/S0160-7383(02)00064-6

Poria, Y., Butler, R., & Airey, D. (2004). Links between tourists, heritage, and reasons for visiting heritage sites. *Journal of Travel Research*, *43*(8), 19–28. doi:10.1177/0047287504265508

Poria, Y., Butler, R., & Airey, D. (2004). How tourists decide which heritage site to visit? *Tourism Review*, *59*(2), 12–16. doi:10.1108/eb058431

Prentice, R. (1993). *Tourism and Heritage Attractions*. London: Rutledge.

Prideaux, B. R., & Kininmont, L. J. (1999). Tourism and heritage are not strangers: A study of opportunities for rural heritage museums to maximise tourism visitation. *Journal of Travel Research*, *37*(3), 299–303. doi:10.1177/004728759903700312

Raivo, P. J. (2002). The peculiar touch of the East: Reading the post-war landscapes of the Finnish Orthodox Church. *Social & Cultural Geography*, *3*(1), 11–24. doi:10.1080/14649360120114116

Reynolds, P. (1994). Culinary heritage in the face of tourism. *Progress in Tourism. Recreation and Hospitality Management*, *6*, 189–194.

Robinson, K. M. (2001). Unsolicited narratives from the Internet: A rich source of qualitative data. *Qualitative Health Research*, *11*(5), 706–714. doi:10.1177/104973201129119398

Romagnoli, G. F. (1995). Rome's Jewish quarter: A rich culinary heritage. *Gourmet*, *55*(8), 90–94.

Ryan, C. (1995). *Researching Tourist Satisfaction*. London: Routledge.

Shackely, M. (2001). Sacred world heritage sites: Balancing meaning with management. *Tourism Recreation Research*, *26*(1), 5–10.

Shackley, M. (1998). Introduction. In Shackley, M. (Ed.), *Visitor Management: Case Studies from World Heritage Sites* (pp. 1–9). Oxford, UK: Butterworth-Heinemann.

Shackley, M. (1999). Visitor management. In Leask, A., & Yeoman, I. (Eds.), *Heritage Visitor*.

Shettel-Neubar, J., & O'Reilly, P. (1981). *Now Where? A Study of Visitor Orientation and Circulation in the Arizona-Sonora Desert Museum*. Jacksonville, FL: Jacksonville State University.

Silberg, T. (1995). Cultural Tourism and Business Opportunities for Museums and Heritage Sites. *Tourism Management*, *16*(5), 361–365. doi:10.1016/0261-5177(95)00039-Q

Smith, H. D. (2001). *The British Maritime Heritage*. Paper presented at the RGS-IBG Annual Conference.

Strange, C., & Kempa, M. (2003). Shades of dark tourism: Alcatraz and Robben Island. *Annals of Tourism Research*, *30*(2), 386–405. doi:10.1016/S0160-7383(02)00102-0

Swarbrooke, J. (1993). The future of heritage attractions. *Insights*, *4*, D15–D20.

Swarbrooke, J. (1995). *The Development and Management of Visitors Attraction*. Oxford, UK: Butterworth-Heinemann.

Teo, P., & Yeoh, B. S. A. (1997). Remaking local heritage for tourism. *Annals of Tourism Research*, *24*(1), 192–213. doi:10.1016/S0160-7383(96)00054-0

Timothy, D., & Boyd, S. (2003). *Heritage Tourism*. London: Prentice Hall.

Voase, R. (1999). 'Consuming' tourist sites/sights: A note on York. *Leisure Studies*, *18*(4), 289–296. doi:10.1080/026143699374862

Wedow, S. (1977). Review of the tourist: A new theory of the leisure class. *Contemporary Sociology*, *6*(2), 200–202. doi:10.2307/2065807

Worden, N. (2003). National identity and heritage tourism in Melaka. *Indonesia and the Malay World*, *31*(89), 31–43. doi:10.1080/13639810304447

This work was previously published in International Journal of Asian Business and information Management, Volume 1, Issue 4, edited by Dasho Karma Ura and Patricia Ordóñez de Pablos, pp. 1-19, copyright 2010 by IGI Publishing (an imprint of IGI Global).

Chapter 7
Embracing Guanxi:
The Literature Review

Jilong Zhang
RMIT University, Australia

Nattavud Pimpa
RMIT University, Australia

ABSTRACT

Guanxi is deeply ingrained in Chinese society and business etiquette; therefore, it is critical to understand and appreciate it to gain a commercial foothold in China. This paper explores the fundamental differences between Chinese and Western approaches to business dealings, to identify the factors that make it difficult for Western managerial practises to adopt guanxi. The authors identify the differences in the perception of trust in both societies and how this ultimately leads to clashes in cultures. However, there is also a need for Western practices to be more flexible and appreciate guanxi and its implications if multinational corporations are to succeed in China. The paper concludes by suggesting a possibility for both approaches to co-operate well, given the relative success of each approach in their respective cultures.

INTRODUCTION

Since 1978, China's progress in economic reform has been one of the great economic success stories of the post-war era. China has become the world's seventh largest economy and second largest recipient of foreign direct investment (OECD, 2008). China's performance is all the more remarkable in that its reforms have been gradual and its development has occurred despite extensive, though declining, state ownership and intervention in the economy. The accession of China to the World Trade Organisation (WTO) marks an important milestone along the reform path.

China has been liberalising its international trade and investment policies since the mid-1980s and is now as open as some present WTO members. Although China stands to gain significantly

DOI: 10.4018/978-1-4666-0276-2.ch007

from the opening of its export markets under the terms of its accession, the depth and breadth of its commitments to liberalise access to its domestic economy are acknowledged to be more extensive than those agreed to by previous adherents to the WTO. This willingness reflects the fact that opening to international markets promotes market discipline, access to technology, and other qualities that have been important goals of domestic economic reforms. In this respect, WTO entry is a complementary aspect of the next phase of China's reforms.

Globalisation has led to opportunities for corporations to expand their businesses across borders and extend their reach into new markets. This makes understanding culture and its implications on business in the international arena vital for success. China is fast becoming a world superpower, but this new land of opportunity is only open to those who fully appreciate the complexities of business culture of the country. In pursuit of deeper understanding of what are the underlying cultural differences between Chinese and Western managerial approaches, this literature review attempts to explore the concept of "Guanxi," its origins, links to the Chinese societal system and criticisms on its implications on business practises – especially in regards to interactions between Chinese and Western firms. The literature attempts to identify the factors that may make it difficult to integrate 'guanxi' into Western managerial practise and in doing so investigates the differences in grassroots culture and business culture between Chinese and Western societies.

FOUNDATIONS OF GUANXI: THE LITERATURE REVIEW

While it is emphasised in the Western concept on "what you know," refers to technological expertise, including the price and quality of tendered product or service, it is emphasized in Confucian societies on "who you know," which refers to personal connections with the appropriate authorities or individuals. These connections are known in Chinese as guanxi, which is believed to be a striking feature that helps to enter a profitable and growing Chinese market. Thus, the issues of what indeed guanxi is and how it works in effect thus have become focal discussions

Guanxi literally means relation or relationships, ties or connections – and this already establishes the foundation on which to understand the Chinese approach to social network building and ultimately – business relationships. According to Pablos (2006) the origin of guanxi relates to the Confucianism heritage of Asian countries and Chinese communities in non-Asian countries. According to it, five major relationships dictate all relationships: Emperor–subject, father-son, husband-wife, elder-younger brothers and senior friend- junior friend. To ensure social harmony, order and stability, appropriate behaviours are needed. Lee et al. (2006) states that "With a history of more than 2500 years, Confucianism has exerted a fundamental influence on the Chinese and East Asian modes of thinking and way of behaving." Thus, by definition, 'Guanxi' embodies this influence, and so it is not surprising that "Guanxi is deeply embedded in the mindset of Chinese and is in every aspect of their personal and organisational interactions" (Park & Luo, 2001). The basis of these relationships also confirms Asian countries such as China, Japan, Korea, Taiwan and Singapore as relatively high power distance countries.

Guanxi is a good example of the development of cultural and economic dimensions. The context of cultural power includes intellectual factors (such as technology, education, etc.); spiritual factors (such as ideal, belief, ethic, value, etc.) and the social-cultural network affect our daily life along in addition to the power of traditional culture (Gia, 1998). Economics and culture are not developing as two parallel lines. They interact and affect each other reciprocally. Based upon the economic development collaboration, productivity increase, and social progress, the power of various cultures integrates and externalizes with the need of overall development from the power of culture subjectively constructed in the process of cultural value and civilization creation, which pushes, guides, condenses, and inspires the economic development (Gia, 1998)

According to Hofstede's (2001), one dimension of culture is power distance – which distinguishes the level of inequality in social and work hierarchies that is tolerated by the people – and their respect of the rights of power and authority of those high social hierarchical standings. Leung and Wong (2001) state that the "social hierarchical theory (that is guanxi's five relationships) has prompted almost all Chinese rulers to adopt Confucianism as a strategic tool to achieve social stability in the vast Chinese continent." Already there is theory of culture in international management that supports the concept of 'guanxi,' which emphasises the importance of understanding this concept.

GUANXI AND CONFUCIANISM

Guanxi interaction has been viewed by many scholars as an idiosyncratic cultural phenomenon (Lee & Wong, 2001) where Confucianism is the major life philosophy. In China, guanxi has been a way of life since time immemorial, thus the personal investment required to develop and maintain good social relations is accepted as an unavoidable fact of life. For each specific relationship, people in China "make out" through interpersonal accommodation and negotiation (Lee & Wong, 2001).

With respect to social context, Chinese culture provides a tightly knitted social framework in which individuals are protective of one another and organizations safeguard their members' interests. It projects a preference for cooperation, group decision-making, and relationships. Thus, the importance of networks lies in their emphasis on collectivism and group harmony. Social status is determined largely by guanxi system to which one belongs – a determination often preordained by the composite membership of one's extended family. The Chinese strive to keep relationships among guanxi members stable and harmonious since guanxi is the basis on which they exchange a lifetime of favors, resources, and business leverage (Pearce & Robinson, 2000).

Social norms, though somewhat are different around the world, mostly are attached to the social status, which is the position where people stand in a social system. And the regulation integrated for a certain status is the "role" (Hui & Graen, 1997). For someone who takes the proper role and plays it accordingly receives high praise for being qualified for his social status. Wu-lune, the basic norms of guanxi, is the main thought of the ideal type of manhood and patriarchy and filial piety and the spirit of trust and justice in regulating different levels of guanxi and setting them in hierarchical order with priorities in Chinese societies. It consists of five levels: the monarch (and his subjects), the father and his son, brothers, friends, and the husband and his wife In this guanxi system, while the three basic guanx*i* of the father and his son, brothers, and the husband and his wife are inside of the family, the political guanxi of monarch and his subjects is up-pushed to the national level, and guanxi of friends contains the rest of those who are outside of the families. From the perspective of Chinese societal progress, the "family – nation – world" is the structure from inside out where the family ethic is the basis of Confucius' human relations, where each one ought to be committed, responsible, and content to his given role. As long as every member in the group follows the rules, the whole society stays in harmony. As time goes by, the Confucius followers elaborate and report Confucius' doctrine to the societies in accordance with the trend of the times. Thus, there is bifurcation of Confucianism, such as neo-Confucianism and others, which states the difference in the implementation of it; nevertheless the core thoughts basically stay in common.

Guanxi is characterized into categories in accordance with its bases (Hui & Graen, 1997). However, no matter, how guanxi is classified and implemented theoretically, in practice there are certain rules for each role played among parties in accordance with Wu-lune. And rules are often borrowed when guanxi cannot be clearly defined. For example, a business relationship built between strangers is seen as most distant unless credentials and a mutual sense of identities are established in terms of insiders, such as the "Zijiaren", then the guanxi can be upgraded and seen either as brothers or friends, and the non-written rules, "Hsin" for brothers or "Yi" for friends is borrowed from Wu-lune as metaphors, with which people involved in business follow along without suspicions. In markets, people see one invisible hand – price, while in China, another hand – guanxi, is seen as a potential solution for most problems of entering and operating in China. While guanxi plays the role as the "protocol," setting the rules governing the communication and the transfer of affection between individuals in Chinese societies, the teaching of Confucianism deeply rooted in effect constructs the norms of guanxi for its establishment and maintenance. Although in this paper we limit our discussion within the phase of guanx*i* and business, we can still see guanxi networks knitted across, reach out broadly, and wide spread into every corner of Chinese society, which has essential influence in multi and inter dimensions.

With respect to social context, Chinese culture provides a tightly knitted social framework in which individuals are protective of one another and organizations safeguard their members' interests. It projects a preference for cooperation, group decision-making, and relationships. Thus, the importance of networks lies in their emphasis on collectivism and group harmony. Social status is determined largely by guanxi system to which one belongs – a determination often preordained by the composite membership of one's extended family. The Chinese strive to keep relationships among guanxi members stable and harmonious since guanx*i* is the basis on which they exchange a lifetime of favours, resources, and business leverage (Pearce & Robinson, 2000).

GUANXI: THE WESTERN AND CHINESE PERCEPTIONS OF ESTABLISHING 'TRUST'

One aspect of guanxi is "xinyong" which directly translates to 'credit' but which is equivalent to what Western cultures know as trust. Although they are similar in definition, the difference in concepts of trust is distinguishable between Eastern and Western cultures. "Trust, in Chinese business relationships, is more based on one's credit of ability to return favours than relied on one's integrity and competency, as in Western relationship context" (Wang, 2005). Wang (2005) also argues that members of a guanxi web are held together by an invisible and unwritten code of reciprocity and members are accountable by the 'threat' of losing face - which embodies an individual's integrity, honour and personal equity. "The '**theory of reasoned action**' assumes intention as a main predictor of behaviour; it suggests that behavioural intentions can drive individual behaviour and intentions could be determined by attitudes (general feelings) and subjective norms" (Fishbein & Ajzen, 1975). In support, Huang et al. (2008) propose that general feelings comprise of salient beliefs including a perceived possibility and judgment of consequences resulting from the behaviour" whereas subjective norms are "an individual's perception about how people important to them judge the behaviour." If the attitudes and subjective norms do not match up within the reasoned action theory, this would result in damaging one's own social functions and interactions within the whole network. This suggests that establishing personal relationships as a foundation for business activities is key in Chinese business culture. The difference is that Western companies focus on a 'transactional strategy' with the supplier organisations; Western business relationships are short term and contracted based. "Once a contract is performed, no further obligations will apply between the buyer and the supplier" (Lee et al., 2006). It can be seen that according to the different approaches and cultural norms between Eastern and Western cultures, trust is also perceived and treated differently.

"Xinyong" often relies on one's oral commitment and ability to fulfil his/her promise and obligation; in certain ways differ from Western concepts of credibility, which largely rely on written agreements and legal contracts (Wang, 2005). Wang (2005) states the difference with the Western system is its reliance and trust placed on the system rather than the individual. In contrast to this idea of individualism and focus on the individual- as proposed by Wang (2005), Farh et al. (1998) argues that with business relationships, a "West-based independent self" is created that differs from an "interdependent self" as with Chinese-oriented social relations. Farh et al. (1998) states that "social relations are accorded great significance, and relationships are often seen as ends in themselves rather than means for realizing individual goals." In support, "this interdependent self would be associated with Hofstede's (2001) construct of collectivistic identity (Song & Werbel, 2007)." In contrast to Chinese culture, according to Michailova and Worm (2003), personal networking in the west is based primarily on individualism.

Edward T. Hall's research into cultural contexts and the theories of "context-space-time" allow the categorisation of counties based on their behaviours and preferences. Characteristics of high-context cultures include long lasting relationships and the preference of agreements between persons being spoken rather than written. Thus, a connection can be drawn between these characteristics and the Chinese concept of 'xinyong (trust and credibility)' and 'guanxi.' Western companies contemplating entering the Chinese market must make it a top priority to understand "guanxi" and exercise the concept to maintain relationships with their business partners. According to Styles and Amber (2003), China is a "relational society" and the low-trust nature of this society, "relationship orientation is only applied to insiders of "guanxi" network but not to outsiders of networks such as a foreign firm." Again this differs from a Western perspective, where most Western countries are high-trust cultures "in which people can trust each other even though they do not have blood relations (Fukuyama, 1995)." Thus because of a general distrust of outsiders in Chinese culture, it cannot be automatically assumed that the Chinese will enter into business freely. "To develop Guanxi is to form the basis for a gradual transition from outsider to an insider so that a long-term relationship can be built" (Haley, Tan, & Haley, 1998). Lee et al. (2006) supports this notion by suggesting that the development of an effective guanxi network has an important role in helping to gain a commercial foothold. Foreign investors most often establish business linkages through a joint venture or wholly owned subsidiary. Under either option, they must be able to gain access to the appropriate guanxi webs, through their Chinese partners, in order to facilitate quick acceptance of the investor.

To oppose this however, Wong (2007) states that in the West, emphasis is put on "what you know," referring to technological expertise, whereas in Confucian societies, emphasis is put on "who you know," which refers to personal connections with the appropriate authorities and individuals." This suggests to me that it is actually the Chinese society that is relatively trusting – because relationships are encouraged on the personal level – whereas in the West, contracts and legal papers are set in place to 'protect' intellectual property.

THE CLASH OF BUSINESS CULTURES

Criticisms of both "guanxi' and Western approaches are applicable when trying to find a link between how the concept of 'trust' is interpreted between cultures and how this can affect establishing business relationships. Yum (1988) suggests that due to the nature and dynamic of guanxi relationships – 'favours' can be paid back in the long run with the expectation that the relationship will last into the unforeseeable future. This suggests that one does not simply pay back his/her 'debt,' but could hold onto it until the right time when the other party needs it. By extension, Hwang (1987) suggests that because of this the degree of reciprocity in guanxi relationships cannot be objectively measured. In contrast, from a Western perspective, the advantage of a set agreement and the use of legal contracts would be that the terms of repayment and method of repayment (cash debts) are established – and are much easier to document and track. The reason why the concept of "guanxi" is so influential in business relationships in China is it provides an alternative dispute resolution mechanism – since formal legal systems and proper distribution channels are still at a relatively immature stage (Lee et al., 2006) – compared to the West. Wang (2005) states that the guiding principles of relational behaviour in guanxi are driven by morality and social norms; This is supported in the literature by Lee et al. (2006) who confirm that concepts such as guanxi have a major influence on both social and business norms."

The conflict would occur when there is a clash of belief systems and differences in what is perceived as social and business norms. "Thus, within the Chinese business context there is this ambiguity between elements of the long term business relationship and the short term transaction" (Lee et al., 2006). Western based business may be reluctant to do business whereby 'debts' are non-measurable and unpredictably outstanding – and Chinese based firms may find it hard to trust Western firms who do not appreciate the value of long term relationships which may imply their untrustworthiness. Thus, the interpretation of trust is one factor that may make it difficult for relationships to form.

Also prominent is how ethical "guanxi" is perceived. The positioning of guanxi may be the factor that leads to the confusion and misinterpretation of the activities involved. "In the process of cultivating guanxi, gifts are normally used (Brunner et al., 1989)." "Gift giving and wine-and-dine are the means of establishing and maintaining guanxi, but the activities are not equivalent to corruption (Leung & Wong, 2001)." Leung and Wong (2001) elaborate by stating that "outright bribery may be good enough to get a business transaction done on a one-off basis, but it cannot produce interpersonal relationships" that are the foundation of healthy long-term relationships. It may be the ethnocentric nature of the West to only take this apparent corruption into account when labelling "guanxi" practices.

EMBRACING GUANXI: A SHIFT IN PERCEPTION

From a buyer-supplier perspective - which may ultimately mirror the relationship between Chinese and Western business partners – relationships are categorised into two types; "adversarial competitive" and "collaborative relationship" (Humphreys et al., 2001). In support of this, the literature by Shapiro (1986) elaborates by arguing that the adversarial approach is dependent on three major activities, one being "the buyer assumes an arms'-length posture and uses only short term contracts." This method forgoes the need or implication of a long-term relationship because the buyer is wary of "not making direct use of the total resources of the supplier." On the other hand, "Collaborative relationships require trust and commitment for long-term co-operation along with a willingness to share risks (Humphreys et al., 2001)" and this is emphasised by the need for "efficient communication at all levels, open information sharing and continuous inter and intra-improvements." Many manufacturers now recognise that their ability to become world-class competitors is based to a great degree on their ability to establish high levels of trust and co-operation with suppliers (Humphreys et al., 2001)." Ohmae (1989) advocates that in a world of converging consumer tastes, rapidly spreading technology, escalating fixed costs and growing protectionism, more collaborative relationships with suppliers are crucial instruments for serving customers in a global environment.

Instead of seeing it as an obstacle to the prosperity to be had in China, Lee et al. (2001) proposed that guanxi "has positive effects on organisational performance and offers good means of facilitating transactions by allowing access to limited resource and information, preferential treatment in business dealings and protection from external threats." In other words, guanxi seems to have a strong positive effect on the collaborative relationship among buyer and suppliers.

The benefits of a guanxi relationship are aligned with the benefits of joint venture, which include the ability for parent country firms to utilise partner's local knowledge of the market. Establishing and maintaining strong guanxi networks can also allow Western firms to bypass host nation government regulations associated with joint ventures as well as address the long lead time to first sale, especially associated with Greenfield wholly-owned subsidiary operations (International Trade Operations, 2007). In relation to this, Lee et al. (2006) note that "given the uncertainty and confusion in China's transactional economy, firms can use guanxi as an entrepreneurial tool to bridge gaps in information and resource flows between unlinked firms and between firms and important outside stakeholders."

With this, it reasonable to suggest that Western business culture can embrace guanxi and use it to their advantage in their ventures in China (or most East Asian and south-East Asian countries) – and this would be a major shift in perception. In fact, the concept guanxi - of utilising personal relationships and connections to facilitate business is not a new idea – and for that matter not restricted by any means to China. In the literature, Walder (1986) has noted that the "term *'blat'* in Russia and *'pratik'* in Haiti refer to the same type of instrumental–personal connections." Spicer (2008) adds that one of the most interesting countries at the moment is Russia too. There are "stories of multinationals being unable to get any idea of what kind of benefits and rewards are paid (in terms of compensation in HR policy), partly because everything is done on such an informal basis... and it is not because of anything illegal, it's just the way things are done there." Thus, although the different forms of guanxi across the world will not be mirroring each other exactly, the use of relationships to facilitate business transactions is common "particularly when the business environment is uncertain and an adequate legal system is not available (Lee et al., 2006)." Park and Luo (2001) establish the link between transactional relationships between both Western and Chinese business cultures by stressing the importance of guanxi "by suggesting that while a relationship follows successful transactions in the West, transactions often (also) follow successful guanxi in China." In support, the "system of guanxi has led to business success in Chinese societies reflecting the ability to be flexible and adapt through the quick reaction within networks based on trust and reciprocity" (Buttery & Wong, 1999). This implies that there is a possibly that the two concepts have enough similarity to both work successfully, and may be transferable between both cultures and serves to confirm the benefits and the need to embrace guanxi or at least appreciate the concept.

IMPLICATIONS OF GUANXI IN THE BUSINESS CONTEXT

The literature confirms a strong relationship among societal links that Guanxi possesses in Chinese society (or society which are strongly influenced by Confucious philosophy). With 2500 years of history, and based on the principles' of Confucianism - guanxi is not just a simple concept;

it is a complex, unwritten rule that binds social and business relationships in Chinese society and is a foundation of Chinese culture. Amongst the major differentials between guanxi and Western business approaches is the perception and context of trust, and how it is expressed in both cultures. In Chinese society, trust is the foundation and guideline of guanxi – and dictates how it is conducted and ensures that it works. This is possible because of the importance of 'face' or personal dignity of each member of a guanxi web. This differs from the Western approach, which is focused on a transactional strategy with business partners – where trust embodies the belief in each party to perform their part of a written contract or agreement. In this approach, the personal attributes are removed, which is consistent with the Western adage, "you don't mix business with pleasure." This suggests that it is the Chinese approach of guanxi that is more 'trusting,' even though Chinese society is a low-trust, relational society. The guanxi approach is flexible enough to accept those within a guanxi web, and to allow those who are outsiders – where as trust is harder to establish (but easier to encapsulate) with the Western approach, which relies of contracts and agreements that advocate short-term commitments and hinder long term relationships. Thus, criticisms can apply from both sides of the spectrum – which the Chinese wondering why Westerners are so restricted and afraid of commitment, whilst Westerners remain uncomfortable with the subjective nature of guanxi, as well as the apparent informal structure of business relationships – that are not held together by legality and written guidelines.

As globalisation progresses and China continues to grow as an economic superpower – the West may have to ultimately succumb to the influence of guanxi, and a major shift in perception is required to do this. Guanxi should not be seen as an obstacle to establishing business relationships across cultures – it should be viewed as a tool to facilitate it. Western managers should accept that the apparent informality of business relationship structure in China is nevertheless effective – and has been across time and other cultures. Although this concept may not be easily comprehended, there is evidence to support that it is the significant step to business success in China.

REFERENCES

Buttery, A. E., & Wong, Y. H. (1999). The development of a Guanxi framework. *Marketing Intelligence & Planning, 17*(3), 147–154. doi:10.1108/02634509910271605

Farh, J. L., Tsui, A. S., Xin, K. R., & Cheng, B. S. (1998). The influence of relational demography and guanxi: the Chinese case. *Organization Science, 9*(4), 471–488. doi:10.1287/orsc.9.4.471

Fishbein, M., & Ajzen, I. (1975). *Belief, attitude, intention, and behavior: An introduction to theory and research.* Reading, MA: Addison-Wesley.

Fukuyama, F. (1995). *Trust: The social virtues and the creation of prosperity.* New York: The Free Press.

Haley, G. T., Tan, C. T., & Haley, U. C. V. (1998). *New Asian emperors.* Burlington, MA: Butterworth Heinemann.

Hall, E. T. (1966). *The Hidden Dimension.* New York: Doubleday.

Hofstede, G. (2001). *Culture's Consequences, Comparing Values, Behaviors, Institutions, and Organizations across Nations.* Thousand Oaks, CA: Sage Publication.

Huang, Q., Davison, R., & Gu, J. (2008). Impact of personal and cultural factors on knowledge sharing in China. *Asia Pacific Journal of Management, 25*, 451–471. doi:10.1007/s10490-008-9095-2

Hui, C., & Graen, G. (1997). Guanxi and professional leadership in contemporary Sino-American joint ventures in mainland China. *The Leadership Quarterly, 8*(4), 451–465. doi:10.1016/S1048-9843(97)90024-2

Humphreys, P. K., Shiu, W. K., & Chan, F. T. S. (2001). Collaborative buyer-supplier relationships in Hong Kong. *Supply Chain Management: An International Journal*, *6*(4), 152–162. doi:10.1108/EUM0000000005708

Hwang, K. (1987). Face and favour: The Chinese power game. *American Journal of Sociology*, *92*(4), 944–974. doi:10.1086/228588

Lee, D. J., Pae, J. H., & Wong, Y. H. (2001). A model of close business relationships in China (guanxi). *European Journal of Marketing*, *35*(1), 51–69. doi:10.1108/03090560110363346

Lee, K. C. P., & Humphreys, K. P. (2006). The Role of Guanxi in Supply Chain Practises. *International Journal of Production Economics*, 450–467.

Michailova, S., & Worm, V. (2003). Personal networking in Russia and China: Blat and Guanxi. *European Management Journal*, *21*(4), 509–519. doi:10.1016/S0263-2373(03)00077-X

Ohmae, K. (1989). The global logic of strategic alliances. *Harvard Business Review*, *67*(2), 143–154.

Pablos, P. (2006). Western and Eastern views on social networks. *The Learning Organization*, *12*(5), 436–456. doi:10.1108/09696470510611393

Park, S., & Luo, D. (2001). Guanxi and organizational dynamics: Organizational networking in Chinese firms. *Strategic Management Journal*, *22*(5), 455–477. doi:10.1002/smj.167

Shapiro, R. (1986). *Towards effective supplier management: international comparisons* (Working Paper No. 9-785-062). Boston, MA: Harvard Business School.

Song, J. L., & Werbel, J. D. (2007). Guanxi as impetus? Career exploration in China and the United States. *Career Development International*, *12*(1), 51–67. doi:10.1108/13620430710724820

Spicer, S. (2008). *Think global, act local: International Employee Benefits*. Retrieved from http://cendoc.esan.edu.pe/paginas/infoalerta/odp/InfoAlertaODP.html

Styles, C., & Ambler, T. (2003). The coexistence of transaction and relational marketing: Insights from the Chinese business context. *Industrial Marketing Management*, *17*(6), 539–557.

Walder, A. G. (1986). *Communist Neo-Traditionalism: Work and Authority in Chinese Industry*. Berkeley, CA: University of California Press.

Wang, L. C. (2005). Guanxi vs. relationship marketing: Exploring underlying differences. *Industrial Marketing Management*, *36*(1), 81–86. doi:10.1016/j.indmarman.2005.08.002

Wong, M. (2007). Guanxi and its role in business. *Chinese Management Studies*, *1*(4), 257–276. doi:10.1108/17506140710828532

Yum, J. O. (1988). The impact of Confucianism on interpersonal relationships and communication patterns in East Asia. *Communication Monographs*, *55*(4), 374–388. doi:10.1080/03637758809376178

This work was previously published in International Journal of Asian Business and Information Management, Volume 1, Issue 1, edited by Dasho Karma Ura and Patricia Ordóñez de Pablos, pp. 10-22, copyright 2010 by IGI Publishing (an imprint of IGI Global).

Section 2
Marketing in Asia

Chapter 8
A Study of Customer Satisfaction of an Automotive Company in Malaysia

Noor Azlina Fakiruddin
International Islamic University Malaysia (IIUM), Malaysia

Hebat Allah A. Wagih
International Islamic University Malaysia (IIUM), Malaysia

Waleed F. Faris
International Islamic University Malaysia (IIUM), Malaysia

ABSTRACT

This paper was conducted to present an analysis of the current level of customer satisfaction on the services offered by an automotive outlet in Malaysia. In an effort to make the company more effective, many have targeted various change programmes to meet customer satisfaction. Customer satisfaction is one of the most important factors in the organizations success, and an important determinant of customer satisfaction is how well the product performs. The objective of this study is to help the organization to improve their customer level of satisfaction through providing continuous efforts to improve their level of service and product productivity.

INTRODUCTION

This outlet mainly handles used cars as well as cars which are traded in when buyers purchase new ones of the same kind. This company currently has 3 centers, in Malaysia and will open 4 new centers this year. It handled 9,000 trade-in cars last year, and 5,000 cars to date this year. This figure is expected to hit 12,000 by year end, and 40,000 by 2010 with a total of 20 centers. The company guides customer and gives tips on choosing and buying the right car. They also show how to take care of the car, so that it will always be in great condition.

The services provided by the center are: 38 Point checks, car body care tips, tires check, start your engines, a lesson before buying, listen to your car, changing your transmission fluid, engine fault

DOI: 10.4018/978-1-4666-0276-2.ch008

finding tips, maintaining your car, the right oil for the right car, check your engine oil.

This outlet is a second hand car dealer for an international brand Malaysia. This outlet company operates as a highly sophisticated Motor Trader car for an international brand with operation throughout Malaysia. A recognized and trusted leader in quality, technology and supply change management Motor Trader, and competitive advantage to their customer by improving time to time and scalability.

The objective of the study is as important as for the organization to recognize and improve their product and service productivity. It will help organization to realize the level of their customer satisfaction.

LITERATURE REVIEW

This paper aims to measure the customer satisfaction of a certain car consumers. We need to identify the customer and the customer satisfaction focusing on their effect or benefit, and their factors. Next, we will relate TQM with customer satisfaction and how it can lead to the satisfaction. Finally, we will discover what the factors that may not lead to repurchase are.

Customer is Group or individual who has a business relationship with the organization--those who receive and use or are directly affected by the products and services of the organization (Interoperability Clearinghouse Glossary of Terms, n.d.). Recently, relationships with customers are vital because they can easily evaluate the cost of various products and easily switch buying habits with minimal inconvenience (Hellriegel & Slocum, 2004). On other hand, they define *Satisfaction* as an "overall customer attitude towards a service provider" (Levesque & McDougall, 1996). Based on the differentiation strategy Hellriegel and Slocum (2004) suggest to provide the customer with something unique and make the organization's product and services distinctive from its competition. The managerial intend behind this strategy is that the customer will pay higher price for the distinctive product. Because the superior value is achieved through higher quality, or some special appeal, Toyota's strategy with Lexus is based on exceptional manufacturing quality, the use of genuine wood paneling, advanced sound system, high engine performance, and comparatively high fuel economy (for luxury cars) (Hellriegel & Slocum, 2004).

Customer satisfaction is a measure of the degree to which a product or service meets the customer's expectations (Industry Canada, n.d.). It is a business term, a measure of how products and services supplied by a company meet or surpass customer expectation. Customer satisfaction is seen as a key performance indicator within business (Wikipedia). It is a psychological and physical variable which associated with satisfaction behaviours. However, it varies from person to person and product/service to product/service. The level of satisfaction can also vary depending on other options that the customer can find and compare it with other organizations' products. Clearly, motorcars' companies as Toyota is one of various field that need to have customer service division and concern about customer satisfaction. Toyota must have consumer care process and customer service should be one of the most important objectives for them.

Research conducted by Athanassopoulos (2000) indicates that product innovations, staff service, price, convenience and business profile are all determinants of customer satisfaction.

BENEFITS OF EFFECTS OF CUSTOMER SATISFACTION

- Build strong customer connections produce *enhanced financial benefits*, a more complete view of customer requirements is needed, which *incorporates an understanding of the emotional dimensions*

Figure 1. Benefits of customer satisfaction (adapted from Fleming & Asplund, 2007)

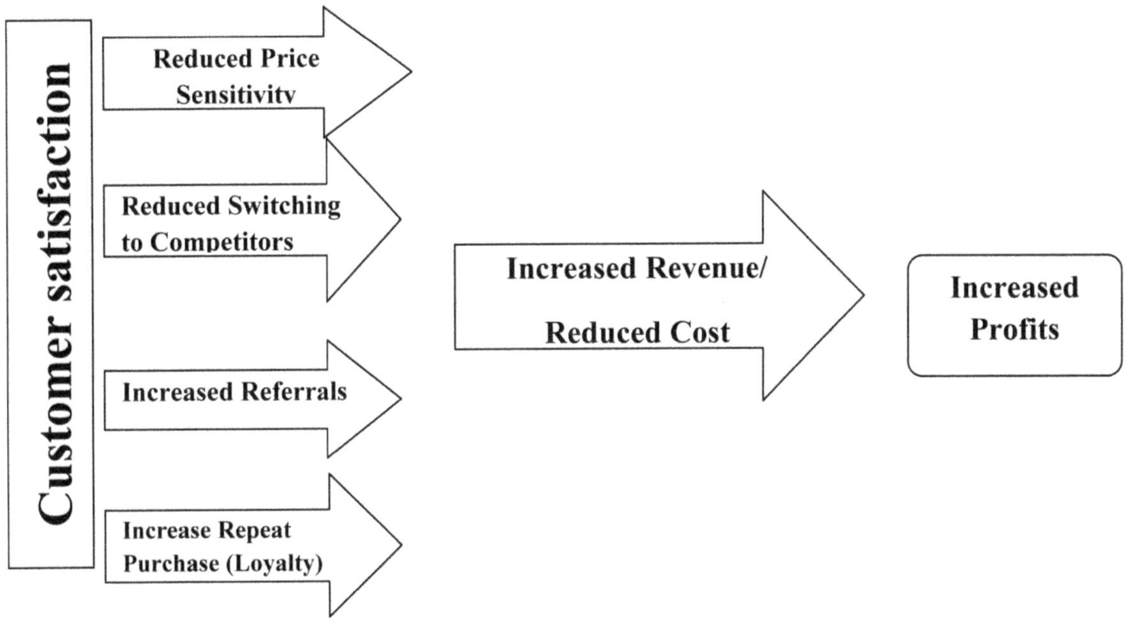

of customer commitment (Fleming, & Asplund, 2007) (Figure 1).

- Satisfaction does have a modest effect on primary store customer share, as predicted; this effect is moderated by shopper characteristics. Specifically, satisfaction is clearly a more important determinant of customer share for consumers with a low economic orientation than for consumers with a high economic orientation (Magi, 2008).

FACTORS LEAD TO CUSTOMER SATISFACTION

At this part we will manifest the factors that had been used in the survey to measure the customer satisfaction of these car outlet customers. These factors are:

1. Shopping for your vehicle.
2. Main entrance, Reception and Vehicle Display Area.
3. *Sales negotiation*. Satisfied customers are less price sensitive, buy additional products, are less influenced by competitors and stay loyal longer (Zineldin, 2000)
4. Sales dealing.
5. *Delivery*, the customers always develop perceptions during the service delivery process and subsequently they compare their perceptions to their expectation in evaluating the outcome of the service encounter
6. Your overall experience.
7. Intention to recommend/ repurchase.

PROBLEM STATEMENT

Customer satisfaction is the main goals for the target Company. Like other organization, it's important to be clear on exactly what's meant by the term. Customer satisfaction is the state of mind that customers have about a company when their expectations have been met or exceeded over the lifetime of the product or

Table 1. Interval scale section of questionnaire

Unacceptable	Dissatisfied	Somewhat Dissatisfied	Somewhat Satisfied	Satisfied	outstanding
1	2 3	4 5	6 7	8 9	10

service. There are some important implications of this definition, first because customer satisfaction is a subjective, no quantitative state, measurement won't be exact and will require sampling and statistical analysis. Second customer satisfaction measurement must be undertaken with an understanding of the gap between customer expectations and attribute performance perceptions. Finally there should be some connection between customer satisfaction measurement and bottom-line results. Measuring and managing customer satisfaction, will takes a continuous effort to maintain a high customer satisfaction levels (Cacioppo, 2007). Organizations in "Sales" are really in the Customer Satisfaction business. Almost every successful sales person will agree that it is much easier to work with a previous customer or a referral than with a new prospect. Studies have shown that any drop from total satisfaction results in a drastic decrease in loyalty--a drop from Very Satisfied to Satisfied results in a decrease of 83% in repeat business (Pearcea, 2001).

The Company goals to get valuable feedback from all its customers in Malaysia that will assist them in continuous efforts to improve product and service.

The survey based on few variables that important or relates to the company nature of business:

1. To identify the level of customer satisfaction on their customer service and vehicle display.
2. To identify the level of customer satisfaction on their sales negotiation
3. To identify the level of customer satisfaction on their sales dealing
4. To identify the level of customer satisfaction on their product and service delivery
5. To identify the level of customer satisfaction on customer overall experience with the company.

THE SURVEY DESIGN

The processes are

- Identify the data information
- Collection of data information
- Analyze the data
- Disseminate the information

The nature of the study is to analyze level of customer's satisfaction based on selected variable. The time horizon in this survey is cross sectional study because the data are gathered just once where the questionnaires are distributed to customers at the outlet, where they purchases their vehicle.

DATA COLLECTION METHOD

The data collection method is an important process in this study. The data are collected through:

Primary Data

Questionnaires were distributed to respondents who are the customers at the outlet. In this survey, the questionnaires consist of close-ended question. Questionnaire used an appropriate scale as a mechanism of answering the questionnaire.

In using the interval scale, the responses were categorized into six categories. The six responses categories are 1-(Unacceptable), 2-3 (Dissatisfied), 4-5 (Somewhat Dissatisfied), 6-7 (Somewhat Satisfied), 8-9 (Satisfied), 10 (outstanding) (Table 1).

Secondary Data

Data gathered can be internal or external to the organization and can be accessed through the computer or giving through recorded or published information. In this survey the external data and sources are from the open literature.

RESPONDENTS TARGET AND SAMPLE SIZE

The population of this study focuses on all customers at the outlet since the survey is to find out the level of customer satisfaction and improve their product and service productivity provided to their customer. In this survey, we used simple random sampling where 288 customers were selected.

Sampling Techniques

The sampling techniques that were chosen are simple random sampling. This method explains about the equal chances of being selected in the sample. In the process, 288 questionnaires were

Table 2. Questionnaire sample

Please Circle the rating for each experience that best describe your satisfaction level					
Unacceptable	Dissatisfied	Somewhat Dissatisfied	Somewhat Satisfied	Satisfied	Outstanding
1	2 3	4 5	6 7	8 9	10

Main entrance, Reception and Vehicle Display Area

1	Customer parking is easily available to you	1	2 3	4 5	6 7	8 9	10
2	Appearance of staff in term of uniform, clean look.	1	2 3	4 5	6 7	8 9	10
3	Cleanliness and appearance of facilities of the outlet (including amenities)	1	2 3	4 5	6 7	8 9	10
4	Vehicles were displayed for easy viewing and access.	1	2 3	4 5	6 7	8 9	10
5	Wide selection of vehicle in term of car make and models	1	2 3	4 5	6 7	8 9	10
6	Vehicles information was readily available and sufficiently displayed on price board	1	2 3	4 5	6 7	8 9	10
7	Comfort of area where sales transaction was conducted	1	2 3	4 5	6 7	8 9	10
8	Attitude of staff when dealing with you in term of courtesy and friendliness.	1	2 3	4 5	6 7	8 9	10
9	Recommended the right vehicle to suit your needs	1	2 3	4 5	6 7	8 9	10

Sales Negotiation

1	The price you paid for the vehicle was relative to your expectation.	1	2 3	4 5	6 7	8 9	10
2	Flexibility of purchase terms & condition during the negotiation	1	2 3	4 5	6 7	8 9	10
3	Fulfillment of all promises/commitment throughout your dealing	1	2 3	4 5	6 7	8 9	10
4	Made clear and understandable negotiation with you	1	2 3	4 5	6 7	8 9	10
5	Provide a hassle-free paperwork/purchase/finance process.	1	2 3	4 5	6 7	8 9	10

Sales Dealing

continued on following page

Table 2. Continued

1	Knowledge about loan/finance/insurance/purchases/payment option	1	2 3	4 5	6 7	8 9	10
2	Knowledge about the vehicles	1	2 3	4 5	6 7	8 9	10
3	Vehicles allowed to be inspected before signing order.	1	2 3	4 5	6 7	8 9	10
4	Cleanliness of vehicle when delivered to you.	1	2 3	4 5	6 7	8 9	10
5	Condition of vehicle when delivered to you was as per the agreement.	1	2 3	4 5	6 7	8 9	10

Delivery

1	Ability to deliver vehicle at the same time where the sales Advisor has promised you.	1	2 3	4 5	6 7	8 9	10
2	Vehicle's features were explained to you	1	2 3	4 5	6 7	8 9	10
3	Cleanliness of vehicle when delivered to you	1	2 3	4 5	6 7	8 9	10
4	Condition of vehicle when delivered to you was as per the agreement.	1	2 3	4 5	6 7	8 9	10

Your overall experience

1	How satisfied was your overall experience with this outlet compared to other used car dealer(s) that you have visited?	1	2 3	4 5	6 7	8 9	10

Table 3. Questions used by SPSS techniques a) Main entrance, Reception and Vehicle display area

Q1	Q2	Q3	Q4	Q5	Q6	Q7	Q8	Q9
8.5	9.0	8.8	8.6	8.3	8.2	9.0	9.1	8.8

b) Sales Negotiation

Q1	Q2	Q3	Q4	Q5
8.3	8.4	8.6	8.7	8.8

c) Sales Dealing

Q1	Q2	Q3	Q4	Q5
8.7	8.6	8.9	8.9	8.9

d) Delivery

Q1	Q2	Q3	Q4
8.8	8.7	8.8	8.8

e) Your Overall Experience

Q1
8.9

Figure 2. Analyses for improvements of services and products

a)

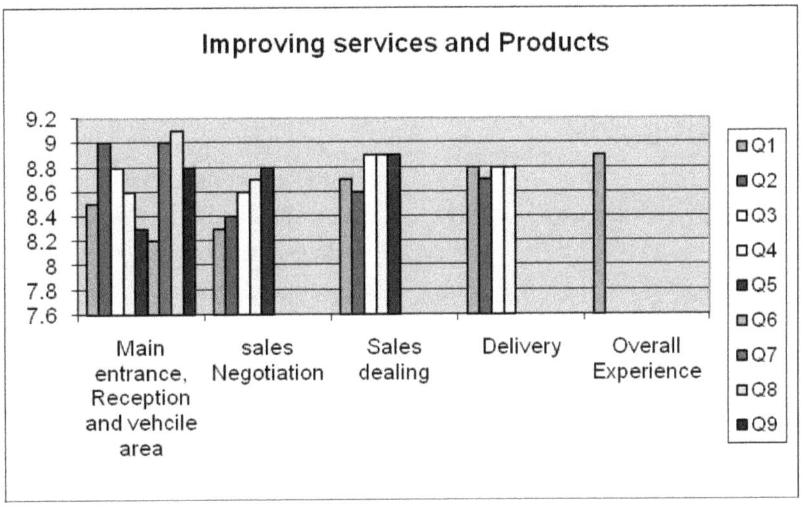

b)

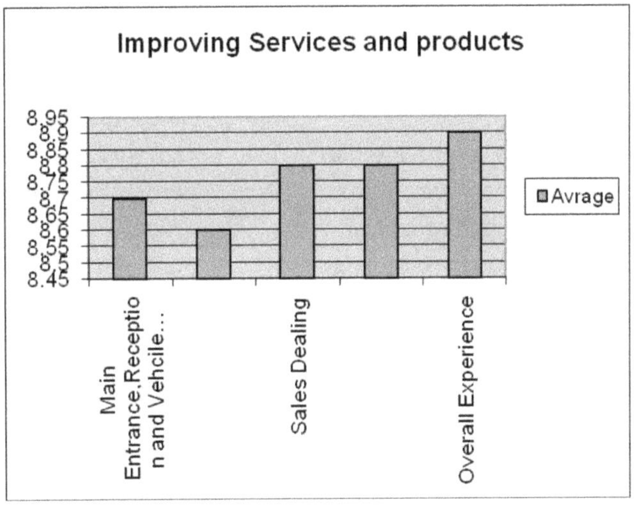

distributed to 288 customers who are selected randomly, with the help of sales advisor at the outlet. The process of distributing the questionnaire took 2 to 3 weeks. Sales advisor request their previous and latest customers to assist in answering the question. With a mission of continuous effort to improves organization service and product productivity (Table 2).

The Analysis

These sample questionnaires were distributed randomly to 288 customers of the outlet with the help of the sales advisor there. The samples were analyzed using SPSS techniques (Table 3 a-e) using the mentioned questions as independent variables that affect the main variables (Main Entrance-Reception and Vehicle display area, Sales Negotiation, Sales Dealing, Delivery and overall Experience) as dependant variable. We got these results as mentioned in the following tables:

And we can describe these analyses through Figure 2 a-b.

RESULTS AND DISCUSSION

From the analysis of the previous data we found that:

- The most effective factor for improving the service at the main entrance is the attitude of staff when dealing with the customer in terms of courtesy and friendliness.
- The most effective factor for improving sales negotiation was provided a hassle-free paper work/ purchase/ finance process.
- The most effective factors for improving sales dealing are vehicle allowed to be inspected before signing order, clearly explained all documents to the customer and honesty when dealing with him.
- The most effective factors for improving the delivery are ability to deliver vehicle at time where the sales advisor has promised, cleanliness of vehicle when deliver to the customer and condition of vehicle when deliver to the customer.
- From the overall experience of the outlet customers, we can see that they are satisfied with the service and the product of the company comparing with the other used car dealers.
- The customers are satisfied with the services and products of the outlet (all the factors got High rating over 8).
- To continue improving the services and products of the outlet, the company should continues focusing on improving these variables especially sales dealing and delivery system because they got higher rating comparing with the other variables. And that of course will continue the satisfied overall experience with the company.

CONCLUSION

Customer satisfaction is seen as a key performance indicator within business.

It is a psychological and physical variable which associated with satisfaction behaviours.

Toyota is one of various fields that need to have customer service division and concern about customer satisfaction. Toyota has consumer care process and customer service is one of the most important objectives for them.

In the literature review we manifest the variables that had been used in the survey to measure the customer satisfaction. These variables are.

1. Main entrance, Reception and Vehicle Display Area.
2. *Sales negotiation.* Satisfied customers are less price sensitive, buy additional products, are less influenced by competitors and stay loyal longer.
3. Sales dealing.
4. *Delivery*, the customers always develop perceptions during the service delivery process and subsequently they compare their perceptions to their expectation in evaluating the outcome of the service encounter
5. Your overall experience.

These factors have been proven through the survey that has been conducted.

Satisfied customers are loyal customers and so high levels of satisfaction lead to repeat business and increased sales.

In a shell nut, customer satisfaction tends involved people in the organization and outside organization as well that give impact on the organization performance and perception of the customers.

REFERENCES

Alomaim, N., Zihni, M., Tunca, S., & Zairi, M. (2003). Customer satisfaction @ virtual organizations. *Management Decision*, *41*(7), 666–670. doi:10.1108/00251740310495595

Bennett, R., & Rundle-Thiele, S. (2004). Customer satisfaction should not be the only goal. *Journal of Services Marketing*, *18*(7), 514–523. doi:10.1108/08876040410561848

Cacioppo, K. (2007, November-December). *Harvard Business Review*.

Customer definition. (n.d.). Retrieved January 24, 2009, from http://www.ichnet.org/glossary.htm

Customer satisfaction. (n.d.). Retrieved January 24, 2009, from http://www.ic.gc.ca/eic/site/stco-levc.nsf/eng/h_qw00037.html

Desatnick, R. L. (1989, April). Long Live the King. *Quality Progress*, 24-26.

Federal Quality Institute. (1991). *Introduction to Total Quality Management in the Federal Government*. Washington, DC: Government Printing Office.

Feigenbaum, A. V. (1986). *Total Quality Control*. New York: McGraw-Hill.

Fleming, J. H., & Asplund, J. (2007, September 13). Customer Satisfaction: A Flawed Measure; Your customers want more than "satisfying" transactions — they want engaging relationships. *Gallup Management Journal*. Retrieved January 25, 2009, from http://findarticles.com/p/articles/mi_6770/is_2007_Sept_13/ai_n28455398/pg_2?tag=artBody;col1

Garvin, D. A. (1987, November-December). Competing on the Eight Dimensions of Quality. *Harvard Business Review*, 101–109.

Magi, A. W. (2008). Share of Wallet in Retailing: The Effects of Customer Satisfaction. *Loyalty Cards and Shopper Characteristics, 79*. Retrieved January 25, 2009, from http://www3.babson.edu/Publications/JR/PastIssues/Volume79Issue2/ShareWalletRetailing.cfm

Pearcea, J. D. (2001). *Unique source of information, training and more*. Retrieved January 13, 2009, from www.jpsalestraining.com

SureVista Solutions. (2007). *Benefits ofCustomer Satisfaction*. Retrieved January 25, 2007, from http://www.surevista.com/benefits-customer-satisfaction.html

This work was previously published in International Journal of Asian Business and Information Management, Volume 1, Issue 2, edited by Dasho Karma Ura and Patricia Ordóñez de Pablos, pp. 24-32, copyright 2010 by IGI Publishing (an imprint of IGI Global).

Chapter 9
Creating a Chinese "Miele"? Asymmetries in the Consumer Perception of Chinese and German Brands

Cora F. Jungbluth
Heidelberg University, Germany

ABSTRACT

This paper analyzes the country-of-origin effects of Chinese brands going international. To approach this topic, the author uses a comparative case study, in which a Chinese manufacturer of white goods, Haier, and a German manufacturer, Miele, serve as examples. Fueled by "China trash" reports in EU and US media, the image of China as country of origin still has a negative connotation. The Chinese government and Chinese companies are working to change this image, since they both regard branding as important means for corporate internationalization. This paper argues that Chinese brands could achieve a similar development as did Japanese and Korean brands in the second half of the last century.

INTRODUCTION

Would German consumers voluntarily exchange their Miele washing-machine for a Haier product? Would Chinese consumers do so?

A few years ago one could clearly have negated both questions. The label "made in Germany" emanating German cardinal virtues such as efficiency and reliability versus the label "made in China" associated with cheap prices but low product quality would have been a truly unequal battle. And even today one can think of a wide range of world-famous German brands in the realm of consumer electronics – Bauknecht, Bosch, Liebherr, Miele, Siemens – to name just a few. But can we name Chinese brands equally renowned beyond the borders of the Middle Kingdom? Of course, Lenovo, formerly Legend, has recently achieved

DOI: 10.4018/978-1-4666-0276-2.ch009

global fame due to its acquisition of the PC division of IBM. The company was thus able to sell a large number of their products under the world famous brand name of IBM. But does this step not prove that Chinese companies cannot survive in the global sphere without an established Western label by their side? *The McKinsey Quarterly* clearly pointed out the issue at question: "Can Chinese brands make it abroad?" (Gao, Woetzel, & Wu, 2003) And Wolfgang Hirn (2005), journalist at German *Manager Magazine*, headlined "Do you know Huawei?" All this suggests that Chinese companies lack an important feature of internationally successful corporations, i.e., strong and renowned brand names trusted by consumers worldwide and well established within a global framework.

Haier, a Chinese manufacturer of white goods, for example, has attained the position of the world's largest provider in its segment in number of pieces. Its brand name, however, has so far not been able to catch up with American or European brands in terms of reputation and recognition. Of course, technological development and shorter tradition play a role, but do not suffice to explain the asymmetry in consumers' perception of these brands, especially when taking into consideration that Haier used to collaborate with German Liebherr in a joint venture. The obvious lack of brand reputation in important consumer markets such as the European Union (EU) represents to Haier's further international development an obstacle that has to be overcome as swiftly as possible. As Benjamin Barber (1992) has quite adequately pointed out, "It is less the goods than the brand names that do the work, for they convey life-style images that alter perception and challenge behavior."

My paper therefore investigates the following questions: Why do German consumers tend to look down upon Chinese brands? Why are German brands well-reputed among Chinese consumers? Where do these attitudes stem from? Which role do country-of-origin effects play in this context? And last but not least, are there any signs indicating a change in consumer perception?

First of all, I will address the question in how far the image of a country influences the consumer perception of brands originating from this country. This section also discusses China's and Germany's reputation as country of origin. Section three and four consecutively introduce a German and a Chinese producer of white goods, which both are market leaders in their respective home country. This is where their resemblance ends, since Miele is a successful brand in China, too, whereas Haier is fighting hard to set foot on the German market. Section five analyzes the influence of country-of-origin effects on the perception of these two brands, followed by concluding remarks on the prospects of Chinese brands on the international markets.

THE INFLUENCE OF COUNTRY-OF-ORIGIN EFFECTS ON BRAND PERCEPTION

The way of how a consumer perceives a certain product is determined by several internal and external factors. Internal factors, which can be influenced by the company itself, include among others advertising or the positioning of a brand. External factors, which cannot be easily influenced by the company, refer for example to the purchasing experience of the consumer or, even more importantly, the image of the country of origin. Bilkey and Nes (1982) have pointed out that "[b]oth empirical observations and experiments indicate that country of origin has a considerable influence on the quality perception of a product" (p. 89). The image of the country of origin therefore becomes a crucial issue, when a company decides to make branding the core of its internationalization strategy (Bell, 2008). Should the consumers be able to easily associate the brand with the company's country of origin? Or would the company be better off with not putting too much emphasis on this issue? The answer to the question depends on the product category as well as on the already existing image of the country of origin (Roth & Romeo, 1992). Han and Terpstra (1988) have shown that "the value of Germany as a country of origin is positive" (p. 248), but it may still be less favorable for certain product categories, such as brown goods. The image of China as country of origin may be negative for electronic appliances due to a perceived lack of quality, but generally positive for green tea or certain medications, given the good reputation of Traditional Chinese Medicine. Han and Terpstra (1988) therefore argue that a "firm may emphasize or downplay the source country depending on the favorableness of the consumers' perception of that country" (p. 250).

One of the factors which considerably influence this perception is the degree of economic development (Han & Terpstra, 1988; Bilkey & Nes, 1982). Especially companies from newly industrialized or developing countries may face difficulties when they start marketing their own brands abroad. In times of an increasing variety and ephemerality of products and an intensifying global competition, however, branding has become an ever more important strategy for companies – be it from industrialized or developing countries – to distinguish their products from those of their competitors and find a way to survive on the international markets (Gao, Woetzel, & Wu, 2003). Summarizing important aspects of the discourse on branding, Rooney (1995) holds that "the purpose of branding is essentially to build the product's image. This image will influence the perceived worth of the product and will increase the brand's value to the customer, leading to brand loyalty" (p. 48).

Brands are designed by enterprises in order to create an outstanding product image and, thereby, achieve higher profit margins as compared to no-name goods. On the one hand, enterprises are able to influence a consumer's perception of a certain product insofar as they can make him or her recognize this product as belonging to a certain brand (Esch, Tomczak, & Kernstock, 2004; Keller, 2003). Through marketing and advertising campaigns, it is further possible to transfer values, emotions or lifestyles to a brand. Consumers thus do not merely purchase a product but also associate it with a range of intangible assets (Rooney, 1995, p. 48). On the other hand, these intangible assets have to be chosen carefully and must be reconsidered regularly according to changing social patterns. Horx and Wippermann (1996) emphasize in this context that "trends of consumption mainly refer to the needs which are reflected in the wishes and desires of consumers, though always against the background of the great processes of value change" (pp. 66-67). This applies to the positioning of brands as one of the internal factors noted above: Brands that were successful with a certain image about ten or twenty years ago will be turned down by their target consumer group and vanish if they do not adapt over the years. Miele's advertising slogan of the 1950s "Mutti macht's mit Miele" (Mummy works with Miele), for example, would probably prevent today's consumers, especially women, from buying these products.

The image of the country of origin, which influences the perception of a brand as an external factor, is determined by neither the enterprise nor the consumer alone. It rather depends on the international reputation of the country itself, its politics, culture and historical background (Roth & Romeo, 1992, pp. 479-480). Kim (1995) describes this effect as such: "[...] a country name has a set of associations which significantly influence consumers' attitude toward products originating from particular countries in a multinational product market. Products sourced from industrialized countries generally tend to have favorable associations, and hence tend to be evaluated favorably compared with those from developing countries" (pp. 66-67). The image of the country of origin implies certain values, customs and also prejudices. These exert an effect on products or brands associated with this country. German brands for example, such as Miele, Siemens or Bosch, are renowned all over the world for the quality and longevity of their products. The label "made in Germany" carries a notion of such "Prussian" characteristics as assiduity and precision. It hence has become a brand itself and is heavily used for marketing German products abroad (Seiwert, 2003, p. 15). In China, Germany as country of origin also enjoys an excellent reputation, with the Chinese proverb *yi si bu gou* 一丝不苟 (be scrupulous about every detail) preferentially used to refer to those "virtues" mentioned above (Xinhua Wang, 2002; Li, 2008, pp. 28-29).

Especially in European countries, Chinese products to the contrary are at best renowned for their inexpensive price, at worst for their lack of quality and dangerous contents. This opinion was held by about 55% of the respondents in a survey conducted by *Synovate* in 2005, for German respondents the percentage was 60. As Yang, Xin, Liu, and Dai (2008) put it, only China's low production costs are world-famous, "but until present China does not yet have world-class brands" (p. 47). This issue combined with the notion of China as a non-industrial country makes it difficult for newly emerging Chinese brands to gain a competitive edge on the US and EU markets. Some Chinese enterprises, including Haier, Lenovo and TCL, have nevertheless chosen branding as core of their internationalization strategy. In this process, they can count on the support of the Chinese government, which has made branding an important part of the so-called "Going-Global-Strategy 走出去战略" (CAITEC, 2005) [1]. Since the government has recognized that the image of China may need a face-lift in order to give Chinese products a better standing abroad, it also has integrated the export of Chinese culture and values into its strategy. This includes, for example, the foundation of an increasing number of Confucius Institutes abroad. The Olympic Games 2008 in Beijing also were supposed to contribute to this goal (Bell, 2008, pp. 124-125). It might be a battle against windmills when considering recent surveys on China's image in the world. These mirror the stark contrast in the perception of China and, for example, Germany or the US as countries of origin. The *Korean Trade-Investment Promotion Agency* found out that consumers were willing to pay 80.2 US-Dollars more for a German-branded product than for the same Chinese-branded product and 66.4 US-Dollars more for the American product (Korea Times, 2009). In 2008, Germany was ranked first by the annual Nation Brands Index (NBI) of *Anholt-GfK Roper* (2008, p. 4), with the US ranking seventh and China 28th out of a total of 50 countries. An unrepresentative survey among German consumers further showed that 51% of all respondents would buy Chinese electrical products such as refrigerators or TV sets only because of their cheap price, 0% would do so because of their brand (Bell, 2008, pp. 122-123). As one can tell from these results, German consumers seem to be particularly prejudiced against Chinese products. This is hardly surprising when considering the increasing number of press reports on "China trash" products that have circulated widely in all kinds of German media in recent years (e.g., Manager Magazin, 2007; Frankfurter Allgemeine Zeitung, September 17, 2008).

Chinese companies that intend to enter the German market with their own brands thus start from a rather disadvantageous position, whereas German brands in China, such as Miele, enjoy a bonus due to the positive image of Germany as country of origin. Only a few companies, Haier among them, have therefore embarked on the hazardous road to establishing their brand in Germany, which they regard as an important step stone to access the EU common market (Deng, 2006, p. 50). In order to discuss the question in how far these different images of China and Germany as countries of origin influence the perception of a German brand in China and vice versa, the two succeeding sections introduce the cases of two manufacturers of white goods already mentioned above: One from Germany, Miele, and one from China, Haier. The latter has successfully completed its branding strategy on the domestic market and belongs to the avant-garde of Chinese enterprises in terms of international branding. Miele belongs to the elite of German manufacturers of white goods in terms of quality and reputation and has been conducting successful branding for several decades at home as well as abroad.

MIELE – A GERMAN STORY OF SUCCESS

In 1899 Carl Miele and Reinhard Zinkann founded the centrifugal factory Miele & Cie close to the German city of Gütersloh. Throughout the years the company marketed a broad range of products, including cars and motor cycles as well as washing-machines, dish-washers and vacuum-cleaners. Today Miele concentrates on its core competence in the consumer goods segment, i.e., household appliances, especially white goods. The company presently has more than 16000 employees worldwide with representations in 42 countries. In 2007/2008 about 73 per cent of the company's turnover, amounting to 2.81 billion Euros, was generated abroad (Miele Pressedienst, 2008). As the only premium brand of household appliances present on all five continents, Miele has marketed this "premium" image not only on its German home market but also on the international markets, thus targeting consumer groups with strong purchasing power. The company places high emphasis on the quality and longevity of its products. For some of these it issued 10-year warranties as early as 1916 (Miele Homepage, no year).

Even though the company, as one the typically German *Familienunternehmen* (family business), belongs to the category of small and medium-sized enterprises, it has been able to compete successfully with large corporations such as Siemens or Whirlpool for brand reputation and trust among consumers and customers: In 2007 Miele won the corporate brand award by the renowned market research company *GfK Market Research* (GfK Marktforschung, 2007). It has attained top positions in the "European Trusted Brands" survey by the *Reader's Digest* (2010) for several years in a row despite fierce competition and an increasing number of brands on the national and international markets. In the category consumer electronics Miele won the first place for five years in a row (2006-2010) in Austria, Germany, the Netherlands, Switzerland, Portugal and Belgium. According to this survey, German consumers especially value product safety and tradition – two features Miele is easily able to offer with its 100-odd years of corporate history and well proven product performance. The Miele commitment to quality, sustainability and longevity also appeals to German consumers, who are found to be especially sensitive to these issues (Ibid.).

However, the image of tradition and commitment would not be sufficient if in the end the products did not keep their promise. Miele – faithful to its corporate slogan *immer besser* (ever better) – lives up to these expectations: Its products have reached the highest scores in the rankings of independent product testing institutes several times (Stiftung Warentest, 2008). This prompted the Austrian *News Agency for the Electronic Industry* to declare "Miele continues 'subscription' to test winner" (Electrojournal.at, 2006). Miele's cross-border success in gaining buyers' confidence does not only rely on excellent quality and a strong corporate name alone, but also on its country-of-origin bonus with consumers worldwide attaching positive associations to German brands. According to one of Miele's CEOs and grandson of the founder, Markus Miele, the company mostly produces in Germany and only to a very small extent in Austria and the Czech Republic in order to market the label *Deutsche Wertarbeit* (German high class workmanship). In an interview with Hamburg-based *Abendblatt* (April 2, 2005), he pointed out that especially in markets such as China, where the company has been developing at a fast pace in recent years, customers highly value German quality. Miele first established a branch in Hong Kong in 1997.

With its brand name translated into Chinese as *Meinuo* 美诺 (beautiful promise) the company laid a sound foundation for introducing its product to the Chinese market. *Nomen est omen* applies to product placement in China even more then in Western countries due to the consumers' affinity to auspicious symbols, figures or meanings (Li & Shooshtari, 2003). Miele representation offices in Shanghai and Beijing, which mainly cooperate with real estate development companies, followed in 2004 and in 2006 respectively (Deguo Meinuo Gongsi, no year). This was a strategically well-placed move, since the target group of these companies mainly consists of members of the urban middle class with relatively strong purchasing power (Farrell, Gersch, & Stephenson, 2006). Finally, in March 2008, the official all-China subsidiary Miele China was founded in Beijing (Yijia Jiaju, 2008).

In China, the company is exclusively targeting the market for high end goods, which is made very clear on the Chinese website as well as in Chinese-language publications on the company (Deguo Meinuo Gongsi, no year; Hu, 2007, p. 55). In a Chinese review on Miele, Markus Miele is quoted saying: "This brand has never intended to be sold to common people [...]. Taking Mercedes Benz as an example, one is not supposed to see this car everywhere on the street. They, too, have never considered producing goods suitable for the need of the masses" (Hu, 2007, p. 55). The trade magazine *Corporate Culture* points out that Miele "has become the first choice of people worldwide pursuing a modern lifestyle and high taste. [...] Bill Gates and the Beckham couple [...] favor the use of Miele appliances" (Chen, 2008, p. 53).

Miele has thus established itself in China as a high end brand that enjoys an excellent reputation and whose products and services are of premium quality. However, as their prices are top end, too, there is only a tiny group of Chinese customers who can actually afford to purchase these products. Even though the brand perception of Miele in China is highly positive, local brands such as Haier still have the advantage of much cheaper prices. It is not clear, though, if this competitive edge will prevail in face of a fast growing Chinese middle class, which "consume[s] globally branded luxury goods voraciously" (Farrell, Gersch, & Stephenson, 2006, p. 60). Even on the domestic market, Chinese companies thus have to reposition themselves in order to survive in an internationalizing environment. A strong brand with the ability to gain access to the premium segment of the market represents one option to achieve this. As for the Chinese market, Haier has already reached this goal. Its progress on the international markets is topic of the following section.

HAIER – THE CHINESE RESPONSE?

In an era of China's growing involvement in the international economy, formally enforced by its WTO access in 2001, Chinese companies cannot count on the domestic market as "safe harbor" any longer. They endure increasing competition from transnational enterprises with well-tried sophisticated branding and marketing strategies. Chinese companies have to establish their own brands, if they do not want to take on the role of original equipment manufacturer forever (Gao, Woetzel, & Wu, 2003, p. 55). Managers and political decision-makers therefore have recognized that they ought to change China's image as mere back factory of the world in order to create a long-term basis for international competitiveness. A "branding" movement has already been evolving in China during the past few years, as Chinese companies try to follow in the footsteps of global players such as Siemens, Samsung and Sony. These owe a large amount of their international success to their strong brand names combined with reliable quality.

As mentioned at the beginning, the Chinese government has integrated the ambition to establish uniquely Chinese brands into its "Going-Global-Strategy", which offers strong political support for outward investment by "enterprises with potential 有实力的企业" (Zhongguo Wang, 2001). Companies, such as Haier, which have made the creation of a world-level brand one of their top priorities, certainly belong to this kind of enterprises and can thus count on government support. Political influence on economic activities in China still ought not to be underestimated. It is of utter importance for Chinese companies involved in ventures abroad to be able to rely on a favorable political situation at home, which facilitates their operations to a great degree. Support by policy-makers, however, does not replace a company's own strategic planning and decision-making. Many Chinese enterprises attempt to expand on the international markets adhering to the notorious principle of "trial and error", which has been applied by the Chinese Communist Party in many fields subject to reform, economic policy being one of them [2]. The case of TCL's initial failure shows how high the costs of this strategy might be: Its random acquisitions of European brands caused an operational deficit for the first time in years (Wirtschaftswoche, 2006). Haier, quite to the contrary, employed a more cautious strategy, which followed the principle of trial and error as well, but – paradoxical as it may sound – in a planned manner. The most important difference between Haier and many other Chinese companies is its desire to go abroad with its own corporate brand. Relying on a sound technological and intellectual basis laid by the former joint venture cooperation with German Liebherr, Haier first established itself as one of the leading manufacturers of white goods *in* China. From the Chinese translation of Liebherr (利勃海尔 Libohai'er) Haier also obtained its corporate name. The story of success started in 1984, when Zhang Ruimin 张瑞敏 took over the management of the virtually bankrupt Qingdao Refrigerator General Company 青岛电冰箱总厂 – the predecessor of Haier – and succeeded to re-establish its profitability within a few years. One story about the initial action of the new CEO is retold over and over again: Accordingly, Zhang was fed up with the bad quality products manufactured in this factory. So he used a hammer to destroy brand-new refrigerators and washing-machines in front of the workers in order to demonstrate that from now on such inferior products would not be accepted any longer – This corporate myth became known not only throughout China but also abroad and contributed to the international reputation of the company (Renmin Wang, 2008; Wang, 2008). Today in China, the brand Haier stands for quality, innovation and customer service (Wu, 2003). The sponsorship of the 2008 Olympic Games has increased its popularity in China even more and probably also contributed to growing brand awareness abroad (Bell, 2008, p. 176). A final appraisal of the influence of the Olympics on Chinese brands abroad is not possible yet due to a dearth of in-depth empirical research on this topic.

Presently, Haier is the world's fourth largest manufacturer of white goods with about 50,000 employees in 30 countries and an annual turnover of about 11.8 billion Euros (Haier Jituan, a). Its domestic market share in white goods presently amounts to about one quarter (Wu, 2003, p. 108). The internationalization of the company represents one of the most important goals of CEO Zhang Ruimin. He puts a major focus on "establish[ing] a brand reputation overseas" (Ibid., p. 109). According to its corporate website Haier therefore started the "strategy of establishing a global brand 全球化品牌战略" in 2005 in order "to create a localized Haier brand on the markets of every single country" (Haier Jituan, b).

Haier began its internationalization with the expansion to the markets of developing countries in South East Asia, the Middle East and Africa, thus "practicing" competition with transnational enterprises from developed countries on third markets (Hu & Yu, 2002, p. 31). Zhang Ruimin summarized this strategy as follows: "In South East Asia and the Middle East, we can make large amounts of foreign exchange. But we won't be able to create a world-famous brand there. If we want to reach the highest spheres of market competition, i.e. marketing a brand, Haier has to enter the European and US-region, where brands stand in great numbers" (Ibid.). The CEO and his top managers in Europe and the United States know too well the challenge they face: Michael Jemal, president of Haier America is fighting hard against the prejudice of "shoddy quality", which is held by many US-consumers when it comes to Chinese products (Businessweek, 2008). The German market is even more difficult to conquer, since – as Zhang Ruimin emphasizes – German consumers can only to be persuaded by high quality products. Allegedly, he has even been told by a German customer that she would only purchase Miele since their products were like artworks. This led the CEO to the conclusion that once Haier's products have conquered the German market, the company will have reached an important strategic milestone (Ibid.; Financial Times Deutschland, 2005).

Up to now, however, Haier has not yet been able to establish a bond with German consumers. A recent investigation of the German product testing institute *StiftungWarentest* (2008) has reported the worst grading possible (5.0) for one of Haier's dish-washers [3]. This result will not contribute positively to this situation: In the same test, Miele occupied the pole position. The independent tests of *Stiftung Warentest* exert a crucial influence on the purchasing decisions of German consumers: These usually forgo products that have received a bad grading (Dradio, 2004). In general, products "made in China" have to cope with reputation problems in Germany similar to those in the United States: As mentioned above, about two third of the German customers associate Chinese products with low quality. The negative image of Haier's country of origin represents a barrier to establishing brand awareness among German consumers, who mistrust a product made in China and sold with a Chinese brand name even more than a product made in China but sold with a Western brand name. These asymmetries in the perception of brands originating from China and Germany, respectively, will be discussed in the following section.

ASYMMETRIES IN BRAND PERCEPTION

The image of Germany as country of origin generally is perceived positively abroad. Chinese customers therefore regard Miele appliances as products of top quality, but can rarely afford to buy them. Haier as domestic brand enjoys an excellent reputation and is highly popular in China. Its prices are lower than those of Western brands in general and "buy China" appeals to the national consciousness of Chinese consumers (Lane, St-Maurice, & Süssmuth Dyckerhoff, 2006, pp. 39-40). Since Miele represents a status symbol comparable to Mercedes Benz and thus only occupies a small market share, the perception of its brand rather depends on hypothetical factors rather than on the practical experience of consumers really using the products. Haier to the contrary offers products affordable for a larger – but still well-situated – part of the Chinese consumers, who assume Haier products to be of high quality as well, especially when compared to other local brands.

The situation is different in Germany where the reputation of China as country of origin has been overshadowed by the "China trash image," which is deeply-rooted in the mind of German consumers. These also prefer to stick to their domestic brands "made in Germany," which they are proud of and which they have trusted for decades. An unrepresentative survey among German academics with Chinese Studies background conducted for this paper in August 2009 found that 80% of the respondents associated "made in Germany" with high quality [4]. Even among this specific target group, which generally hold a more differentiated image of China than the average German consumer, more than 60% of the respondents think of "made in China" in terms of bad quality and cheaply produced goods. This percentage corresponds with that revealed by the *Synovate* survey mentioned above.

The brand Miele was known to all and Haier to 78% of the participants. The high degree of brand awareness Haier achieved in this survey can be ascribed to the specialized background of this target group. In a survey among academics with little or none expertise on China, only 28% of the respondents were familiar with Haier (Bell, 2008, p. 122). As for the image of the two brands, the participants favored Miele above Haier: 94% thought that Miele products were of good or very good quality, whereas 47% did so for Haier. This shows, however, that Haier's brand image among this specific consumer group is much better than one would assume when considering their opinion on "made in China".

If respondents were to choose a brand for the purchase of a washing machine, the majority would still opt for Miele (58%; Haier: 14%) and would be willing to pay 1000 Euros or more for this appliance (65%; Haier: 5%). Presently, more than one third of the respondents actually have a Miele washing machine at home (36%; Haier: 3%). Even those participants below the age of 30 hold the opinion (80%) that Miele is a symbol of traditional German virtues and quality standards. Miele thus has a walk-over on the German market, since it has managed to maintain the bond between brand and consumer despite on-going changes in society and between different generations. Up to now, it has never disappointed its customers in terms of quality and service. Haier has had a comparatively difficult start in Germany, the negative image of China as country of origin being one of the reasons. The company initially also neglected the organization of an efficient and well-structured marketing campaign to make German consumers more familiar with its brand and thus convince them that Chinese products, too, may be of excellent quality: Only in 2007 did the company finally hire a local agency to develop a marketing strategy for Germany (Die Brandenburgs, 2007). Moreover, the bad test result that one of Haier's dish washers achieved in 2008 will not contribute to building up a relation of trust with potential consumers. Even though it was only one single product out of a myriad of others and has been withdrawn from the market, it still left a negative impression: One of Haier's first products to be tested by *Stiftung Warentest* once more confirmed the "China trash image".

One can see from these two cases that the country of origin plays an important role in the initial consumer perception of a brand, be it on the domestic or a foreign market. Enterprises from positively perceived countries can make use of this advantage in the marketing of their products (Han & Terpstra, 1988). Enterprises from negatively perceived countries must find ways to overcome the initial disadvantage they face. For them, it is vital to gain consumers' trust and prove that they are worth it in terms of quality and service. A marketing strategy which is well adapted to local consumption patterns is an important step to achieve this. Relocating a part of the production to a country with a more favorable image, in the case of Haier Germany, may also be a means to tackle this issue, since Haier products might thus profit from the positive image of "made in Germany" [5].

MIELE AND HAIER – RETHINKING CONSUMER PERCEPTION?

For Western brands, the initial steps on the Chinese markets have been easier due to their (industrialized) country-of-origin bonus, which often gained them a leap of faith with the consumers [6]. Additionally, they were more experienced in strategic marketing and branding. Chinese enterprises are latecomers which are only just starting to gather experience in this field. Moreover, they have to fight against the bad reputation of "made in China," putting them into a disadvantageous position abroad at the beginning.

This is true even on the domestic market, especially if one regards recent scandals about local products in China (Xinhua Wang, 2008). Nevertheless have the Chinese started to readjust their pattern of consumption, revealing a strong tendency towards the preference of domestic brands. This makes it easier for local brands to win market shares, but more difficult for foreign companies to hold their position and to market their products. They will have to increasingly sinicize their advertisement and marketing, if they want to keep up with this new trend in Chinese brand perception. Brands like Miele, which are mainly targeting the luxury market, have to pay attention to this development, too. Haier and other Chinese manufacturers of white goods are on the verge of entering this margin-rich fast growing market segment and they learn quickly how to manage their brands.

For Chinese brands, it may become easier to establish themselves abroad with increasing experience in international business. Further support stems from the Chinese government's ambition to change the image of China merely being the factory of the world. Brands like Lenovo or Haier are already trying to attach a more international touch to their brand with focus on modern management strategies and advanced technology. Haier has for example tried to gain the attention of German consumers via a new modern and innovative technology, as it presented a washing machine working without any detergent to the public in 2007 (The Brandenburgs, 2007). The recruitment of a German marketing agency familiar with the local marketing patterns also was supposed to be helpful for the establishment of a trusting relationship with German consumers. Since the lack of knowledge about country-specific characteristics is regarded as one of the general problems of Chinese companies going global (Hemerling, Michael, & Michaels, 2006, p. 15), these should clearly keep in mind the well-known strategy "think globally, act locally". Wang Jing (2008, p. 176) also holds that the adaptation to local conditions is extremely important for Chinese companies abroad, especially when it comes to communication with consumers. Otherwise they will repeat the mistakes made by Western companies before, which – in spite of renowned brands – had to learn that long-term success on Asian markets cannot be achieved without a marketing strategy adapted to regional cultures (Ibid., p. 105).

The introduction to this paper quoted Gao, Woetzel & Wu, who brought up the question "Can Chinese brands make it abroad?" They themselves give the answer immediately: "[...] yes, but it won't be easy" (Gao, Woetzel, & Wu, 2003, p. 55). If those Chinese companies that have chosen branding as strategy of internationalization are able to better adapt to a foreign environment and persuade through well-priced products of good quality, while at the same time offering excellent customer service, their native brands have the chance to successfully establish themselves in the international brand scenery in the future. Much also depends on the development of China's image as country of origin, since until present the majority of European and American consumers associate negative aspects with this image.

I would still like to conclude with a rather optimistic statement of Hans Joachim Fuchs, CEO of *China Brand Consulting Shanghai*: "It will not take ten years until Chinese brands will enjoy the same recognition and the same good reputation in [Germany] as Bosch, Miele or Volkswagen" (Fuchs, no year, p. 58). It would not be the first time for brands from East Asian countries to achieve this kind of transformation: In the 1960s, Japanese and, later, Korean companies and their brands were laughed at by their Western incumbents. Today, the latter have stopped laughing or have even ceased to exist. In this respect, Chinese brands might well be able to contribute to the repetition of history.

ACKNOWLEDGMENT

The research for this paper was conducted within the project *B12 Rethinking Trends: Transcultural Flows in Global Public(s)* at the Cluster of Excellence "Asia and Europe in a Global Context" at Heidelberg University. A first draft was presented at the 3rd China Goes Global Conference, Sep. 30 - Oct. 2, 2009, Harvard University, Cambridge, USA. The author would like to thank Prof. Dr. Barbara Mittler and Dr. Lena Henningsen (University of Heidelberg) as well as Dr. Felix Boecking (University of Edinburgh) and two anonymous conference reviewers for their highly useful and valuable comments.

REFERENCES

Abendblatt. (2005, April 2). *Herr der Waschmaschinen (Master of Washing Machines)*. Retrieved October 5, 2010, from http://www.abendblatt.de/daten/2005/04/02/ 416504.html

Anholt-GfK Roper. (2008). *2008 Anholt-GfK Roper Nation Brands Index℠ Report Highlights*. Retrieved October 5, 2010, from http://www.gtai.com/uploads/media/Report_Highlights_Roper_Nation_Brands_Index.pdf

Barber, B. R. (1992, March). Jihad vs. McWorld. *The Atlantic*. Retrieved October 5, 2010, from http://www.theatlantic.com/doc/199203/barber

Bell, S. (2008). *International Brand Management of Chinese Companies: Case Studies on the Chinese Household Appliances and Consumer Electronics Industry Entering US and Western European markets*. Heidelberg, Germany: Physica Verlag.

Bilkey, W. J., & Nes, E. (1982). Country-of-Origin Effects on Product Evaluations. *Journal of International Business Studies, 13*(1), 89–99. doi:10.1057/palgrave.jibs.8490539

Brandenburgs, D. (2007). *Haier will Higher*. Retrieved October 5, 2010, from http://www.brandenburgs.net/site/DE/int/03_presse/03_02_berichte/44_container_datum.php?we_lv_start_self=23&PulldownListviewCat=

Businessweek. (2008, July 31). *Haier: Taking a Brand Name Higher*. Retrieved October 5, 2010, from http://www.businessweek.com/magazine/content/06_31/b3995009.htm

Chen, Y. (2008). Meinuo 109 Nian de Zhuanzhu (Miele: Watching 109 Years). *Zhongguo Xin Shidai, 8*, 53–55.

Deguo Meinuo Gongsi. (n.d.). *Miele Zai Zhongguo Miele (Miele in China)*. Retrieved October 5, 2010, from http://www.miele.cn/cn/domestic/ 5727.htm

Deng, L. (2006). Deguo – Zou Chuqu de Xin Wotu (Germany – The New Fertile Soil of (Chinese Enterprises') Going Global). *Da Jingmao, 6*, 50–52.

Dradio. (2004). *Das Gewissen der Verbraucher (The Conscience of Consumers)*. Retrieved October 5, 2010, from http://www.dradio.de/dlf/sendungen/hiwi/324537

Electrojournal.at. (2006). *Miele verlängert 'Abo' als Testsieger: 'Hier offenbart sich Güte der Geräte-Konstruktion' (Miele Continues, Subscription' to Test Winner: Displaying the Quality of Construction)*. Retrieved October 5, 2010, from http://www.elektrojournal.at/ireds-44767.html

Esch, F.-R., Tomczak, T., & Kernstock, J. (2004). *Corporate Brand Management – Marken als Anker strategischer Führung von Unternehmen* (Corporate Brand Management – Brands as Anchor of Strategic Corporate Governance). Wiesbaden, Germany: Gabler.

Farrell, D., Gersch, U. A., & Stephenson, E. (2006). The Value of China's Emerging Middle Class. *The McKinsey Quarterly*, 60–69.

Financial Times Deutschland. (2005, May 25). *Haier and Higher*. Retrieved October 5, 2010, from http://www.karma-news.com/archives/56

Frankfurter Allgemeine Zeitung. (2008, September 17). *Verseuchte Milch: Die Profiteure gehen über Kinderleichen (Poisoned Milk: Profit-makers Walk Across Children's Bodies)*. Retrieved October 5, 2010, from http://www.faz.net/s/Rub8E1390D3396F422B869A49268EE3F15C/Doc~E9E86EA9FF47D4C1B81C2D818094AE143~ATpl~Ecommon~Scontent.html

Fuchs, H. J. (n.d.). Vor einer grossen Schlacht (Before a Great Battle). *Denkpausen – Inspirationen zu Management und Leadership, 5*, 52-58.

Gao, P., Woetzel, J. R., & Wu, Y. (2003). Can Chinese Brands Make it Abroad? *The McKinsey Quarterly*, 55–65.

Haier Jituan (a). (n.d.). *Gongsijianjie (Company Profile)*. Retrieved October 5, 2010, from http://www.haier.cn/about/about.shtml

Haier Jituan (b). (n.d.). *Fazhan Zhanlüe (Developing Strategy)*. Retrieved October 5, 2010, from http://www.haier.cn/about/vision_index.shtml

Han, M. C., & Terpstra, V. (1988). Country-of-Origin Effects for Uni-National and Bi-National Products. *International Business Studies, 19*(2), 235–255. doi:10.1057/palgrave.jibs.8490379

Heilmann, S. (2008). Policy Experimentation in China's Economic Rise. *Studies in Comparative International Development, 43*, 1–26. doi:10.1007/s12116-007-9014-4

Hemerling, J., Michael, D. C., & Michaels, H. (2006). *China's Global Challengers - The Strategic Implications of Chinese outbound M&A (BCG Rep.)*. Boston: Boston Consulting Group.

Hirn, W. (2005). Kennen Sie Huawei (Do You Know Huawei)? *Manager Magazine*. Retrieved October 5, 2010, from http://www.managermagazin.de/unternehmen/mittelstand/0,2828,344730,00.html

Homepage, M. (n.d.). *Das Unternehmen – Zeitreise (The Company –Journey through Time)*. Retrieved October 5, 2010, from http://www.miele.de/de/haushalt/ unternehmen/2335_4859.htm

Horx, M., & Wippermann, P. (1996). *Was ist Trendforschung* (What is Trend Research). Düsseldorf, Germany: Econ Verlag.

Hu, T., & Yu, Q. (2002). Zhongguo Qiye Hexin Jingzhengli Peiyang de Liang Zhong Guoji Qudao - 'Haier Moshi'he 'Huali Moshi' de Bijiao Yanjiu. *Hubei Shangye Gaodeng Zhuanke Xuexiao Xuebao, 14*(4), 30–32.

Hu, Z. (2007). Miele - Baise Jiadian zhong de Benchi Miele (Miele – Mercedes-Benz among the White Goods). *Qiye Wenhua, 6*, 55–57.

Keller, K. L. (2003). *Strategic Brand Management – Building, Measuring, and Managing Brand Equity*. Upper Saddle River, NJ: Pearson Education.

Kim, C. K. (1995). Brand Popularity and Country Image in Global Competition: Managerial Implications. *Journal of Product and Brand Management, 4*(5), 21–33. doi:10.1108/10610429510103818

Lane, K., St-Maurice, I., & Süssmuth Dyckerhoff, C. (2006). Building Brands in China. *The McKinsey Quarterly*, 34–41.

Li, F., & Shooshtari, N. H. (2003). Brand Naming in China - Sociolinguistic Implications. *Multinational Business Review,* 1-17.

Li, J. (2008). Strict and Tenacious German Style – German Top Brand for Electronic Appliances Miele. *Jiadian Keji, 4,* 26-27, 30-31.

Magazine, M. (2007). *Gefährliche Exporte: EU kritisiert China (Dangerous Exports: EU Criticizes China)*. Retrieved October 5, 2010, from http://www. manager-magazin.de/unternehmen/artikel/0,2828,496155,00.html

Marktforschung, G. (2007). *Best Brands - Die Gewinner (Best Brands – The Winners)*. Retrieved October 5, 2010, from http://www.bestbrands.de/index_fs.html

Pressedienst, M. (2008). *Miele im Geschäftsjahr 2007/2008 erfolgreich bei der Internationalisierung vorangekommen (Miele advanced successfully in its internationalization in 2007/2008)*. Retrieved October 5, 2010, from http://www.miele.de/de/presse/artikel/artikel_184_2008.aspx

Reader's Digest. (2010). *Comparison of winning brands 2006 - 2010: Results for: Kitchen Appliances*. Retrieved October 5, 2010, from http://www.rdtrustedbrands.com/trustedbrands/results/tables/Comparison%2520of%2520winning%2520brands%25202006%2520to%25202010.category.Kitchen%2520appliances.shtml

Rooney, J. A. (1995). Branding: A Trend for Today and Tomorrow. *Journal of Product and Brand Management*, *4*(4), 48–55. doi:10.1108/10610429510097690

Roth, M. S., & Romeo, J. B. (1992). Matching Product Category and Country Image Perceptions: A Framework for Managing Country-Of-Origin Effects. *Journal of International Business Studies*, *23*(3), 477–497. doi:10.1057/palgrave.jibs.8490276

Schramm, M., & Staack, T. (2005). *Evolving Brand Preferences: The Case of China - An Empirical Analysis*. Retrieved October 5, 2010, from www.thinkdesk.biz/news-english/images/IAMA2005_THINKDESK-Paper.pdf

Seiwert, M. (2003). Geopolitische Spannungen: Müssen die Marken in Zukunft ihre Herkunft verleugnen? (Geo-political tensions: Will Brands Have to Deny their Origins in the Future). *Absatzwirtschaft*, *5*, 14–20.

Synovate. (2006). *Western Consumers still Skeptical about Quality of 'Made in China' and 'Made in Korea' Labels*. Retrieved October 5, 2010, from http://www.synovate.com/ insights/infact/issues/200601/

Times, K. (2009, January 21). *Korean Brand Value Up, Closer to US, Japan*. Retrieved October 5, 2010, from http://www.koreatimes.co.kr/www/news/biz/2009/01/123_ 38278.html

Wang, J. (2008). *Brand New China: Advertising, Media, and Commercial Culture*. Cambridge, MA: Harvard University Press.

Wang, R. (2008, August 13). *Zhang Ruimin: Dachui zachu shijie mingpai (Zhang Ruimin: A Big Hammer Helps a World-class Brand to Come into Existence)*. Retrieved October 5, 2010, from http://finance.people.com.cn/GB/8215/126457/7661094.html

Wang, X. (2002, September 24). *Deguo zhizao de xinyu conghe er lai (Where Does the Reputation of ,made in Germany' come from)*? Retrieved October 5, 2010, from http://news.xinhuanet.com/fortune/2002-09/24/content_572868.htm

Wang, X. (2008, September 12). *Sanlupai yingyou'er naifen shijian gundong baodao (Reports on the Issue of Sanlu Milk Powder for Children and Infants)*. Retrieved October 5, 2010, from http://news.xinhuanet.com/newscenter/2008-09/12/content_9935963.htm

Wang, Z. (2001). *Zhonghua Renmin Gongheguo Guomin Jingji he Shehui Fazhan di shi ge Wunian Jihua gangyao (Outline of the 10. Five-Year- Plan on Economic and Social Development of the People's Republic)*. Retrieved October 5, 2010, from http://www.china.com.cn/ch-15/plan8.htm

Warentest, S. (2008). Die Langsamspüler (The Slow Washers). *Test*, *4*, 64–68.

Wirtschaftswoche. (2006). Werte vernichtet (Values destroyed). *Special Issue on China,* 68-71.

Wu, Y. (2003). The Refrigerator Magnat. *The McKinsey Quarterly*, *3*, 106–111.

Yang, G., Xin, R., Liu, S., & Dai, H. (2008). *Xianhua yu jingji: Tanxun Zhongguo qiye quanqiuhua zhi lu* (Fresh Flowers and Sharp Thorns: In Search of a Road to Globalization for Chinese Enterprises). Beijing, China: Shangwu Yinshuguan.

Yijia, J. (2008). *Meinuo (Miele) Zhongguo gongsi xuanchuan chengli (Announcement of the Foundation of Miele China)*. Retrieved October 5, 2010, from http://www.e-jjj.com/Article/HTML/0063/63484.html

ENDNOTES

1 The Going-Global-Strategy has been part of the national Five-Year-Plan since 2001. Its most important goal is the encouragement of able Chinese enterprises to invest overseas in order to strengthen their overall international competitiveness as well as to spur China's integration into the world economy.

2 As for the method of trial and error see Heilmann (2008).

3 The grading system of *Stiftung Warentest* works similar as the grading system in German schools, which uses 1.0 as the best and 6.0 as the worst grade. *Stiftung Warentest* only employs the scale from 1.0 to 5.0, the latter equaling F in the US grading system.

4 The survey was conducted online in August 2009 among 36 scholars and students of Chinese Studies in order to examine the attitude of highly educated consumers with over-average China experience towards Chinese and German brands and products with focus on Haier and Miele.

5 Han and Terpstra (1988, p. 250) suggested this move for Korean automobile manufacturers, arguing that their image would profit from relocating production to the US.

6 It is much harder to be successful on the Chinese market from a long-term perspective, since these advantages might fizzle out quickly (Schramm & Staack, 2005; Fuchs, n.d.).

This work was previously published in International Journal of Asian Business and Information Management, Volume 1, Issue 4, edited by Dasho Karma Ura and Patricia Ordóñez de Pablos, pp. 30-42, copyright 2010 by IGI Publishing (an imprint of IGI Global).

Section 3

Economic Development and Investments

Chapter 10
Chinese Investments in Italy:
Is the Wave Arriving?[1]

Francesca Spigarelli
University of Macerata, Italy

ABSTRACT

This paper focuses on the effects of the Go Global policy in Italy. The main type of Chinese investments, their strategic motivations, as well as the role of ethnic networks are analyzed. The phenomenon is still not significant in quantitative terms, but trends are impressive. Italy can provide an access to western markets and strategic logistic services, as well as to a wide array of distinctive skills/intangible assets in manufacturing industries. At the moment, there is a prevalence of greenfield initiatives, but acquisitions are rising sharply. In geographical terms, the locations chosen by Chinese investors favor areas that offer a wealth of distinctive skills (typically, but not only, district areas), but are not limited to industries in which Chinese ethnic groups are involved.

1. INTRODUCTION

One of the most interesting phenomena to have emerged in the international economic and financial scenario of late is the leading role played by various emerging economies as global investors (Unctad, 2006). New multinational enterprises (MNEs), primarily those tied to what is known as the "BRIC" area (Brazil, Russia, India and China), have embarked upon rapid globalization processes based on investment transactions not only of the South-South type, but also targeting industrialized economies, particularly Europe and the USA (Wilson & Purushothaman, 2003; Sauvant, 2005, 2006).

China is certainly the most striking example of this phenomenon, due in part to the strong role its government plays in the creation of a vast network to aid and support the country's enterprises. Aid consists mainly of information services, financial

DOI: 10.4018/978-1-4666-0276-2.ch010

and fiscal incentives, insurance coverage, and consulting support (Shi, 2002). Several cases of acquisitions of well known Western companies (Wu, 2005; Anzkiewicz & Whalley, 2006) have drawn the attention of scholars and the media to the current and prospective effects of this policy, known as *Go Global*, especially considering the significant acceleration shown by Chinese initiatives in recent times and the rapid pace at which these enterprises are expanding on a global scale (Hong & Sun, 2004; Unctad, 2006, pp. 7-8).

The characteristics and effects of the *Go Global* phenomenon have yet to be studied in depth, chiefly due to the fragmentary nature and limited extent of available data. The recent scientific interest in this issue has produced several studies describing trends and drivers of Chinese investments (Taylor, 2002; Deng, 2003, 2004; Wong & Chang, 2003; Buckley et al., 2008b; Morck et al., 2008; Yeung & Liu, 2008; Alon & McIntyre, 2008), quantitative analyses of the reasons behind Chinese investments (Buckley et al., 2007, 2008b; Morck et al., 2008; Gao, 2008), and case studies of individual internationalization experiences (Liu & Li, 2002; Warner et al., 2004; Antkiewicz & Whalley, 2006; Bonaglia et al., 2007; Rui & Yip, 2008; Liu & Tian, 2008).

The goal of this paper is to explore the initial effects and primary characteristics of the *Go Global* phenomenon in Italy. The purpose is to spur reflection on how to best organize further study aimed at outlining and predicting the possible medium- and long-term effects of this phenomenon on the Italian economy. Passive internationalization, driven by Chinese investment flows, could give rise to new fields of study, new questions regarding the process of Italian economic and industrial development, and the reconfiguration of knowledge and local value allocation processes.

The paper is divided into two parts. The first part consists of a reconstruction of the theoretical framework on which the analysis is based. The second part is focused on the Italian situation, in an attempt to draw a map of the phenomenon and verify several initial hypotheses.

2. MNEs FROM EMERGING COUNTRIES AND GO GLOBAL: AN OVERVIEW OF THE LITERATURE

In the broader context of internationalization studies, new literature has recently developed (Meyer, 1998; Bonaglia et al., 2007; Mathews, 2002, 2006; Ramamurti, 2008; Rabellotti & Sanfilippo, 2008; Sauvant, 2008), centered on the appearance in the global scenario of new multinationals, late entrants to the market (Bartlett & Ghoshal, 2000) from emerging economies. Following several limited cases of MNEs from developing nations in the early 1980's (Lecraw, 1977; Kumar & McLeod, 1981; Wells, 1983; Lall, 1983), in the early 1990's a growing number of enterprises from Chile, China, Egypt, Malaysia, Mexico, Russia, South Africa, Thailand, and Turkey (Aykut & Goldstein, 2006, p. 8) began to enter the world scene (Aykut & Goldstein, 2006, pp. 11-12; Unctad, 2006, pp. 18-36). These enterprises, driven by the search for competitive advantages, follow atypical patterns of development: they internationalize in order to grow and make ample use of alliances and acquisitions in order to acquire distinctive resources such as brands, technology, and knowledge (Gammeltoft, 2008). These resources are functional to the rapid creation of an international competitive advantage capable of expanding and diversifying their basic competencies.

In this context, the traditional interpretation based on Dunning's approach – the OLI theory (Dunning, 1977, 1979, 1980, 1988) does not seem capable of explaining the outreach towards international markets. According to several authors, the approach taken by MNEs from emerging economies is the opposite of the traditional one (Bonaglia et al., 2007, p. 3; Dunning et al., 2008). They internationalize in order to achieve competitive advantages rather than to affirm their current advantages on a global scale. Their success appears to be tied to their ability to exploit the opportunities afforded by global competition and their position as late entrants to the market (Ramamurti, 2008): by developing a geocentric mentality (Perlmutter, 1969) they are able to project their product on a global level before having achieved global scale themselves (Bartlett & Ghoshal, 2000). In addition, they specifically leverage strategic and organizational innovation in order to internationalize rapidly (Matthews, 2002, 2006).

By focusing attention on firm-specific elements, albeit with differences related to the differing contexts (Ramamurti, 2008)[2], the following triggers of international action by MNEs from emerging countries may be identified (Bertoni et al., 2008):

- push factors, related to the increasing costs of resources available on domestic markets, in addition to the need to follow the internationalization policies of clients, suppliers and competitors;
- pull factors, related to opportunities for growth and expansion abroad, the availability of natural resources or strategic assets, the investment attraction policies of the host countries, and government incentives offered by home countries;
- managerial factors such as the search for skills and knowledge needed for business growth through internationalization.

In the context of the new MNEs from emerging countries, China represents one of the cases most analyzed and discussed by scholars and experts (Zwig, 2002; Alon & McIntyre, 2008). Box 1 contains a summary of the characteristics of the *Go Global* policy and the results of the main studies conducted in this area. Those recent studies emphasize the peculiar nature of an internationalization process strongly promoted by the government, but also provide evidence of critical issues and weaknesses associated with this process.

Box 1 – The *Go Global* Policy

As early as the late 1970's, in parallel with the Open Door policy, which transformed exports and foreign investment into the drivers of Chinese growth (Yao, 2006; Long, 2005; Branstetter & Lardy, 2006), the government launched an additional phase of international integration, based on encouraging its enterprises to make transnational investments.

The *Go Global* policy, formally introduced in the 10th Five-Year Plan (2001-2005) and marked as strategic in the 11th Plan (2006-2010), led the government to make a gradual transition. From acting as a mere approver and selector of globalization initiatives, the government started serving as a vehicle for providing fiscal and financial aid, and, lastly, in recent years, becoming a concrete provider of support to enterprises in terms of operating services and assistance (Bellabona & Spigarelli, 2007).

In keeping with the government's goals, all sectors have been interested in the push towards internationalization (from the primary sector, in the form of natural resources and energy, to the secondary sector, particularly mechanics, the automotive industry, and electronics, and the tertiary sector, such as financial services, research and development, and ICT). Many enterprises, and not only large, publicly controlled companies, are globalizing (Bellabona & Spigarelli, 2007) through greenfield initiatives, and especially through acquisitions (Hess, 2006; Rui & Yip, 2008).

The very justifications for internalization are evolving, as desired by the authorities (Yeung & Liu, 2008). Chinese interests are increasingly moving beyond the initial search for sources of energy and raw materials (Lawrence, 2002). Recent empirical analyses (Zhang, 2005; Battat, 2006; Jiang, 2006) have shown that enterprises frequently internationalize in order to conquer new markets (Cross & Voss, 2008; Liu & Tian, 2008) due to the strong competition and saturation of their domestic markets (Taylor, 2002; Zhang, 2003; Deng, 2004; Buckley et al., 2008b; Morck et al., 2008). The new multinationals seek to reduce production costs and overcome commercial barriers. By acquiring Western companies, they secure advanced technologies and strategic competencies, including managerial skills (Cross, Voss, 2008), as well as internationally recognized brands, in order to avoid long market acceptance times (Morck et al., 2008; Rui & Yip, 2008).

Go Global is taking a strategic role among Chinese development policies, due in part to the high degree of fragmentation shown by the internal market. The international outreach of enterprises is seen as a tool to bridge the highly significant development gap shown in many national areas (Meyer, 2008).

Despite the significant impact that *Go Global* is having at the international level, there are also signs of weakness and fragility inherent in an "induced" internationalization process. Firstly, the organizational structure of conglomerated enterprises is limiting the efficacy of their actions in the global context and their advantages of scale frequently prove a matter of mere appearance. For historical and governance-related reasons, subsidiaries tend to operate as quasi-independent entities, often pursuing goals that conflict with the strategies outlined by the government or on behalf of the government by holding companies (Nolan, 2001; Nolan & Zhang, 2002). There have been numerous cases of bankruptcy or severe difficulties encountered on international markets (Accenture, 2005).

Another critical issue, which has been the subject of extensive debate in both academic circles and among the Chinese management, is the effective ability of Chinese enterprises operating abroad to absorb technology, especially where not accompanied by an adequate training process for the managerial and professional class (Tan, 2005)[3].

Another problematic issue associated with the *Go Global* policy is that government support for the internationalization of Chinese enterprises is an integral part of an economic diplomacy program the strategic and economic objectives of which are frequently confused (Yeung & Liu, 2008). During the most recent phases of expansion, primarily focused on intangible assets, government intervention has been viewed with suspicion, slowing or preventing agreements, alliances and acquisitions, as had previously occurred elsewhere in Asia (Goldstein, 2008)[4]. The strong bonds between government and enterprises are having deleterious effects on both managerial culture, as mentioned above, and the perception of the risk of foreign investments (Yeung & Liu, 2008).

A recurring issue underlined in recent scientific contributions in this area is the need for a revision of the general theory of foreign investment. This theory must be adapted to account for the many specific factors that characterize the investment strategies of emerging economies, and of China in particular (Buckley et al., 2007; Li, 2007). Imperfections in capital markets, specific advantages of ownership and institutional factors are argued to have such a significant effect on investment expansion patterns that a new *ad hoc* key theory is required to interpret international investment flows.

In the case of China in particular, imperfections in capital markets mean that investing companies have access to capital at relatively advantageous non-market conditions. The imperfections in question pertain to the fact that public companies have access to subsidized loans aimed at supporting highly ambitious foreign acquisition policies and proposing a "light" assessment of the associated risk (Lardy, 1998; Scott, 2002; Warner et al., 2004; Morck et al., 2008). In addition, the existence of an undeveloped, inefficient banking system is argued to favor the contracting of loans at non-market conditions (Warner et al., 2004; Child & Rodrigues, 2005; Antkiewcz & Whalley, 2006). Furthermore, the presence of conglomerated enterprises in China is argued to result in a sort of internal capital market that is also capable of providing significant support for foreign investments (Liu, 2005). Lastly, family enterprises are argued to have facilitated access to financial resources by drawing on the liquidity possessed by the families that own them (Tsai, 2002; Child & Pleister, 2003; Erdener & Shapiro, 2005).

The advantages of ownership that condition foreign expansion processes consist of the qualities and aptitudes that Chinese enterprises have developed over time, and which allow them to compete more effectively in the international arena. Various factor include: flexibility, economic efficiency in the use of certain key resources – such as labor – (Morck et al., 2008; Rui & Yip, 2008), familiarity with operating in the situations typical of emerging and developing economies (Cuervo-Cazurra & Genc, 2008; Deng, 2003; Hong & Sun, 2006; Kalotay, 2008; Klein & Wocke, 2007; Pananond, 2007; Rui & Yip, 2008; Bertoni et al., 2008), and the ability to establish inter-company relations or relations with other players in local target markets. One example of this latter aspect is the presence of ethnic networks spread throughout Asia in particular, but also present in the rest of the world, created by the Chinese diaspora (Buckley et al., 2007, 2008; Zhou, 2006; Zanier, 2006)[5]. These human and social relationship assets (Lecraw, 1977; Wells, 1983; Lau, 2003) could represent a key resource explaining the contours of Chinese foreign investment growth (Luo, 1997; Standifird & Marshall, 2000; Tong, 2003; Gao, 2008). A positive relationship between investment activity and the establishment of ethnic Chinese in the host countries has been demonstrated in a number of international contexts (Buckley et al., 2007). Cultural affinity is argued to decrease psychological distances, reduce transaction costs and create a positive network effect capable of attracting capital[6].

Lastly, several international factors contribute to explaining the trend in Chinese investment, as suggested by the emerging literature in this field – the institution-based view of strategy (North, 1990; Peng, 2002; Meyer & Nguyen, 2005; Wright et al., 2005). More specifically, these factors consist of the formal and informal rules of play dictated by the institutional context of reference. The presence of public support, aid and subsidies is also argued to allow Chinese enterprises to overcome many difficulties associated with access to foreign markets (Hong & Sun, 2006). At the same time, the degree of bureaucratization of authorization procedures is argued to condition the amount and direction of investment flows.

3. ITALY AND THE *GO GLOBAL* EFFECT

The foregoing paragraph examined the main theoretical positions among which to situate analyses of emerging economies' new role as global investors. By driving significant investment flows towards Western countries, enterprises from countries undergoing rapid development, and first and foremost from China, are giving rise to new models of internationalization, requiring a revision of consolidated theoretical approaches (Aulakh, 2007; Goldstein, 2007).

This part of the paper focuses on the impact that the Chinese government's policy of providing a strong stimulus for foreign investment is having on Italy.

This exploratory analysis, as mentioned above, aims to identify possible areas for further academic inquiry and points of departure for additional courses of study.

An analysis of the extension of *Go Global* in Italy could prove interesting, both in connection with the well known debate as to the future of Italian districts (Carabelli et al., 2006), as well as the phenomenon known as onsite delocalization, which consists of the establishment of foreign entrepreneurs in various district chains (Fubini, 2006).

The changes underway in local contexts (Corò et al., 2005), including from the standpoint of the transnational fragmentation or re-composition of knowledge (Rullani, 2006), may be conditioned by the effects of *Go Global*, albeit within certain limits. By acquiring brands, knowledge and technology, increasingly often through acquisitions of local companies, Chinese enterprises could lead to "alterations" or "anomalies" in current economic development patterns, especially in certain geographical areas and sectors.

Furthermore, reflections on *Go Global* in Italy appear highly interesting in the light of the Chinese government's recent decision to involve small and medium enterprises in the international expansion process to a significant degree. To this end, it is interesting to consider the possible consequences of internationalization initiatives that lead to a "contamination" of two industrial models which share various similarities. Much like Italy (Fortis, 2006), China also has a strong presence of small and medium enterprises specialized in traditional industries (manufacturing operations, typically in the fields of textiles, apparel, footwear, and mechanics) and organized in district areas (Amighini & Chiarlone, 2003).

Considering the literature and the results of previous studies on the *Go Global* phenomenon, several "raw" research hypotheses (key research points) may be identified and analyzed through preliminary exploratory inquiries into the Italian context.

Hp1- In its initial stages, the *Go Global* policy has driven large public enterprises to internationalize in search of natural resources. For this reason, countries devoid of raw materials and energy sources, such as Italy, still benefit to a limited extent from Chinese capital inflows.

Hp 2 - At the current "mature" stage of the *Go Global* policy, as in the case of MNEs from other emerging nations, Chinese enterprises are internationalizing in search of high-value assets (technology, knowledge, brands, and strategic locations) in order to achieve an active role in international competition. Countries with a wealth of such resources, such as Italy, become privileged targets.

Hp 3 - In their search for strategic assets that are difficult to replicate or internally realizable only in the long term, Chinese enterprises view acquisitions as their favored method of international expansion.

Hp 4 - Ethnic networks play a key role in identifying target countries for investment. The presence of Chinese communities may represent an advantage of ownership capable of facilitating the penetration of Western countries.

A variety of information sources are used in order to analyze these hypotheses and lay down an overview of the Chinese presence in Italy that is as up-to-date and complete as possible. Specifically, these are:

- Chinese direct investment flows into Italy, as surveyed by Banca d'Italia[7];
- a map of Italian companies with Chinese ownership, whether incorporated in Italy ex novo or controlled through the acquisition of Italian companies;
- the number and geographical location of individual business initiatives, as surveyed by Unioncamere (Association of Italian Chambers of Commerce), with respect to persons of Chinese ethnicity.

3.1 Chinese Investment Flows into Italy

The first source of information helpful to reflecting on the impact of *Go Global* in Italy is related, as discussed above, to the surveys conducted by Banca d'Italia concerning direct investment flows on the statistical basis of the balance of payments. Although, as amply emphasized by the literature (Mori & Rolli, 1998; Lipsey, 2001; Federico & Minerva, 2007), this figure is not significant if viewed in isolation, it may nonetheless provide several interesting points of departure for further reflection[8]. See w.

At present, inflows of Chinese resources into Italy are certainly negligible in extent, especially

Table 1. Chinese FDIs in Italy represented by region (per thousand €)

2001		2004	
Emilia Romagna	22	Lazio	1,215
Lazio	107	Tuscany	250
Lombardy	1,197	Trentino Alto Adige	70
Piedmont	225	Veneto	659
Tuscany	536	Not specified	612
Veneto	167		
Not specified	100		
Total PRC	2,354	Total PRC	2,806
Total from World	37,186,823	Total from World	106,482,743

Source: calculation based on data provided by Banca d'Italia

when compared to total foreign investments (Rabellotti & Sanfilippo, 2008). Even compared to other Asian countries, China's role is decidedly limited: it is the origin of 2.9% of Asian resources brought into Italy.

However, the growth rate of inflows in recent years is striking. Between 2001 and 2007, investment flows increased by 303%, and although they have slowed after the 2006 boom (+136% compared to the previous year), expansion continues to show a significant uptrend. In other words, the phenomenon is highly dynamic and is developing rapidly.

Although the data are not directly comparable, Italy's performance is in line with the general increase in Chinese investment broad: Chinese investment flows throughout the world increased by 172% between 2001 and 2007 (Spigarelli, 2008; Wei, 2008), whereas Chinese investments destined for Europe, including Russia, increased by 319% between 2003 and 2006 (Tarmidi & Gammeltoft, 2008).

Focusing the analysis at the geographical level shows a considerable increase in inflows to Lazio and Lombardy in recent years, the continuing absorption of resources by Tuscany, and strong, constant expansion in Veneto. In 2007 Veneto absorbed 24% of Chinese inflows into Italy, catalyzing resources primarily in the provinces of Vicenza and Treviso.

One noteworthy phenomenon is the fact that new, and perhaps unexpected, regions began to receive resources in 2006. Inflows to Calabria, Marche, Molise, Puglia and Umbria began to be reported, as opposed to the typical textile manufacturing regions, such as Tuscany and Veneto, which are closely tied to *made-in-Italy* products and show a high presence of residents of Chinese ethnicity. It appears that Chinese investments are differentiating and spreading throughout Italy, pushing into new areas, though always where a strong "traditional" industrial base exists. Further discussion of these issues is provided in a subsequent section of this paper.

Turning to an analysis by sector (Table 2), it should be noted that the study is not fully significant if it is considered that the primary share of resources was absorbed by the "Family" class throughout the entire period analyzed (2001-2007)[9]. The percent weight of "Other non-classifiable branches" is also high.

A comparison with the 2001-2006 figures shows a decrease in investment flows in commercial services, probably due to significant past penetration that has tended to saturate the various markets. Nonetheless, the uptrend in the resources allocated to this segment resumed in 2007.

The differentiation of investment targets has increased over the years: new business segments are attracting investment. In particular, in 2006 an unusually high percentage of resources was allocated to the tertiary sector, specifically hotel, restaurant and bar services. In 2007, however, 13% of resources was allocated to the electrical materials and supplies segment, whereas 6.8% was allocated to maritime and air transportation services. These trends are perhaps further evidence that Chinese investments continue to evolve and mature, and not only in geographical terms.

Table 2. Chinese FDIs in Italy represented by sector (thousand €)

2001		%	2004		%
Family	841	36	Family	1,724	61%
Services for trade and commerce	634	27	Minerals and iron/non iron metals	414	15%
Constructions and public works	595	25	Constructions and public works	250	9%
Other non-classifiable branches	123	5	Services for trade and commerce	130	5%
Industrial and agricultural machinery	90	4	Industrial and agricultural machinery	198	7%
Other services	61	3	Other services	90	3%
Minerals and iron/non iron metals	10	0			
Total PRC	2,354	100	Total PRC	2,806	100

Source: calculation based on data provided by Banca d'Italia

3.2 Chinese-Owned Italian Enterprises

In order to achieve a more detailed understanding of the situation of Chinese investments in Italy, it would be helpful to move beyond the overview provided by Banca d'Italia's survey of investment flows to a review of Chinese-owned enterprises incorporated or acquired in Italy. As emphasized in the literature, the analysis of this information is even more fundamental to capturing a country's international expansion policy than that of investment flow figures (Cantwell & Bernard, 2008).

The resulting map, based on various sources[10], indicates that there are 28 companies owned or co-owned by Chinese entities in Italy at the end of 2007[11]. These companies consist primarily of units incorporated ex novo as greenfield initiatives and located in northern Italy (Table 3). Several recent cases, as highlighted by the media, show that the acquisition of enterprises, typically small or medium-sized enterprises, in a wide range of business segments, is becoming increasingly common.

In terms of business segments, it is interesting to note that Chinese enterprises are present in Italy both in mature sectors (mechanics, textiles, apparel) and high-technology sectors (electronics, automotive, communications). Leading *Go Global* players in Italy include large groups, some of which are public in origin, with a worldwide reputation for the speed and breadth of their internationalization initiatives. From Cosco and China Ocean Shipping Company in the logistics segment, to Haier in the home appliances industry, to Huawei in telecommunications and ICT, and Anhui Jianghuai Automobile Co. Ltd, ChangAn Automobile Group and Nanjing Motor Corporation in the automotive sector. Transactions are commonly financed by drawing on the extensive financial reserves possessed by Chinese enterprises. Nonetheless, there is no lack of support from the government and the Chinese banking system, in the form of the internationalization support measures provided under the *Go Global* policy (Bellabona & Spigarelli, 2006).

After delving into these cases in further detail and reviewing the primary activities undertaken by Chinese-controlled Italian enterprises, it may be stated that these enterprises commonly act as commercial branch offices and are assigned tasks such as market analysis, positioning and product studies, and group marketing. Twelve of the 28 cases surveyed match this description.

The market-seeking objective seems clear. However, positioning commercial companies on the Italian market could also be interpreted as the first step in a broader internationalization strategy, which at the current stage aims to explore the Italian and European situation.

Table 3. Italian companies with Chinese ownership

Italian Company	*Sector/activity*	Chinese owner	Year of establishment/ acquisition
Transport and logistics			
Air China	Commercial offices	Air China	1986
China Shipping Agency Co. srl	Shipping and logistics	China Shipping Co.	1999
Co.Na.Te.Co.	Shipping and logistics	Cosco Group	2002
Cos Cos srl	Shipping and logistics	Cosco Group	1996
Penta Shipping Italia srl	Shipping and logistics	Cosco Group	1995
ICT			
Huawei Techologies Italia	Commercial and marketing services for the group (TLC, ICT)	Huawei Techologies	2004
Primary sector			
Baosteel Italia Distribution Center spa	Commercial and marketing services for the group (steel)	Baosteel group	2001
Temi Group	Commercial and marketing services for the group (copper products)	Tongling Nonferrous Metals Inc	n.d.
Manufacturing			
Anhui Jianghuai Automobile Co.	R&D for the group (automotive)	Anhui Jianghuai Automobile Co. Ltd	2004
Benelli QJ srl	Motorcycle	Quinjiang Group	2005
Cemate Machinery - Technology srl	Industrial machinery	Cemate Machinery	1988
Changan Automobile Group	Commercial and marketing services for the group (automotive)	Chang An Automotive Co. Ltd.	2005
Chunlan	Commercial and marketing services for the group (white goods)	Chuanlan Group Co.	n.a.
Elios spa	Electric and electronic components	Feidiao Electrics Co. Ltd	2006
Haier Appliances (Italy) spa	White goods production	Haier Group Corp.	2001
Haier Europe Trading	Commercial, marketing and R&D services for the group (white goods)	Haier Group Corp.	2000
Haitian Europe	Machinery and molds for plastic products	Ningbo Haitian Group Co. Ltd	n.a.
Hengdian Group Europe	Commercial and marketing services for the group (pharmaceutical, chemical, electronics)	Hengdiang Group	n.a.
HPM Europe	Wholesale of contructor	Hunan Sunward Intelligent Machinery	2007
Hisense Italy	Commercial services for the group (ICT, white goods)	Hinsense Group	2001
Nanjing Motor Corporation	Commercial, marketing and R&D services for the group (automotive)	Nanjing Motor Corporation	1986
Peaktop Technologies srl	Plastic products	Peaktop Technologies	2000
Univent spa	Mechanical tools	Xi'an Electric Motor Works	n.a.

continued on following page

Table 3. Continued

Italian Company	*Sector/activity*	Chinese owner	Year of establishment/ acquisition
Wilson	Design and production of shoes	Wenzhou Hazan	2004
Trading			
China Long Co. srl	Wholesale trade	China National Arts And Crafts I&E	n.a.
China Silk	Commercial and marketing services	China Silk	n.a.
Temax Italia srl	Commercial and marketing services for the group (trading)	Genertec Group	1991
Financial services			
Bank of China	Banking	Bank of China	1998

Source: map based on data provided by Sviluppo Italia, CCPIT, specialist economic press

On occasion, such units are also assigned research and development activities, as is the case in the ICT and automotive industries in the province of Turin. In this case, the strategic asset-seeking objective of the investment becomes readily apparent, given the presence of consolidated, locally based knowledge in both industrial and academic capacities. Chang An Automotive Co. Ltd and Anhui Jianghuai Automobile Co. Ltd. preferred Turin to other possible locations around the world precisely in order to acquire, through a process of "contagion", planning and design knowledge and specializations capable of improving their products and making them attractive in the eyes of demanding Western consumers.

Other cases of the search for high value-added assets include the non-greenfield acquisitions aimed at obtaining brands. Particularly relevant in this regard is the case of Benelli QJ Srl, a motorcycle manufacturer acquired in a state of severe crisis by Qianjiang Group Co. Ltd (QJ Group), which also manufactures motorbikes and scooters (Bellabona & Spigarelli, 2006, p. 157). A brand with historical resonance in motorcycle racing, Benelli is intended to provide a new image for Chinese products to be distributed in the West, in addition to high-level design skills (Spigarelli et al., 2009).

Brand-related concerns also characterized the acquisition by Fediao Electrics of Elios S.p.A., which operates in the electric and electronic components, lamp socket, and interior and exterior lighting decoration segment (Ridella, 2007). It is interesting to note that in this case the brand is required not so much in order to establish the company in the eyes of a vast public of image-conscious consumers, but rather to emerge in a "technical" business-to-business segment.

The search for strategic assets by Chinese investors is also related to the acquisition of the advantages of localization and the access to specialized know-how and skills capable of interpreting the European market's needs and technical standards in highly competitive sectors. This is the case of Haier, which has become one of the world's leading players in the home appliance sector. In 2001 Haier acquired from Meneghetti Srl, based in Padua, 100% of the business unit specialized in the manufacture of built-in refrigerators. The decision to acquire an Italian facility is part of a broader strategy also related to the localization of Haier Europe in Varese. The latter, in addition to coordinating the marketing and sale of home appliances in Europe, manages all logistical activities on the Continent[12].

Another significant sector in which to acquire high-value intangible resources is logistics. In accordance with a trend shown at the global and European level, Italy also seems to be becoming a destination of increasing interest for Chinese entrepreneurs aiming to dominate the commercial maritime shipping business, which is expanding rapidly due to the intensification of trade between China and the rest of the world (Morelli, 2004, pp. 8-10). Italy is viewed as a gateway to the Mediterranean, the vast European market, and the Balkans, in particular. The acquisition of port infrastructure and logistical networks is typically achieved through joint ventures with local partners who have long-term experience in the successful operation of Italy's primary hubs. In this case, Chinese players are large multinationals such as Cosco and China Ocean Shipping, as cited above, which boast dense international networks of firms and partnerships.

3.3 Ethnic Networks and Chinese Entrepreneurism

Another area of interest which requires further inquiry to understanding the *Go Global* phenomenon in Italy is the relationship between the enterprises that act as protagonists in the internationalization process and the broad network of Chinese micro-entrepreneurs with deep roots in district areas.

Although individual entrepreneurism is symptomatic of a type of internationalization that differs from that sought under the *Go Global* policy, its strong presence in Italy ought to be compared and linked to the phenomenon of the internationalization of Chinese companies examined in the foregoing paragraph. It is sufficient to consider that the historical presence of Chinese micro entrepreneurs in Italy could become a resource for large and medium enterprises to exploit, by forming relationships with business initiatives that have already been launched and established in district areas. As mentioned above, several recent studies (Buckley et al., 2007) demonstrate that the presence of an ethnic network, consisting of strong social relationships that also extend to business, creates a key ownership asset for investor enterprises.

One way of understanding the phenomenon in Italy is to analyze the data on the nationality of natural individual persons who act as entrepreneurs, as surveyed by Unioncamere[13].

An analysis of the ethnic composition and distribution by industry of non-EU entrepreneurs was conducted drawing on the most recent available data (2004-2007).

The results indicate, firstly, that entrepreneurial activity is highly dynamic: the presence of Chinese entrepreneurs increased by 45% between 2004 and 2007, compared to the 22% increase in total non-EU entrepreneurs (Table 4).

The numbers are particularly significant if compared with the corresponding percentage calculated on national data: active Italian enterprises grew at rates near 1%. In terms of geographical distribution, at the end of 2007, the regions in which Chinese entrepreneurs were prevalent were Lombardy and Tuscany (20%), Veneto (11%), Emilia Romagna and Lazio (10%). In these regions, the high incidence of Chinese entrepreneurs is also supported by the significant presence of Chinese citizens, who represent a significant percentage of the population compared to other ethnic groups and are particularly concentrated in the areas mentioned above (Istat, 2007).

Polarization in geographical areas that are home to specialized districts[14] does not prevent new Chinese business initiatives from expanding throughout the territory. The preferred locations for new establishments are unexplored regions, including those of southern Italy, that show the same tendency towards specialized districts[15]. ISTAT data on foreign citizens confirm significant growth of new Chinese residents in southern Italy in the three-year period 2005-2007 (Istat, 2007).

Another significant phenomenon that comes to light from an interpretation of the data is the concen-

Table 4. Chinese entrepreneurs in Italy (number of individual entrepreneurs)

	2004		2005		2006		2007	
	Chinese entr.	Total non UE entr.	Chinese entr.	Total non UE entr.	Chinese entr.	Total non UE entr.	Chinese entr.	Total non UE entr.
Lombardy	6,222	70,140	6.222	70,140	7,790	84,864	8,355	83,911
Tuscany	6,164	28,837	6,164	28,837	7,705	36,288	8,475	35,588
Lazio	3,021	33,970	3,021	33,970	3,773	41,497	4,112	39,129
Emilia Romagna	2,996	28,666	2,996	28,666	3,798	37,260	4,255	37,761
Veneto	2,614	30,241	2,614	30,241	3,800	37,992	4,576	37,586
Campania	1,607	17,140	1,607	17,140	1,976	20,559	2,241	21,548
Piedmont	1,341	23,268	1,341	23,268	1,795	30,073	1,959	28,113
Sicily	1,056	14,420	1,056	14,420	1,595	16,991	1,759	17,682
Marche	656	8,898	656	8,898	1,011	11,272	1,163	11,339
Puglia	621	9,737	621	9,737	87	11,067	948	11,424
Liguria	618	10,241	618	10,241	764	12,392	810	12,869
Abruzzo	596	9,018	596	9,018	807	10,443	907	10,489
Sardinia	481	5,320	481	5,320	663	6,274	721	6,408
Calabria	378	7,590	378	7,590	597	8,794	630	8,939
Friuli V. Giulia	354	9,790	354	9,790	515	11,119	620	11,123
Trentino A. Adige	165	4,082	165	4,082	251	5,191	303	5,388
Umbria	144	4,517	144	4,517	230	5,704	292	5,509
Basilicata	45	1,513	45	1,513	92	1,655	98	1,672
Molise	33	1,387	33	1,387	50	1,543	51	1,520
Aosta Valley	6	525	6	525	16	629	15	612

Source: calculation based on data provided by Unioncamere

tration of Chinese entrepreneurs in specific business segments (ethnic specialization) and the tendency of this ethnic group to saturate business initiatives in the sector of "initial establishment" (industry ethnicization). Consider that in 2007, overall, Chinese business initiatives represented 81% of the non-EU total in the manufacture of articles of apparel and fur articles, 78% in the preparation and tanning of leather and the manufacture of travel goods, and 45% in the textiles industry. As already shown in previous studies (Borsari et al., 2006), Chinese entrepreneurs, and especially in the textiles and apparel segment, take a place in the district production chain by offering competitive advantages such as processing costs, the flexibility of production capacity, and the rapidity of execution of orders (Borsari et al., 2006, p. 5).

Another interesting trend is the gradually increasing positioning of Chinese entrepreneurs in retail distribution of personal and household articles (+73% of initiatives between 2004 and 2007) leveraging products imported directly from their home country. Production, ordered in the homeland, is adapted to European specifications, thereby maintaining highly competitive prices and avoiding the problems and bureaucratic issues typical of the Italian market (Borsari et al., 2006, p. 16).

4. CLOSING REMARKS

On the whole, Italy appears to be feeling the effects of the *Go Global* policy to a progressively increasing extent. The phenomenon is still not significant in quantitative terms. What is interesting, however, is the trend shown in recent years and the rapid pace at which the situation is evolving.

The reason behind the delay of the arrival of the wave of Chinese investments in Italy is most likely due to the gradual maturity and evolution of China's international penetration strategy.

As emphasized above, Chinese authorities have progressed from initial resource-seeking expansion aimed at the procurement of scarce raw materials, to direct stimulation of the entry into foreign markets, for market-seeking purposes in search of new commercial markets. Only in recent times the government has enacted specific measures and taken action aimed at stimulating investments in strategic assets. Italy, which offers few natural resources, yet provides a deep market outlet and a wide array of distinctive skills in manufacturing industries, has only become an interesting target for Chinese internationalization over the last few years. As a matter of fact, in addition to the 28 enterprises surveyed in this paper, the press and media have reported a number of cases in which negotiations are ongoing or nearing the closing stages. The uptrend in investments over time, as also reported in other studies, bears witness to the increasing interest shown in Italy[16]. In this regard, research hypotheses 1 and 2 may be confirmed.

The recent expansion of Chinese investments in Italy may also be linked to the gradual internationalization of medium-size enterprises. Such enterprises, to a much greater extent than public groups or large multinationals, tend to find fertile ground for alliances and acquisitions in Italy. Indeed, they have a greater focus on high value-added and intangible resources (image, brands, research, and innovation) in the traditional industries that are so well represented in Italy. Medium-sized Chinese enterprises, attracted by the quality and brands of many exemplars of the *made-in-Italy* tradition, could increasingly come to view the acquisition of the resources of Italian companies as a gateway to rapid "external" growth capable of offsetting their image as low-cost, lower-quality producers. In addition, the availability of financially "accessible" target companies of limited size in Italy could tend to facilitate penetration and entry into the European market.

The scope of this paper offers a general overview highlighting areas of interest that require further analysis. Firstly, the scarce presence of Chinese enterprises in Italy could also be linked to the symmetrical weakness of the Italian presence in China in the form of enterprises, institutional representation, and university exchanges. An equally interesting issue is attempting to understand the extent to which managerial, cultural and bureaucratic difficulties are affecting the decision as to whether to invest in Italy, in a manner similar to the obstacles encountered by Italian companies in China.

The survey of production and commercial investments in Italy by Chinese enterprises shows that there is currently a prevalence of greenfield initiatives. However, acquisitions have risen sharply in recent months. Acquisitions seem to have become the dominant strategy, as opposed to the opening of new establishments or commercial units. It can consequently be stated that the third research hypothesis is substantially, but not fully, confirmed. In this case as well, further pursuit of the analysis would be of interest, in an attempt to understand fully which factors (institutional, situational, localization-related, and firm-specific) tend to stimulate, and which tend to hamper, M&A transactions. Scouting and the search for information on potential targets, as well as the availability of consulting and information support in the homeland or Italy, also represent subjects worthy of investigation.

In geographical terms, the locations chosen by Chinese investors tend to favor areas that offer a wealth of distinctive skills (typically, but not only, district areas), but are not limited to industries in which Chinese ethnic groups are involved. At present, the mechanics segment is that most affected by both acquisitions and greenfield initiatives.

The logistics and IT industries are also common targets. Textiles, apparel and footwear, industries in which the Chinese ethnic group has historically been concentrated in Italy, have still not emerged as target segments for the internationalization processes of Chinese enterprises. Nonetheless, Banca d'Italia figures confirm that the rapid growth of Chinese investments in Italy also involves segments and regions that manufacture typical *made-in-Italy* products and are organized into specialized districts. These are areas in which the data on non-EU entrepreneurs show a strong Chinese ethnic presence.

On the whole, the phenomenon is probably still in the "embryonic" stage and therefore does not allow for an accurate evaluation of the fourth research hypothesis, which at present, given the limited data available, may not be confirmed. In addition, it should be emphasized that, as compared to other countries, in which Chinese MNEs may join other large Chinese companies created and developed in the West, Italy offers a wealth of small entrepreneurs. The data currently available do not allow it to be determined whether, and to what extent, enterprises created by Chinese immigrants serve as effective drivers to draw MNEs to Italy. This is yet another interesting issue open to further inquiry through specific analyses in the field[17].

In conclusion of this paper, it would be pertinent to shed light on its principle limitations. In particular, the observations refer to a set of data that is very limited in quantitative and temporal terms. *Go Global* is both recent, especially in regards to its impact on Italy, and therefore difficult to survey. On the one hand, the sources are fragmentary and heterogeneous, and on the other, a part of the phenomenon is not covered by the available official statistics and do not allow econometric and more sophisticated analysis. Consider, for example, the problems encountered in the attempt to survey companies acquired or incorporated by Chinese entities. In addition, even the data provided by Banca d'Italia are inadequate to an exhaustive overview of foreign investments due to their lack of timeliness and the methods by which they were collected.

In view of these considerations, what may be the positive aspect of this study is its attempt to shed light on a very recent phenomenon by building a framework of inquiry helpful to organizing additional courses of study. In addition, the topic of MNEs from emerging countries is meeting with increasing interest from the academic community, given the need to adjust consolidated internationalization theories and understand the effects of the actions of the new global investors on Western nations. Continuing to investigate this phenomenon also seems an interesting course of action considering its significant policy implications. There is an urgent need to understand the current extent and primary drivers of the phenomenon, in order to assess the expedience of specific attraction policies, as implemented by the main Western countries, which view passive internationalization as an opportunity to turn mature segments around or achieve growth in strategic sectors. The action taken by the Italian central government and regions appears not only sluggish and lagging behind that taken by other European countries, but also often uncoordinated and heterogeneous. In addition, there are numerous issues for further study and hypotheses to be tested, beyond those already mentioned, in order to understand *Go Global*'s possible effects on Italy's distinctive skills, local aggregations, and the strategies of national companies. The foregoing is in relation to the potential effects on local knowledge (contamination, appropriation, fragmentation, reinforcement) and national and local economic development patterns.

REFERENCES

Accenture. (2005). *China spreads its wings – Chinese companies Go Global*. Retrieved from www.accenture.com

Alon, I., & McIntyre, J. R. (2008). *Globalizing of Chinese Enterprises*. New York: Palgrave Macmillan.

Amighini, A., & Chiarlone, S. (2003). Rischi e opportunità dell'integrazione commerciale cinese per la competitività internazionale dell'Italia (Working Paper, 149, CESPRI).

Anzkiewicz, A., & Whalley, J. (2006). Recent Chinese buyout activity and the implications for global architecture (Working Paper, 12072, NBER).

Aulakh, P. S. (2007). Emerging multinationals from developing economies: motivations, paths and performance. *Journal of International Management, 13*(3), 235–402. doi:10.1016/j.intman.2007.05.001

Aykut, D., & Goldstein, A. (2006). Developing countries multinationals: south-south investment comes of age (Working Paper, 257, OECD).

Bartlett, C. A., & Ghoshal, S. (1989). *Management Across Borders. The Transnational Solution*. Boston, MA: Hutchinson Business Books.

Bartlett, C. A., & Ghoshal, S. (2000). Going global: Lessons from late movers. *Harvard Business Review, 78*(2), 133–142.

Battat, J. (2006). *China's Outward Foreign Direct Investment*. Washington: FIAS, World Bank Group.

Bellabona, P., & Spigarelli, F. (2006). La dimensione transnazionale dell'economia cinese. In Abbiati, M. (Ed.), *Propizio è intraprendere imprese* (pp. 147–165). Venice: Cà Foscarina.

Bellabona, P., & Spigarelli, F. (2007). Moving from Open Door to Go Global: China goes on the world stage. *International Journal on Chinese culture and management, 1*(1), 93-107.

Bertoni, F., Elia, S., & Rabbiosi, L. (2008, October 9-10). *Drivers of acquisition from Brics to advanced countries: firm level evidence*. Paper presented at Emerging Multinationals: Outward Foreign Direct Investment from Emerging and Developing Economies Conference, Copenhagen, Denmark.

Bonaglia, F., Goldstein, A., & Mathews, J. A. (2007). Accelerated internationalization by emerging markets' multinationals: The case of the white goods sector. *Journal of World Business, 42*(4), 369–383. doi:10.1016/j.jwb.2007.06.001

Borsari, S., Ceccagno, A., Gallo, A., & Gaudino, S. (2006). *I cinesi e gli altri. L'imprenditoria extracomunitaria in Italia*. Bologna, Italy: Consorzio Spinner.

Branstetter, L., & Lardy, N. (2006). *China's embrace of globalization* (Working Paper, 12373, NBER).

Buch, C. M., Kleinert, J., & Toubal, F. (2006). Where enterprises lead, people follow? Links between migration and FDI in Germany. *European Economic Review, 50*(8), 2017–2036. doi:10.1016/j.euroecorev.2005.11.002

Buckley, P. J. (2002). Is the International Business research agenda running out of steam? *Journal of International Business Studies, 33*(2), 365–373. doi:10.1057/palgrave.jibs.8491021

Buckley, P. J., Clegg, J., Cross, A., Voss, H., Rhodes, M., & Zheng, P. (2008a). Explaining China's Outward FDI. In Sauvant, K. P. (Ed.), *The Rise of Transnational Corporations From Emerging Markets: Threat or Opportunity?* Willinston, VT: Edward Elgar Publishing.

Buckley, P. J., Clegg, L. J., Cross, A. R., Liu, X., Voss, H., & Zheng, P. (2007). The determinants of Chinese outward foreign direct investment. *Journal of International Business Studies, 38*(4), 499–518. doi:10.1057/palgrave.jibs.8400277

Buckley, P. J., Cross, A. R., Tan, H., Liu, X., & Voss, H. (2008b). Historic and Emergent trends in Chinese Outward Direct Investments. *Management International Review, 48*(6), 715–748. doi:10.1007/s11575-008-0104-y

Cantwell, J., & Barnard, H. (2008). Do Firms from Emerging Markets have to Invest Abroad? Outward FDI and the Competitiveness of Firms. In Sauvant, K. P. (Ed.), *The Rise of Transnational Corporations From Emerging Markets: Threat or Opportunity?* Willinston, VT: Edward Elgar Publishing.

Carabelli, A., Hirsch, G., & Rabellotti, R. (2006). *Italian SMEs and Industrial District on the move: Where are they going?* (Working Paper, 3). Bologna, Italy: University of Bologna, Department of Economic Science.

Child, J., & Pleister, H. (2003). Governance and management in China's private sector. *Management international, 7*(3), 13-24.

Child, J., & Rodrigues, S. B. (2005). The internationalization of Chinese firms: a case of theoretical extension? *Management and Organization Review, 1*(3), 381–410. doi:10.1111/j.1740-8784.2005.0020a.x

Chittoor, R., & Ray, S. (2007). Internationalization paths of Indian pharmaceutical firms – A strategic group analysis. *Journal of International Management, 13*(3), 338–355. doi:10.1016/j.intman.2007.05.008

Corò, G., Volpe, M., & Bonaldo, S. (2005, May 27). *Local production systems in Italy between fragmentation and international integration.* Paper presented at the Economia Internazionale e Sviluppo conference, University of Urbino, Urbino, Italy.

Cross, A. R., & Voss, H. (2008, February 1-2). *Chinese direct investment in the United Kingdom: An assessment of motivations and competitiveness.* Paper presented at the Corporate Strategies in the New Asia Conference, University of Bremen, Bremen, Germany.

Cuervo-Cazurra, A. (2007). Sequence of value-added activities in the multinationalization of developing country firms. *Journal of International Management, 13*(3), 258–277. doi:10.1016/j.intman.2007.05.009

Cuervo-Cazurra, A., & Genc, M. (2008). Transforming disadvantages into advantages: developing-country MNEs in the least developed countries. *Journal of International Business Studies, 39*(6), 957–979. doi:10.1057/palgrave.jibs.8400390

d'Italia, B., & Cambi, U. I. (2004). *Manuale della Bilancia dei Pagamenti e della posizione patrimoniale dell'Italia sull'estero.* Roma. Retrieved from http://www.bancaditalia.it/statistiche/quadro_norma_metodo/metodoc/manuale_bilpag.pdf

De Felice, A. (2006). Investimenti Diretti Esteri in Romania: nuove prospettive per il Mezzogiorno. In Del Prete, F. (Ed.), *Prossimità e Sviluppo*. Milano, Italy: Franco Angeli.

Deng, P. (2003). Foreign direct investment by transnationals from emerging countries: the case of China. *Journal of Leadership & Organizational Studies, 10*(2), 113–124. doi:10.1177/107179190301000211

Deng, P. (2004). Outward investment by Chinese MNCs: motivations and implications. *Business Horizons, 47*(3), 8–16. doi:10.1016/S0007-6813(04)00023-0

Di Minin, A., & Zhang, J. (2008, October 9-10). *Preliminary Evidence on the International R&D strategies of Chinese companies in Europe*. Paper presented at the Emerging Multinationals: Outward Foreign Direct Investment From Emerging and Developing Economies Conference, Copenhagen, Denmark.

Dunning, J. H. (1977). Trade, Location of Economic Activity and the Multinational Enterprise: A Search for an Eclectic Approach. In Ohlin, B., Hesselborn, P. O., & Wijkman, P. M. (Eds.), *The International Allocation of Economic Activity*. London: Macmillan.

Dunning, J. H. (1979). Explaining Changing Patterns of International Production: In Defence of the Eclectic Theory. *Oxford Bulletin of Economics and Statistics, 41*(4), 269–296.

Dunning, J. H. (1980). Toward an Eclectic Theory of International Production: Some Empirical Tests. *Journal of International Business Studies, 11*(1), 9–31. doi:10.1057/palgrave.jibs.8490593

Dunning, J. H. (1988). The Eclectic Paradigm of International Production: A Restatement and Some Possible Extensions. *Journal of International Business Studies, 19*(1), 1–31. doi:10.1057/palgrave.jibs.8490372

Dunning, J. H., Kim, C., & Park, D. (2008). Old Wine in New Bottles: a Comparison of Emerging Market TNCs Today and Developed Country TNCs Thirty Years Ago. In Sauvant, K. P. (Ed.), *The Rise of Transnational Corporations From Emerging Markets: Threat or Opportunity?* Willinsont, VT: Edward Elgar Publishing.

Erdener, C., & Shapiro, D. M. (2005). The internationalization of Chinese family enterprise and Dunning's ecletic MNE paradigm. *Management and Organization Review, 1*(3), 411–436. doi:10.1111/j.1740-8784.2005.00021.x

Fatiguso, R. (2006, November 14). Investire in Italia? Che fatica. *Il Sole24ore.*

Federico, S., & Minerva, G. A. (2007). Outward FDI and local employment growth in Italy. *Temi di Discussione del Servizio Studi* (Economic Working Papers, 613). Banca d'Italia.

Flisi, S., & Murat, M. (2009). Immigrants links, Diasporas and FDI. An empirical Investigation on Five Eureopean Countries. *Materiali di discussion* (608). Università degli Studi di Modena e Reggio Emilia. Retrieved from http://merlino.unimo.it/web_dep/materiali_discussione/0608.pdf

Fortis, M. (2006). I distretti industriali e la loro rilevanza nell'economia italiana: alcuni profili di analisi. In Quadrio Curzio, A., & Fortis, M. (Eds.), *Società, Stato e Mercato in Italia* (pp. 477–483). Bologna, Italy: Sussidiarietà, Laboratori, Distretti, Il Mulino.

Fubini, F. (2006, July 7). Il distretto di Prato? Salvato dai cinesi. *Il Corriere della Sera.*

Gammeltoft, P. (2008). Emerging multinationals: outward FDI from the BRICS countries. *Int. J. Technology and Globalisation, 4*(1), 5–22. doi:10.1504/IJTG.2008.016184

Gao, L. (2008, April 1-2). *Determinants of China's Outward Foreign Direct Investment.* Paper presented at the 19th CEA, UK Conference – China's Three Decades of Economic Reform (1978-2008), University of Cambridge, Cambridge, UK.

Gao, T. (2003). Ethnic Chinese networks and international investment: evidence from inward FDI in China. *Journal of Asian Economics, 14*, 611–629. doi:10.1016/S1049-0078(03)00098-8

Garg, M., & Delios, A. (2007). Survival of the foreign subsidiaries of TMNCs: the influence of business group affiliation. *Journal of International Management, 13*(3), 278–395. doi:10.1016/j.intman.2007.05.007

Goldstein, A. (2007). *Multinational Companies from Emerging Economies. Composition, Conceptualization and Direction in the Global Economy.* New York: Macmillan, Palgrave.

Goldstein, A. (2008). Who's Afraid of Emerging Market TNCs? Or: Are Developing Countries Missing Something in the Globalization Debate? In Sauvant, K. P. (Ed.), *The Rise of Transnational Corporations From Emerging Markets: Threat or Opportunity?* (pp. 183–203). Willinston, VT: Edward Elgar Publishing.

Goldstein, A., & Pananond, P. (2008). Singapore Inc. goes shopping abroad: profits and pitfalls. *Journal of Contemporary Asia, 38*(3), 417–438. doi:10.1080/00472330802078493

Hernandez, Z. (2008). Outward foreign direct investment from emerging markets: annotated bibliography. In Sauvant, K. P. (Ed.), *The Rise of Transnational Corporations From Emerging Markets: Threat or Opportunity?* (pp. 339–373). Willinston, VT: Edward Elgar Publishing.

Hess, W. (2006). Going Outside, Round tripping and dollar diplomacy: an introduction to outward direct investment. *World Market Analysis*, January. Retrieved from www.globalinsight.com

Hong, E., & Sun, L. (2004). *Go overseas via Direct Investment. Internationalization strategy of Chinese Corporation in a comparative prism* (Discussion Paper 40). London, UK: SOAS University of London, Center for financial and management studies.

Hong, E., & Sun, L. (2006). Dynamics of Internationalization and Outward Investment: Chinese Corporations Strategies. *The China Quarterly*, *187*(September), 610–634. doi:10.1017/S0305741006000403

ICE. (2008). Rapporto ICE 2007-2008 – L'Italia nell'economia internazionale. Retrieved from www.ice.it/statistiche/rapporto.htm

International Monetary Fund. (1993). *Balance of Payments Manual* (5th ed.). Washington, DC. Retrieved from http://www.imf.org/external/pubs/ft/bopman/bopman.pdf

ISTAT. (2007). Cittadini Stranieri, Bilancio demografico anni 2005-7 e popolazione residente al 31 Dicembre. Retrieved from http://demo.istat.it/

Jiang, W. (2006, June 14). Survey: Firms achieve goals abroad. *China Daily*.

Kalotay, K. (2008). Russian transnational and international investments paradigms. *Research in International Business and Finance*, *22*(2), 85–107. doi:10.1016/j.ribaf.2006.12.002

Klein, S., & Wocke, A. (2007). Emerging global contenders: the South African Experience. *Journal of International Management*, *13*(3), 319–337. doi:10.1016/j.intman.2007.05.002

Kumar, K., & McLeod, M. G. (1981). *Multinationals from developing countries*. Lexington, MA: Lexington Books.

Kynge, J. (2006). *China Shakes the World*. Boston: Houghton Mifflin Company.

Lall, S. (1983). *The new multinationals: The spread of third world enterprises*. Chichester, UK: Wiley.

Lardy, N. R. (1998). *China's unfinished economic revolution*. Washington, DC: Brookings Institution.

Lau, H. F. (2003). Industry evolution and internationalization process of firms from a newly industrialized economy. *Journal of Business Research*, *56*(10), 847–852. doi:10.1016/S0148-2963(02)00472-1

Lawrence, S. V. (2002). Going global. *Far Eastern Economic Review*, *165*(12), 30–33.

Lecraw, D. (1977). Direct Investment by Firms from Less-Developed Countries. *Oxford Economic Papers*, *29*(3), 442–457.

Li, P. P. (2007). Toward an Integrated Theory of Multinational Evolution: the Evidence of Chinese Multinational Enterprises as Latecomers. *Journal of International Management*, *13*(3), 296–318. doi:10.1016/j.intman.2007.05.004

Linder, B. S. (1961). *An Essay on Trade and Transformation*. Stockholm, Sweden: Almqvist & Wiksell.

Lipsey, R. E. (2001). Foreign Direct Investment and the Operations of multinational firms: concepts, history and data (Workinng Paper, 8665, NBER).

Liu, H., & Li, K. (2002). Strategic implications of emerging Chinese multinational: the Haier case study. *European Management Journal*, *20*(6), 699–706. doi:10.1016/S0263-2373(02)00119-6

Liu, L. (2005). *China's industrial policies and the global business revolution: the case of the domestic appliance industry*. London: Routledge Curzon.

Liu, L., & Tian, Y. (2008). The internationalization of Chinese enterprises: the analysis of the UK case. *Int. J. Technology and Globalisation*, *4*(1), 87–102. doi:10.1504/IJTG.2008.016189

Long, G. (2005). China's Policies on FDI: Review and Evaluation. In Moran, T. H., Graham, E. M., & Blomstrom, M. (Eds.), *Does Foreign Direct Investment Promote Development?* (pp. 315–336). Washington, DC: Peterson Institute for International Economics.

Luo, Y. (1997). Guanxi: principles, philosophies and implications. *Human Systems Management*, *16*(1), 43–51.

Mathews, J. A. (2002). *Dragon multinationals: A new model of global growth.* New York: Oxford University Press.

Mathews, J. A. (2006). Dragon multinationals: New players in 21st century globalization. *Asia Pacific Journal of Management*, *23*(1), 5–27. doi:10.1007/s10490-006-6113-0

Meyer, K. (1998). *Direct Investment in Economies in Transition*. Cheltenham, UK: Edward Elgar.

Meyer, K. (2008). China's second economic transition: building national markets. *Management and Organization Review*, (1): 3–15. doi:10.1111/j.1740-8784.2008.00094.x

Meyer, K., & Nguyen, H. V. (2005). Foreign investment strategies and sub-national institutions in emerging markets: evidence from Vietnam. *Journal of Management Studies*, *42*(1), 63–93. doi:10.1111/j.1467-6486.2005.00489.x

MOFCOM. (2008). File (2008) 529 issued by the Department of Commerce National Bureau of Statistics State Administration of Foreign Exchange. Retrieved from http://tradeinservices.mofcom.gov.cn/upload/2009/04/10/1239346168141_20005314.pdf

MOFCOM. (2009). Statistical Bulletin of China's Outward Foreign Direct Investments. Retrieved from http://tradeinservices.mofcom.gov.cn/upload/2009/04/10/1239346168141_20005314.pdf

Morck, R., Yeung, B., & Zhao, M. (2008). Perspectives on China's outward foreign direct investment. *Journal of International Business Studies*, *39*(3), 337–350. doi:10.1057/palgrave.jibs.8400366

Morelli, A. (2004). Cina alla conquista degli oceani. *Altraeconomia*. Retrieved from www.altraeconomia.it

Mori, A., & Rolli, V. (1998). Investimenti diretti all'estero e commercio: complementi o sostituti? *Temi di discussione*, 337.

Murat, M., & Pistoresi, B. (2006). Emigrants and immigrants networks in FDI. forthcoming in *Applied Economics Letter.* Retrieved from http://merlino.unimo.it/web_dep/materiali_discussione/0546.pdf

Nolan, P. (2001). *China and the Global Economy: National Champions, Industrial Policy and the Big Business Revolution*. New York: Palgrave.

Nolan, P., & Zhang, J. (2002). The challenge of globalization for large Chinese firms. *World Development*, *30*(12), 2089–2107. doi:10.1016/S0305-750X(02)00148-1

North, D. C. (1990). *Institutions, Institutional change and Economic Performance*. Cambridge, MA: Cambridge University Press.

OECD. (2006). *Benchmark Definition of Foreign Direct Investments* (3rd ed.). Retrieved from http://www.oecd.org/dataoecd/10/16/2090148.pdf

OECD. (2008). *China encouraging responsible business conduct.*

Pananond, P. (2007). The changing dynamics of Thai multinationals after the Asian economic crisis. *Journal of International Management*, *13*(3), 356–375. doi:10.1016/j.intman.2007.05.005

Peng, M. W. (2002). Towards an insitution-based view of business strategy. *Asia Pacific Journal of Management*, *19*(2), 251–267. doi:10.1023/A:1016291702714

Perlmutter, H. V. (1969). The tortuous evolution of the multinational corporation. *The Columbia Journal of World Business*, *4*(1), 9–18.

Posner, V. (1961). International Trade and Technical Change. *Oxford Economic Papers*, *13*(3), 323–341.

Rabellotti, R., & Sanfilippo, M. (2008, October 9-10). *Chinese FDI in Italy.* Paper presented at Emerging Multinationals: Outward Foreign Direct Investment from Emerging and Developing Economies Conference, Copenhagen, Denmark.

Ramamurti, R. (2008). What Have We Learned About Emerging-Market MNEs? In Ramamurti, R., & Singh, J. V. (Eds.), *Emerging Multinationals from Emerging Markets. Cambridge, UK: Cambridge University Press*. Cambridge.

Rauch, J. (2001). Business and social networks in international trade. *Journal of Economic Literature*, *39*, 1177–1203.

Ridella, R. (2007, November 8). *La presenza diretta delle aziende cinesi in Italia*. Paper presented at China meets Italy Conference, Osservatorio, Turin.

Rui, H., & Yip, G. S. (2008). Foreign acquisitions by Chinese firms: a strategic intent perspective. *Journal of World Business*, *43*(2), 213–226. doi:10.1016/j.jwb.2007.11.006

Rullani, E. (2006). L'internazionalizzazione invisibile. La nuova geografia dei distretti e delle filiere produttive. *Sinergie*, *24*(69), 3–32.

Sauvant, K. P. (2005). New sources of FDI: the BRICs. Outward FDI from Brazil, Russia, India and China. *Journal of World Investment & Trade*, *6*(5), 639–709.

Sauvant, K. P. (2006). Inward and outward FDI and the BRICs. In Subhash, C. J. (Ed.), *Emerging Economies and the Transformation of International Business: Brazil, Russia, India and China* (pp. 313–409). Cheltenham, UK: Edward Elgar.

Sauvant, K. P. (2008). The Rise of TNCs from emerging markets: the issues. In Sauvant, K. P. (Ed.), *The Rise of Transnational Corporations From Emerging Markets: Threat or Opportunity?* (pp. 3–14). Willinston, VT: Edward Elgar Publishing.

Scott, W. R. (2002). The changing world of Chinese enterprises: an institutional perspective. In Tsui, A. S., & Lau, M. (Eds.), *Management of Enterprises in People's Republic of China* (pp. 59–78). Boston: Kluwer Academic Press.

Shi, G. (2002, November 13). Speech in the 16th National Congress of the Communist Party of China. *People's Daily*.

Spigarelli, F. (2008). *Nuovi investitori globali: le imprese cinesi in Italia* (Working Paper, 12). Macerata, Italy: University of Macerata, Department of Studies on Economic Development.

Spigarelli, F. (2009). Multinazionali dai paesi emergenti: il caso delle imprese cinesi. *Economia e Politica Industriale*, *2*, 131–159.

Spigarelli, F., Alon, I., & Wei, W. (2009). *A speed race: Benelli and QJ competing on the motorbike international arena. Richard School of Business*. London, Ontario: Ivey Publishing.

Standifird, S. S., & Marshall, R. S. (2000). The transaction cost advantage of Guanxi-based business practices. *Journal of World Business*, *35*(1), 21–42. doi:10.1016/S1090-9516(99)00032-2

Sun, Y., Von Zedtwitz, M., & Simon, D. F. (2007). Globalization of R&D and China: An Introduction. *Asia Pacific Business Review*, *13*(3), 311–319. doi:10.1080/13602380701291867

Tan, W. (2005, June 23). Culture Conflicts in Sino-Foreign Ventures. *Beijing Review.*

Tarmidi, L., & Gammeltoft, P. (2008, October 9-10). *The characteristics of China's multinationals in the manufacturing sector operating in Indonesia.* Paper presented at Emerging Multinationals: Outward Foreign Direct Investment From Emerging and Developing Economies Conference, Copenhagen, Denmark.

Taylor, R. (2002). Globalization strategies of Chinese companies: current development and future prospects. *Asian Business & Management, 1*(2), 209–225. doi:10.1057/palgrave.abm.9200011

Tong, S. J. (2005). Ethnic Networks in FDI and the Impact of Institutional Development. *Review of Development Economics, 9*(4), 563–580. doi:10.1111/j.1467-9361.2005.00294.x

Tong, S. Y. (2003). *Ethnic Chinese Networking in cross-border investment: the impact of economic and institutional development* (Working Paper). Hong Kong, China: University of Honk Kong, Hong Kong Institute of Economics and Business Strategy.

Tsai, K. S. (2002). *Back Alley Banking: Private Entrepreneurs in China.* Ithaca, NY: Cornell University Press.

UNCTAD. (2005). *Prospects for Foreign Direct Investment and the Strategies of Transnational Corporations. 2005-2008.* New York: United Nations.

UNCTAD. (2006). World Investment Report. *Overview FDI from Developing and Transition Economies, United Nations.* New York: Geneva.

UNCTAD. (2007). World Investment Report. *Transnational Corporations, Extractive Industries and development, United Nations.* New York: Geneva.

Valdani, E., & Bertoli, G. (2006). *Mercati internazionali e marketing.* Milano, Italy: Egea.

Vernon, R. (1966). International Investment and International Trade in the Product Cycle. *The Quarterly Journal of Economics, 82*(2), 190–207. doi:10.2307/1880689

Vinciguerra, L. (2007, September 28). Sergio Tacchini, la missione salvataggio arriva dalla Cina. *Il Sole24ore.*

Wagner, D., Head, K. C., & Reis, J. C. (2002). Immigration and the Trade of provinces. *Scottish Journal of Political Economy, 49*(5), 507–525. doi:10.1111/1467-9485.00245

Warner, M., Hong, N. S., & Xu, S. (2004). Late developments experience and the evolution of transnational firms in the People's Republic of China. *Asia Pacific Business Review, 10*(3-4), 324–345. doi:10.1080/1360238042000264397

Wei, J. (2008, April 23). Non-financial investment rises 6%. *China Daily.*

Wells, L. (1983). *Third world multinationals: The role of foreign direct investment from developing countries.* Cambridge, MA: MIT Press.

Wilson, D., & Purushothaman, R. (2003). *Dreaming With BRICs: The Path to 2050* (Global Economics Paper 99). Retrieved from http://www2.goldmansachs.com/ideas/brics/book/99-dreaming.pdf

Wong, J., & Chan, S. (2003). China's outward direct investment: expanding worldwide. *China. International Journal (Toronto, Ont.), 1*(2), 273–301.

Wright, M., Filatotchev, I., Hoskisson, R. E., & Peng, M. W. (2005). Strategy Research in emerging economies: challenging the conventional wisdom. *Journal of Management Studies, 42*(1), 1–33. doi:10.1111/j.1467-6486.2005.00487.x

Wu, F. (2005). The Globalization of Corporate China. *NBR Analysis, 16*(3). Retrieved from http://www.nbr.org/publications/analysis/pdf/vol16no3.pdf

Yao, S. (2006). On economic growth, FDI and exports in China. *Applied Economics*, *38*(3), 339–351. doi:10.1080/00036840500368730

Yeung, H. W., & Liu, W. (2008). Globalizing China: the rise of mainland firms in the global economy. *Eurasian Geography and Economics*, *49*(1), 57–86. doi:10.2747/1539-7216.49.1.57

Zanier, V. (2006). Il sistema economico. In Abbiati, M. (Ed.), *Propizio è intraprendere imprese* (pp. 79–105). Venice, Italy: Cà Foscarina.

Zhang, K. (2005). *Going Global. The Why, When, Where and How of Chinese Companies' Outward Investment intention*. Asia Pacific Foundation of Canada, Vancouver. Retrieved from www.asiapacific.ca

Zhang, Y. (2003). *China's Emerging Global Business: Political Economy and Institutional Investigations*. Basingtoke, UK: Palgrave Macmillan.

Zhao, W., & Liu, L. (2008, October 9-10). *Outward Direct Investment and R&D spillovers: the China's case*. Paper presented at Emerging Multinationals: Outward Foreign Direct Investment From Emerging and Developing Economies Conference, Copenhagen, Denmark.

Zhou, M. (2006). The Chinese Diaspora and International Migration. In Bian, Y., Kwok-bun, C., & Tak-sing, C. (Eds.), *Social Transformations in Chinese Societies*. Leiden, The Netherlands: Brill Academic Publishers.

Zweig, D. (2002). *Internationalizing China: Domestic Interests and Global Linkages*. Ithaca, NY: Cornell University Press.

ENDNOTES

1 A preliminary version of this paper was published in the Italian Review "Economia e Politica Industriale" (n. 2/2009) as "Le multinazionali dei paesi emergenti: gli investimenti cinesi in Italia". See Spigarelli, 2009.

2 For further exploration of the Indian case, see Chittoor & Ray, 2007; Garg & Delios, 2007. On the Brazilian case, see Cuervo-Cazurra, 2007; on the Russian case, refer to Kalotay, 2008. For a broad bibliographical review of the primary scientific contributions concerning MNEs, broken down by emerging nation of origin, see Hernandez, 2008.

3 Furthermore, the recognition of the limits of the Open Door policy in the attempt to the transfer know-how through joint ventures with Western companies was one of the key reasons for implementing a strategy based on foreign acquisitions and expansion (Di Minin & Zhang, 2008; Zhao & Liu, 2008). For a discussion of issues associated with technological transfer and the ability of Chinese enterprises to capture value through the establishment of R&D centers and partnerships, see Sun, *et al.* 2007.

4 In connection with the Asian situation, the case of the government of Singapore and the company Temasek, particularly as regards problems associated with expansion in Thailand, is relevant (Goldstein & Pananond, 2008). As pertains to China, the failed acquisition of the U.S. Unocal by Chinese National Offshore Oil Corporation in 2005 is one of the most striking examples of how the presence of potential political and military public interests may impede international integration transactions (Kynge, 2006).

5 It is estimated that there are over 33 million people of Chinese ethnicity beyond the borders of the PRC (including Hong Kong, Macau, and Taiwan). This population is distributed throughout 151 countries: 80% lives in Asia and 15% in the Americas. The countries with the largest Chinese presence are Thailand (9.2 million), Indonesia (8 million), Malaysia (5.8 million), Singapore (2.7 million), the USA (2.9 million) and Canada

(1 million). Cfr. Zanier, 2006, p. 90; Zhou, 2006, p. 162.

6 For empirical studies on the impact of transnational ties on FDIs see: Gao, 2003; Tong, 2005; Buch et al. 2006, Murat & Pistoresi, 2006; Flisi & Murat, 2009. For reviews on this topic see: Rauch, 2001; Wagner et al., 2002.

7 In this paper, data from Banca d'Italia were preferred to those released by Mofcom because only since 2006 the Chinese statistical systems has been adopting the internationally accepted methodology (OECD, 2006, p. 8). It changed from recording "all occurred in the activities of foreign direct investment enterprises, public, etc..." to recording "all foreign direct investment occurred in the territory of the activities of institutions and individuals" (Mofcom, 2008, p. 1). So since 2006 Mofcom has been keeping track not only of data on foreign direct investments performed by companies, but also by other private and public bodies and individuals. This fact can explain the differences in data released by the two institutions, especially before 2006. For problems related to measurements of China's OFDI see also OECD, 2008, p. 71. Banca d'Italia could provide a complete set of data for the whole period of analysis (2001-2007), built following the internationally accepted methodology and split by sectors and regions of investments. For data on Chinese OFDIs in Italy (flows and stocks) released by Mofcom, see Mofcom, 2009, p.62 and p. 67.

8 In addition to the issues of the completeness and timeliness of the data, of specific relevance to the Chinese case is the consideration that investments are often made by groups that hold companies abroad, sometimes in tax havens, through which the transactions are organized. In such circumstances, transactions are not considered as Chinese in origin and are therefore not recognized in the Banca d'Italia survey.

9 In accordance with the national accounting definition, this item includes artisans, sole proprietorships, and consumers (De Felice, 2006, p. 11). The contents of this highly heterogeneous item do not allow a significant review of the effective destination of investments at the national level.

10 The sources referred to are: Sviluppo Italia, based on Reprint data; CCPIT database; and the specialist economic press. The fact that the information provided by the various sources does not overlap is a symptom of the non-exhaustive nature of this information.

11 Companies based in Hong Kong, Taiwan, Macao or tax havens have been excluded from the map. The analysis is limited to companies based in mainland China, considering that these companies are the privileged beneficiaries of the action taken by the government to support active internationalization. For this reason, for example, the analysis of the Italian case does not consider the acquisition of Sergio Tacchini by Hong Kong Hembley International Holdings, a leading apparel outsourcer and distributor (Vinciguerra, 2007). Likewise, the acquisition of Omas, a manufacturer of luxury pens, by the Xinyu Hengdeli Holdings group, based in the Cayman Islands, in 2007, has also been excluded (ICE, 2008).

12 The choice of Varese and Padua is related not only to the strategic location for Europe, but also to the general tradition in these areas of Italy of the manufacture of home appliances, the professional skills and abilities available, the network of sub-suppliers specialized in industrial components and technologies, and the proximity to the main Western competitors (Bellabona & Spigarelli, 2006c, pp. 159-161).

13 For further information and clarification concerning Unioncamere figures, as well as the data issued by the Register of Companies maintained by Chambers of Commerce and

the REA Archive, see the detailed methodological note in Borsari et al. (2006), pp. 31-33. In the paper cited above, the conclusions drawn refer to data originating from the same source as that used in this paper.

14 Worthy of note is the limited presence of Chinese entrepreneurs in the Marche region (3% of total Chinese entrepreneurs in Italy in 2007), even though 61% of their business initiatives were concentrated in textiles, apparel and footwear in 2007.

15 Consider, for example, the case of the area of San Giuseppe Vesuviano (Napoli), in the textiles-apparel sector.

16 In a recent analysis, Rabellotti & Sanfilippo (2008), p. 17, demonstrate that of all Chinese enterprises that have invested in Italy, including those based in Hong Kong or in foreign tax havens, 19% first became established in Italy in the period from 1986 to 1999, 34% from 2000 to 2004, and 47% from 2005 to the present day.

17 To assess the role of ethnic network in influencing Chinese OFDIs in Italy, data on education level of Chinese immigrants should also be used in the future, as soon as available. In this way, the effect of different levels of skills of Chinese citizens - living in different Italian geographical areas - on investments performed by Chinese companies in the country could be verified.

This work was previously published in International Journal of Asian Business and information Management, Volume 1, Issue 1, edited by Dasho Karma Ura and Patricia Ordóñez de Pablos, pp. 54-76, copyright 2010 by IGI Publishing (an imprint of IGI Global).

Chapter 11
Entry Strategies into the P.R.C. for Post Recession Wealth Management Service

Andrew Chan
City University of Hong Kong, China

Carlson Chan
City University of Hong Kong, China

ABSTRACT

Trillions of dollars were evaporated worldwide as a result of the financial tsunami that shocked the world in 2007 (Soros, 2008). This paper argues that in a time of global recession, banks seeking to expand wealth management service in the People's Republic of China (P.R.C.) need to evaluate their entry strategies thoroughly to gain a competitive edge. Through applying the theory of planned behavior, a sequential mixed methods study was conducted post recession with the aim to identify wealth management service entry strategies into the P.R.C. A sample of 227 Taiwanese nationals based in the P.R.C. indicated three managerial implications consisting of the bank entry identity, deployment of expatriates, and brand-image position.

INTRODUCTION

The financial tsunami and global recession posted great business pressure to banks. Sharp drop in revenue along with the number and wealth of their high net-worth customers stimulated the bank management in searching for opportunities globally (Capgemini & Merrill Lynch, 2007, 2008, 2009; Posner, 2009). The P.R.C.'s accession to the World Trade Organization created unprecedented opportunities for businesses around the world to gain a share of this huge and growing market. This research analyses the causes and aftermaths of the financial tsunami and proposes that the banking sector exercises discretion to device their entry strategies into new markets like China by

DOI: 10.4018/978-1-4666-0276-2.ch011

taking into consideration customers' attitude and preferences that influence their choices of WMS (see Isdale, 2006; Maude, 2006). Our study has attained the following three objectives: first it evaluates the global recession to suggest ways to restore confidence in its aftermaths; second our Mainland China survey tests the factors in attitudinal preference that influence bank customers' choice of WMS by applying the 'Theory of planned behavior' (Ajzen, 1991, 2008), and finally the study proposes strategies for overseas Chinese banks if they were to flagship a foothold in China beyond 2010 (Cousin, 2007; EIU, 2009).

LITERATURE REVIEW

Summarizing the extant literature, four dynamics originating and radiating from USA and Europe have contributed to the global financial crisis in 2008 (Lane, 2009). They are, first, the *laxity of monetary policy and housing bubble*. The low treasury rates and long term real interest rates created an ease of monetary environment (Calomiris, 2009; Ritholtz & Task, 2009). Speculations (Baker, 2009) coupled with aggressive mortgage packaging (Taylor, 2009; Schwartz, 2009; Muolo & Padilla, 2008; Sowell, 2009) were the factors leading to a housing bubble (Bonner & Wiggin, 2009). Second, *financial innovation and risk-taking*. Credit default swaps along with over the counter trading platform were central to the explosion of structured finance (Unterman, 2009; Chorafas, 2009; Leopold, 2009; Taub, 2009; Barbera, 2009). Third, *misguided incentives*. The compensation schemes in the finance industry and substantial fees garnered by the rating agencies were the contributing factors to the crisis (Barth et al., 2009; Immergluck, 2009; Steil, 2009). And last, securitization and the legal structure of special purpose vehicles facilitated the *regulatory arbitrage* to excel (Mason, 2009; Vines, 2009; Wessel, 2009). Altogether these policy blunders (Baker, 2009), corporate misbehavior (Leopold, 2009) and ricochets of human greed (Ritholtz & Task, 2009) have caused the 2008 HNWI Wealth Distribution total global amount to shrink by US$7.9 trillion, compared with that of 2007 (Capgemini & Merrill Lynch, 2009).

In the literature of overseas Chinese business practices, culture plays a vital role in building relationships (e.g., Lee & Dawes, 2005; So & Walker, 2006; Reuvid, 2008). It generally indicates an imperative in very prudent strategy execution in relationship-oriented high-context 'Chinese business culture', with guanxi as a vital ingredient in successful Chinese business activities (Orru, Biggart, & Hamilton, 1997; So & Walker, 2006) especially in personal financial services (Huang, 2000; Pompian & Longo, 2004; Li, 2000; Zhou, 2000). Language forms part of the cultural components and affects the relationship in doing business (Singh, 2004a, 2004b). Hence, the following hypotheses are expected:

H1: Behavioral Intention of home nationals to choose home country banks' WMS is positively related to Attitude toward the Behavior.

H2: There is a positive relationship between the Behavioral Intention and the Subjective Norms of home nationals to give business to home country banks which provide WMS when available.

H3: Perceived Behavioral Control *is positively related to the Behavioral Intention of the PRC based home nationals to choose their home country's WMS products when made available to them.*

H4: There is a positive relationship among the Behavioral Intention and Attitude toward Behavior Subjective Norms, and Perceived Behavioral Control.

METHODS

Data were collected through survey questionnaires distributed through the Taiwanese Businessmen's

Association in May, 2009 targeted at home nationals who were resident in China at the time. The sampling frame consists of 1060. To ensure the level of significance and the eligibility of our total number of responses, we employed Cohen's (1988) statistical power analysis to secure robustness among power for significance criterion (alpha), sample size (N), and population effect size (ES). A minimum sample size of at least 195 is required and we have obtained 227 valid returned questionnaires. The survey consists of three pilot tests and a main study between December 2008 and July 2009. An elicitation study plus three pilot tests were administered for questionnaire development, refinement and finalization. The main study employs a cross-sectional study of Taiwan home nationals resident in China. It was carried out in May 2009 with businesspersons in the PRC. High reliability constructs at range between 0.801 and 0.890 confirm a third pilot to test temporal stability of all direct and indirect measures. A test-retest score of 0.706 and 0.817 showed acceptable test-retest reliability for each measure.

Three independent variables of *Attitude towards behaviors*, *Subjective norms* and *Perceived behavioral control* to measure the behavioral construct of *Intention* are built into our survey research instrument. They are developed following Ajzen (1991, 2008) and Ajzen and Fishbein (2005), and then adapting measurement scales suggested by Ajzen (2006) and Francis et al. (2004). Under each construct, there are direct and indirect measures except for *Intention* which is the dependent variable that can only be measured directly.

Direct measure of all constructs is measured by Likert scales of 1-7 in generic form: *Intention* is measured by 3 items, *Attitude toward behavior* (4 items), *Subjective norms* (3 items), and *Perceived behavioral control* (4 items). Indirect measure is measured by a pair of items in specific form in Likert scales of 1-7 (unlikely=1/likely=7) multiply by a bipolar scale of -3 to 3: *Attitude toward behavior* is measured by 4 pairs, *Subjective norms* (by 3 pairs), and *Perceived behavioral control* (by 3 pairs). Correlations of direct and indirect measures under each construct were analyzed by computing the Spearman's rho. Simultaneous consideration of the predictability of the three independent variables and the dependent variable was completed through a computation of the multiple correlation coefficients using stepwise multiple regressions. In testing the theory, weights were obtained for the three independent variables as indicators of the relative importance of each component in the prediction of intention.

DATA ANALYSIS AND HYPOTHESES TESTING

From the elicitation study, a list of modal salient beliefs was constructed after qualitative data analysis. The beliefs under the *construct of attitude* toward the behavior were:

1. Feeling of cultural affection (AF1),
2. Facility with communicating in cultural language (AF2),
3. Provision of flexible services (AF3), and
4. Feeling of trustworthiness (AF4).

Three beliefs of important others under *the subjective norms construct* were:

1. Taiwanese friends (SF1),
2. Family members (SF2), and
3. Friends in business (SF3).

As for the *construct of perceived behavioral control*, the three beliefs were:

1. Brand name of Taiwanese banks (CF1),
2. Being served by Taiwanese (CF2), and
3. Taiwanese banks to enter P.R.C. via a third regulated location (CF3).

Figure 1 shows the correlation matrix of the salient beliefs and the behavioral intention (BI).

Figure 1. Correlation matrix of salient beliefs and behavioral intention

	BI	AF1	AF2	AF3	AF4	SF1	SF2	SF3	CF1	CF2	CF3
BI	1										
AF1	.615**	1									
AF2	.592**	.661**	1								
AF3	.731**	.547**	.544**	1							
AF4	.736**	.671**	.675**	.659**	1						
SF1	.497**	.510**	.486**	.422**	.441**	1					
SF2	.506**	.569**	.490**	.432**	.472**	.614**	1				
SF3	.511**	.529**	.466**	.456**	.474**	.763**	.722**	1			
CF1	.207**	.123	.045	.040	.145*	.207**	.252**	.263**	1		
CF2	.246**	.173**	.145*	.248**	.159*	.284**	.326**	.325**	.516**	1	
CF3	.513**	.418**	.542**	.487**	.497**	.408**	.400**	.386**	.047	.172**	1

** Correlation is significant at the 0.05 level (2-tailed).*

*** Correlation is significant at the 0.01 level (2-tailed).*

The results reveal that all salient beliefs were positively correlated with the behavioral intention. Most of the beliefs were moderately to highly correlated the behavioral intention except for two: 'brand name of Taiwanese banks' (CF1) and 'being served by Taiwanese' (CF2) were both weakly correlated with the behavioral intention.

A summary of the correlation coefficients and beta weights of the data analysis is presented in Figure 2. The direct and indirect measures should demonstrate positive correlations between each other as both aims to measure the same construct. The results reveal that the direct and indirect measures of attitude toward behavior ($r = 0.791$) and subjective norms ($r = 0.672$) indicate a high correlation whereas a moderate correlation was recorded for perceived behavioral control ($r = 0.445$).

Hypothesis 1 Testing

High correlations coefficients recorded between intention and attitude toward behavior direct (ABD) at 0.852 and that between intention and attitude toward behavior indirect (ABI) at 0.780 indicate that ABD and ABI were highly correlated with the behavioral intention. Stepwise regression analysis reveals that it was significantly beneficial to include the attitude toward behavior construct in the model in terms of estimating the behavioral intention with a p-value 0.000. The first hypothesis was not rejected.

Hypothesis 2 Testing

The correlation coefficients between intention and subjective norms direct (SND) was at 0.802 and between intention and subjective norms indirect (SNI) was at 0.562 exhibiting that both SND and SNI were highly correlated with the behavioral intention. In the stepwise regression, the results indicate that the subjective norms were significantly beneficial to be included in the model for estimating the behavioral intention with a p-value of 0.000. The second hypothesis was not rejected.

Hypothesis 3 Testing

The high correlation coefficients between intention and perceived behavioral control direct (PBCD) at 0.793 indicate that PBCD was highly correlated with the behavioral intention. The correlation coefficients between intention and perceived behavioral control indirect (PBCI) indicate that PBCI was moderately correlated with the behavioral intention. The third hypothesis was not rejected.

Figure 2. Correlation coefficients and beta weight summary

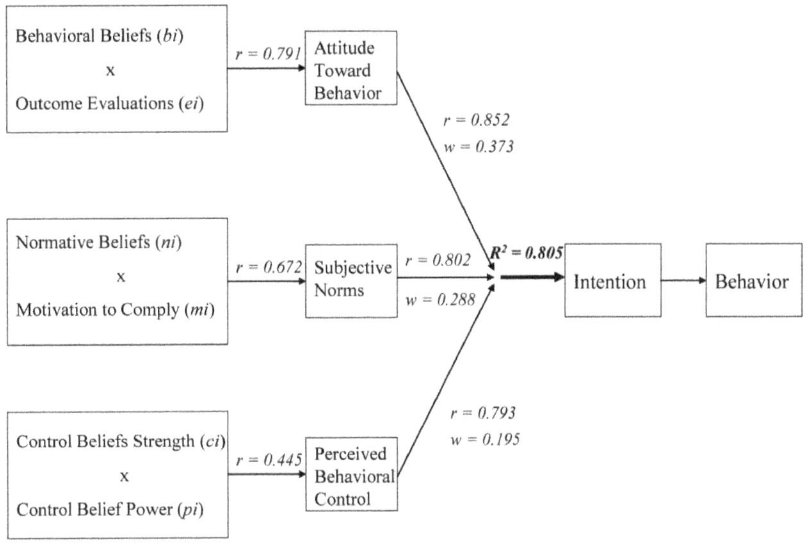

**Correlation is significant at the 0.01 level (2-tailed)*

***r = Spearman's rho; w = Beta weight; R^2 = Adjusted R-square*

Hypothesis 4 Testing

Further analysis was performed to put all the three variables into the same regression model. For the model which include all the direct measure of the three variables (attitude toward behavior, subjective norms, and perceived behavioral control), the adjusted R-square was found to be 0.805 The F value of 312.843 with a p-value of 0.000 indicates that in the F-test, we reject the null hypothesis which states that all coefficients of the independent variables are zero. To further test the significance of each independent variable in the model with the significance level set at alpha = 0.05, the T-value were found at 8.136 for attitude toward behavior, 5.167 for subjective norms, and 4.335 for perceived behavioral control with all of them getting a p-value of 0.000. The last hypothesis was not rejected.

A summary of the hypotheses testing was illustrated in Figure 3. The beta value of the three independent variables: attitude toward behavior (0.373), subjective norms (0.288), perceived behavioral control (0.195) indicate that there is a positive relationship between the dependent variable (behavioral intention) and the three independent variables with the strongest from attitude toward behavior and the least from perceived behavioral control.

DISCUSSION AND CONCLUSION

This research has built a predictive model, based on the *Theory of Planned Behavior* (Ajzen, 1991, 2008), to investigate the behavioral beliefs of the attitude towards the behavior, the normative beliefs of the subjective norms and the control beliefs of perceived behavioral control, in order to explain the correlation among these three constructs affecting the behavioral intention of the respondents for formulating post recession wealth management service entry strategies in the P.R.C.

Building on prior researches and our survey, we have identified entry strategies useful for overseas banks to provide post-recession WMS in China. We suggest the need for banks offering WMS in China to gravitate effort in: acquisition

Figure 3. Summary of hypotheses testing

Hypothesis	Result	Variable	Mean	*s.d.*
H1: Behavioral Intention of home nationals to choose home country banks' WMS is positively related to Attitude toward the Behavior.	Not Rejected	ABD ABI	5.70 12.37	2.009 12.290
H2: There is a positive relationship between the Behavioral Intention and the Subjective Norms of home nationals to give business to home country banks which provide WMS when available.	Not Rejected	SND SNI	5.35 7.55	1.834 12.053
H3: Perceived Behavioral Control is positively related to the Behavioral Intention of the PRC based home nationals to choose their home country's WMS products when made available to them.	Not Rejected	PBCD PBCI	5.68 4.51	1.838 10.428
H4: There is a positive relationship among the Behavioral Intention and Attitude toward Behavior, Subjective Norms, and Perceived Behavioral Control.	Not Rejected			

s.d.=standard deviation; ABD=attitude toward behavior direct; ABI=attitude toward behavior indirect; SND=subjective norms direct; SNI=subjective norms indirect; PBCD=perceived behavioral control direct; PBCI=perceived behavioral control indirect

of existing branch networks to establish corporate entity (CBRC, 2008); deploying expatriates versus indigenous staff to China; and leveraging brand-image positioning (Clow & Baack, 2010; Gallagher, 2004; Gray, 2004). Figure 4 depicts the ten salient beliefs solicited under each construct as ten forces that impinge on business entry strategies.

Firstly, the research findings facilitate effective decisions to be made by the management when deciding on the 'entry mode identity'. Figure 5 depicts four entry identity options. Banks may consider direct entry but need to assess their own internal strengths and weaknesses in terms of their financial status and legal impacts before deciding their appropriate entry mode strategies. Results indicate that the entry mode identity significantly affect the customer preferences on choosing post recession wealth management services.

In evaluating the entry mode to the P.R.C., bank management must pay special attention to the options of establishing branches directly or through a 3rd regulated location for independent jurisdiction. Choosing the later may provide more confidence for those customers who are hesitated to have a connection with the home country for privacy. In this research, respondents indicated that their intention to choose their home country banks that offer wealth management services would not be impeded if it establishes branches directly and not through a subsidiary or regulated in a third location. Another advantage of entering the P.R.C. through a 3rd regulated location through merger and acquisition is the change of identity for those banks with lower reputation in the home country. The new identity provides the bank an opportunity in projecting another image of the bank riding on the existing reputation or specialties of the bank being acquired. An example is a bank with specialty of wealth management strength and reputation. The results in this research showed that when the home country banks do not have good brand names in the home country, there was a weak correlation to the behavioral intention to choose their wealth management services. Banks in such positions must pay special attention when assessing their entry strategies to the P.R.C. targeting this customer segment.

Figure 4. Ten forces on post recession entry strategies

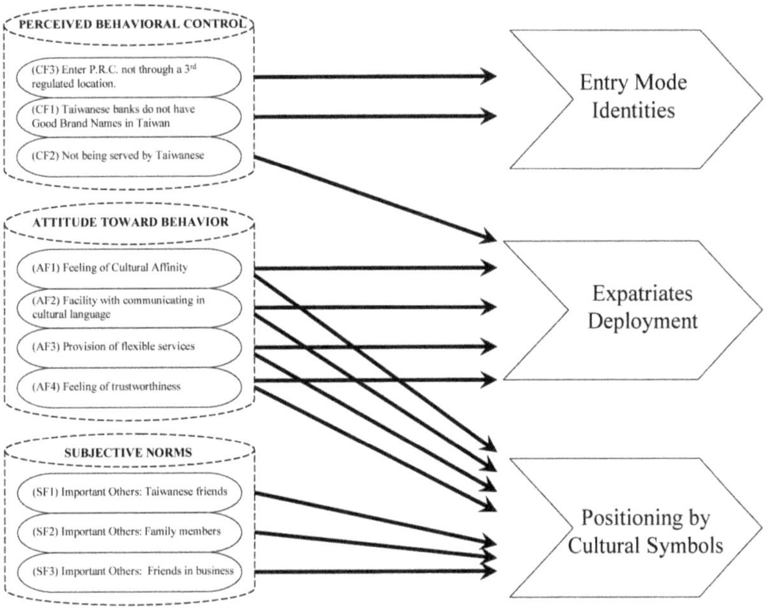

Secondly, in the entry phase, there will be a need for the home country banks to transfer expatriates to the P.R.C. to serve the target segment. The ratio of expatriates to local Chinese staff must be high enough to provide the expected services. These expatriates should be allocated to four major units. At the branch level, a special desk stationed by expatriates to meet with walk-ins or branch visits from the home nationals is needed. Wealth management or portfolio consultants are

Figure 5. Entry identity options

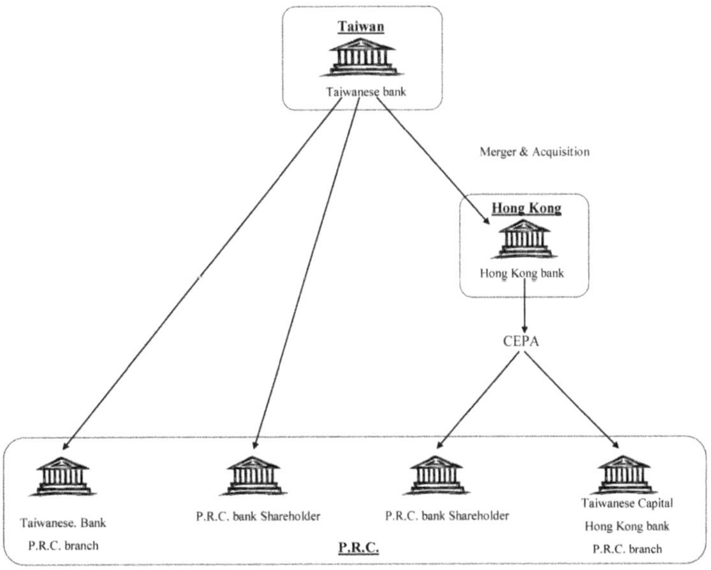

Figure 6. Deployment of expatriate at the entry phase

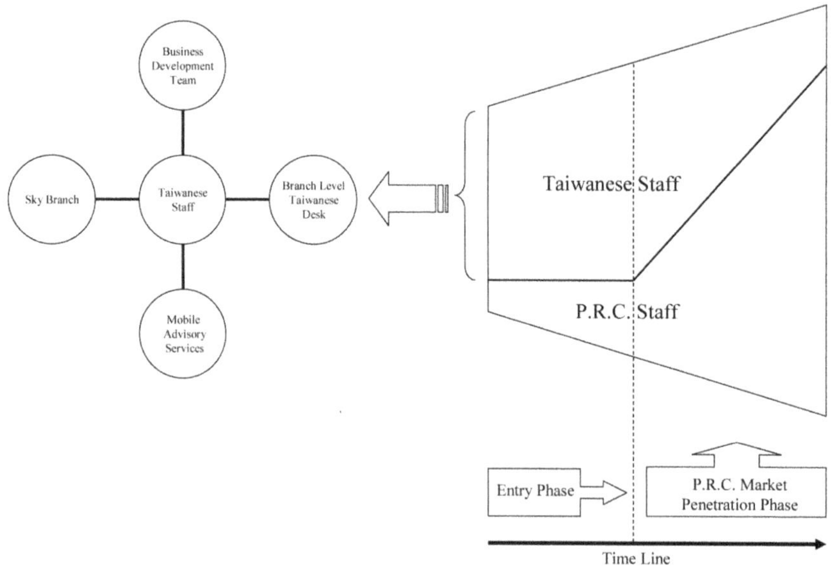

required to provide advisory services. As portfolio advice plays a vital role in wealth management services, we recommend a mobile advisory service team consisting of expatriate advisors to offer more flexible services. When customer needs portfolio review or advice, this team may provide outbound services upon request. A business development team making sales visits can also help to build up business proactively. Another function of the business development team is to coordinate with the referent groups on organizing marketing related activities to reach the target segment. Furthermore, a sky branch using advanced technology will give customers the advantage of communicating with the expatriate over the phone or through video conferencing to provide wealth management services to those customers who are unable to visit mortar and bricks branches. The recommendations are illustrated in Figure 6. It is reaffirmed that in the entry phase of entering into the P.R.C., deployment of expatriate is essential to meet the expectations of the target customer segment in the P.R.C.

Thirdly, in considering building of brand image, the results of this study also support consolidating the bank positioning through the use of cultural symbols which will create a competitive advantage. From this study, the four factors of cultural affection (AF1), facility with communicating in cultural language (AF2), provision of flexible services (AF3), and feeling of trustworthiness (AF4) provided very relevant and useful evidence demonstrating the importance of establishing market positioning through a clear appreciation of the target segment's beliefs. It is vital that banks considering post recession entry strategies therefore develop a brand image based on cultural symbols for promotions and marketing events to capture the wealth management services in the P.R.C.

It is recommended to establish a business development team formed by the expatriates to coordinate with the home country trade associations in the P.R.C. in the organization of promotional activities for referral marketing as presented in Figure 7. Those communication channels of the trade associations such as newsletters are effective means to reach the target segment. This team can sponsor events held by these organizations, such as annual dinners, commercial talks, and seminars

Figure 7. Positioning by cultural symbols

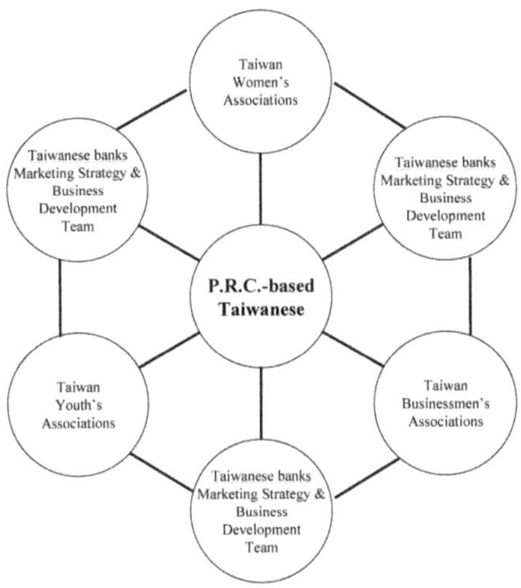

as a means to further help in targeting the customer segment by leveraging the referent groups to influence their choice of wealth management services in the P.R.C.

REFERENCES

Ajzen, I. (1991). The theory of planned behavior. *Organizational Behavior and Human Decision Processes*, *50*(2), 179–211. doi:10.1016/0749-5978(91)90020-T

Ajzen, I. (2006). *Constructing a Theory of Planned Behavior Questionnaire: Conceptual & Methodological Considerations*. Retrieved from www.unix.oit.umass.edu/~aizen/pdf/tpb.measurement.pdf

Ajzen, I. (2008). Consumer attitudes and behavior. In Haugtvedt, C. P., Herr, P. M., & Kardes, F. R. (Eds.), *Handbook of Consumer Psychology* (pp. 525–548). New York: Lawrence Erlbaum.

Ajzen, I., & Fishbein, M. (2005). The influence of attitudes on behavior. In Albarracín, D., Johnson, B. T., & Zanna, M. P. (Eds.), *The Handbook of Attitudes* (pp. 173–221). Mahwah, NJ: Erlbaum.

Baker, D. (2009). *Plunder and Blunder: The Rise and Fall of The Bubble Economy*. Sausalito, CA: Polipoint Press.

Barbera, R. J. (2009). *The Cost of Capitalism: Understanding Market Mayhem and Stabilizing Our Economic Future*. New York: McGraw-Hill.

Barth, J. R., Li, T., Lu, W., Phumiwasana, T., & Yago, G. (2009). *The Rise and Fall of the U.S. Mortgage and Credit Markets: A Comprehensive Analysis of the Market Meltdown*. New York: John Wiley.

Bonner, W., Incontrera, K., & Wiggin, A. (2009). *Financial Reckoning Day Fallout: Surviving Today's Global Depression*. New York: Wiley.

Bonner, W., & Wiggin, A. (2009). *The New Empire of Debt: The Rise and Fall of an Epic Financial Bubble*. New York: Wiley.

C. B. R. C. (2008). *Annual Report 2008*. Beijing, China: China Banking Regulatory Commission.

Calomiris, C. (2009). Financial innovation, regulation, and reform. *The Cato Journal*, *29*(1), 65.

Capgemini & Merrill Lynch. (2007, 2008, 2009). *World Wealth Reports 2007, 2008, 2009. Capegemini Consulting Technology & Merrill Lynch Wealth Management*. Retrieved from http://www.capgemini.com

Chorafas, D. N. (2009). *Financial Boom and Gloom: The Credit and Banking Crisis of 2007-2009 and Beyond*. New York: Palgrave Macmillan.

Clow, K. E., & Baack, D. (2010). *Integrated Advertising, Promotion and Marketing Communication*. Upper Saddle River, NJ: Prentice Hall.

Cohen, J. (1988). *Statistical Power Analysis for the Behavioral Sciences* (2nd ed.). Hillsdale, NJ: Lawrence Erlbaum.

Cousin, V. (2007). *Banking in China*. New York: Palgrave Macmillan.

E. I. U. (2009). New York: Economist Inteligence Unit. *Country Report: Taiwan*, *6*, 1–24.

Francis, J. J., Johnston, M., Eccles, M. P., Grimshaw, J., & Kaner, E. F. S. (2004). *Measurement Issues in The Theory of Planned Behavior: A Supplement to The Manual for Constructing Questionnaires Based on The Theory of Planned Behavior & Theory of Planned Behavior Questionnaires* (Discussion Paper, pp. 43-76). Newcastle upon Tyne, UK: University of Newcastle, Centre for Health Services Research.

Gallagher, G. J. (2004). Great expectations: Challenge and opportunity for today's wealth manager. *The Journal of Wealth Management*, *7*(2), 87–91. doi:10.3905/jwm.2004.434570

Gray, L. (2004). *The New Family Office: Innovative Strategies for Consulting to The affluent*. London: Institutional Investor Books.

Huang, Y. (2000). The banking market system. In Norton, J. J., Li, C. J., & Huang, Y. (Eds.), *Financial Regulation in the Greater China Area: Mainland China, Taiwan and Hong Kong SAR* (pp. 25–63). The Hague, The Netherlands: Kluwer Law International.

Hudson, I. (2009). From deregulation to crisis. In Guard, J., & Antony, W. A. (Eds.), *Bankruptcies and Bailouts*. Halifax, Nova Scotia: Fernwood.

Immergluck, D. (2009). *Foreclosed: High-risk Lending, Deregulation and The Undermining of America's Mortgage Market*. Ithaca, NY: Cornell University Press.

Isdale, M. H. (2006). Strategies for simplifying and increasing the effectiveness of wealth management. *The Journal of Wealth Management*, *9*(2), 14–18. doi:10.3905/jwm.2006.644215

Lane, S. (2009). Financial services in Europe before, during and after the crisis. *Accountancy Ireland*, *41*(2), 16.

Lee, D. Y., & Dawes, P. L. (2005). Guanxi, trust, and long-term orientation in Chinese business markets. *Journal of International Marketing*, *13*(2), 28–56. doi:10.1509/jimk.13.2.28.64860

Leggett, S. (2009). A single global regulatory framework: Solution, or problem? *Corporate Finance Review*, *14*(1), 11.

Leopold, L. (2009). *The Looting of America: How Wall Street's Game of Fantasy Finance Destroyed Our Jobs, Pensions, and Prosperity, and What We Can Do About It*. White River Junction, VT: Chelsea Green Publications.

Li, C. J. (2000). China's financial reform and its implications. In Norton, J. J., Li, C. J., & Huang, Y. (Eds.), *Financial Regulation in the Greater China Area: Mainland China, Taiwan and Hong Kong SAR* (pp. 319–344). The Hague, The Netherlands: Kluwer Law International.

Mason, P. (2009). *Meltdown: The End of the Age of Greed*. New York: Verso.

Maude, D. (2006). *Global Private Banking and Wealth Management: The New Realities*. New York: Wiley.

Muolo, P., & Padilla, M. (2008). *Chain of Blame: How Wall Street Caused the Mortgage and Credit Crisis*. New York: Wiley.

Orru, M., Biggart, N. W., & Hamilton, G. G. (1997). *The Economic Organization of East Asian Capitalism*. Thousand Oaks, CA: Sage Publications.

Osenton, T. (2009). *Boomer Destiny: Leading the U.S. Through The Worst Crisis Since The Great Depression*. Westport, CT: Praeger.

Pompian, M. M., & Longo, J. M. (2004). A new paradigm for practical application of behavioral finance: Creating investment programs based on personality type and gender to produce better investment outcomes. *The Journal of Wealth Management*, *7*(2), 9–15. doi:10.3905/jwm.2004.434561

Posner, R. A. (2009). *A Failure of Capitalism: The Crisis of '08 and The Descent into Depression*. Boston: Harvard University Press.

Reuvid, J. (2008). China and the WTO. In Reuvid, J. (Ed.), *Business Insights: China: Practical Advice on Entry Strategy and Engagement*. Philadelphia: Kogan Page.

Ritholtz, B., & Task, A. (2009). *Bailout Nation: How Greed and Easy Money Corrupted Wall Street and Shook The World Economy*. New York: Wiley.

Schwartz, H. M. (2009). *Subprime Nation: American Power, Global Capital, and The Housing Bubble*. Ithaca, NY: Cornell University Press.

Singh, I. (2004a). Language and ethnicity. In Singh, I., & Peccei, J. S. (Eds.), *Language, Society and Power an Introduction* (2nd ed., pp. 94–111). New York: Routledge.

Singh, I. (2004b). Language, thought and representation. In Singh, I., & Peccei, J. S. (Eds.), *Language, Society and Power an Introduction* (2nd ed., pp. 18–34). New York: Routledge.

So, Y. L., & Walker, A. (2006). *Explaining Guanxi: The Chinese Business Network*. London: Routledge.

Soros, G. (2008). *The New Paradigm for Financial Markets: The Credit Crisis of 2008 and What It Means* (1st ed.). New York: Public Affairs.

Sowell, T. (2009). *The Housing Boom and Bust*. New York: Basic Books.

Steil, B. (2009). *Lessons of the Financial Crisis*. New York: Council on Foreign Relations, Center for Geo-economic Studies.

Taub, J. (2009). *Enablers of Exuberance: Legal Acts and Omissions That Facilitated the Global Financial Crisis* (SSRN Working Paper Series). SSRN.

Taylor, J. B. (2009). *Getting Off Track: How Government Actions and Interventions Caused, Prolonged, and Worsened the Financial Crisis*. New York: Hoover Institution Press.

Unterman, A. (2009). Perverse incentives: Risk taking and reform. *Banking and Financial Services Policy Report*, *28*(6), 11.

Vines, S. (2009). *Market Panic: Wild Gyrations, Risks, and Opportunities in Stock Markets* (2nd ed.). New York: Wiley.

Wessel, D. (2009). *In Fed We Trust: Ben Bernanke's War On The Great Panic*. New York: Crown Business.

Zhou, Z. (2000). The regulation of foreign banks in China. In Norton, J. J., Li, C. J., & Huang, Y. (Eds.), *Financial Regulation in the Greater China Area: Mainland China, Taiwan and Hong Kong SAR* (pp. 65–97). The Hague, The Netherlands: Kluwer Law International.

This work was previously published in International Journal of Asian Business and Information Management, Volume 1, Issue 2, edited by Dasho Karma Ura and Patricia Ordóñez de Pablos, pp. 1-11, copyright 2010 by IGI Publishing (an imprint of IGI Global).

Chapter 12
Chinese Outward Foreign Direct Investment:
In Search of a New Theory

Fernanda Ilhéu
ISEG/Technical Universal of Lisbon, Portugal

ABSTRACT

In past years, China recorded a fast sustainable economic growth with an estimated average GDP growth rate of 9.7% in the period of 1980-2008, turning China into the world's second largest economy. With an export oriented economic model, China is the most attractive developing country for FDI flows, both short and long term. In this regard, China has been able to achieve a foreign exchange reserve of US$ 2.2 trillion, the world's largest reserve currency. Around 50% of this huge reserve is being applied in American bonds, while the remaining supports Chinese health and social security systems, bank solvability, internationalization of their economy, investment in geostrategic positioning, and making foreign aid available to other developing countries. During the 2008 global crisis, China was able to resist better than other major world economies, benefitting from this downturn to implement policies to reduce its economic imbalances. One of these imbalances is the gap between Chinese FDI and OFDI, which is now progressively narrowing. In the near future, OFDI is expected to be larger than FDI, and in this paper, the authors research whether Chinese OFDI can be explained by existing theories or if a new theory is required.

INTRODUCTION

Due to the Reform and Open Door Policies initiated in 1978, China recorded a fast sustainable economic growth with an estimated average GDP growth rate of 9.8% in the period of 1980-2009, turning China, in 2009, into the world's second largest economy, just after USA. With an export oriented economic model, highly supported by Foreign Direct Investment (FDI), mostly from developed countries, China is, since 2002, the most attractive developing country for FDI flows, both at short and long terms, becoming not only the world's factory, but also its number one exporter,

DOI: 10.4018/978-1-4666-0276-2.ch012

after surpassing Germany in 2009. With the biggest current account surplus balance of US$ 253.3 billion in 2009, China has been able to achieve a foreign exchange reserve of US$ 2.3 trillion, the world´s largest reserve currency. Around 50% of this huge reserve is being applied in American bonds, while the remaining supports Chinese health and social security systems, Chinese banks' solvability, internationalization of the Chinese economy, investment in geostrategic positioning to guarantee energy independence and foreign aid to other developing countries.

During 2008's global crisis, China was able to resist better than the major world economies, even benefitting from this downturn to implement policies to reduce its economic imbalances. One of these imbalances is the gap between FDI in China and Outward Foreign Direct Investment (OFDI) of China which is now progressively narrowing. In fact, in the near future, OFDI is expected even to be larger than FDI, Davis (2009) refer that "*China OFDI accounts for not much more than 1% of the global total, far below the country's share of the world trade. However, this total is rising fast and the country will eventually become a major source of global FDI.*"

Mostly two types of Chinese OFDI can be distinguished: trade-oriented investment and resource–seeking investment. Governmental backing has been crucial for the resource–seeking investment. Although the Chinese investment is nowadays more oriented to mature economies, its bulk is mainly directed to the other developing countries mainly to Asian countries, Latin America, and now also to African countries.

LITERATURE REVIEW

FDI is presently one of the most important variables of the international business flows. FDI is defined as an investment, involving a long-term relationship and reflects the objective of establish a lasting interest and control by an individual or organization the foreign direct investor of one country in an enterprise of a foreign country (Foreign Invested Enterprise - FIE). FDI implies that the investor exerts a significant degree of management and control in the FIE. This can be done either by the transfer of capital flows to create a new venture (green field investment) or by the acquisition of equity capital in an enterprise already existing in a foreign country. Reinvested earnings in the FIEs or short or long-terms loans between the foreign direct investor and the FIE is also consider FDI (UNCTAD, 2009).

OECD (2008) benchmarking definition of FDI, adds that the lasting interest implying a long-term relationship and a significant degree of influence, management and control requires at least the direct or indirect ownership of 10% of the voting power. Basically OECD defines FDI as an investment of an entity resident in one country that has acquired, either directly or indirectly at least 10% of the voting power of a corporation or equivalent for an incorporated company resident in another country. The entity that invests abroad can also contribute with other assets either than equity capital, like technology and production knowledge, that although being intangible assets can be evaluated and considered in the capital invested in the venture or can be considered non-equity forms of investment.

Yan (2005) resumes FDI to two types of economic activities the firm´s equity based investment, including the purchase of equity shares in foreign firms in foreign countries and the establishment of firms operations and management in foreign countries the equity based production.

International Business (IB) and Marketing theories, consider FDI (new ventures, joint ventures and acquisitions) as one mode of entry into foreign markets, and in fact as the one which reflects the higher degree of commitment to those markets, allowing the higher involvement in the management and control of the ventures. Bradley (2002) refers that FDI in new ventures, joint ventures or acquisitions can only be consider when the firm is ready for heavy commitment of resources, accepts higher risks and has enlarged capacity of control.

Meyer (2004) also consider that foreign investors establish their operations using three types of modes, normally classified as joint venture, acquisition and greenfield investments.

The mode of entry which represents the higher committed involvement on the international market is the foreign direct investment, with the direct ownership of 100% foreign capital, known in the literature as Wholly Owned Enterprises (WFOEs), which represents the direct ownership of foreign based assembly manufacturing or services facilities (Terpstra & Sarathy, 1994; Kotler, 2003). Only in this entry mode, the firm has a full control over the investment; it is fully independent to develop a long-term international market strategy, without conflict with local partners and can protect the company from local protectionism given to domestic companies. With this entry mode the firm acquires greater knowledge of the market and is able to appropriate local advantages without having to share it, getting the "*possibility of integrating various national operations into a synergistic international system*" (Terpstra & Sarathy, 1994, p. 401).

Normally the company who choose this mode of entry might strengthen its image, and get the goodwill of local government, workers, customers, suppliers, distributors and the local society in general (Kotler, 2003), but has to support an higher investment normally and the political or economic risks involved. The firm can also face higher exit barriers if the local government fear employment and economic lost with the firm withdraw from the country.

The firm can obtain wholly owned foreign production in two ways; developing its own facilities from the ground or through the merger and acquisition of a foreign based firm. In some cases the structure of market can justify the Merger & Acquisitions (M&As) process as the only way to enter the market. One of most common reasons to choose the M&As mode of entry, is the quickness of company´s establishment and of the internalization of knowledge, local market experience, market share, labor force, institutional support, network of suppliers, intermediaries and customers and contacts with local market operators and government, or acquiring a well-known brand with loyal clients, which allows the promptness of returns (Terpstra & Sarathy, 1994; Bradley, 2002).

But this can be a very expensive and risky way to enter a market due to the lack of transparency of the real market and poor finance situation of the local firm. Accordingly with Woodcock et al. (1994, p. 260) the acquisition mode of entry has a cost associated with the risk of paying too much for the target firm, "*the cost of this risk is associated with the asymmetric information problem confronting the acquiring firm due to the firm's inferior knowledge of the resources being purchased*". Also the integration process can be difficult due to the difficulty of integrate two management teams with different language, culture and business practices. These authors also concluded that "*strategic implementation and organizational cultural differences make it very difficult for organizations to merge efficiently and effectively*" and that "*post-acquisitioned performance was negatively influenced by the two firm's top manager's divergent views of organizational culture*" see (p. 255).

So the foreign firm may not be able or want to buy a local operation and might prefer to build up a new venture from the beginning. The firm may want to introduce the latest technology and equipment, may want to avoid heritage problems, and may want to shape the local firm with its own culture practices and image (Terpstra & Sarathy, 1994). Smaller firms normally prefer new ventures, not only because they are frequently less expensive than acquisitions, but also the size of investment and firm's commitment can be controlled and adjusted to the capacity of the foreign firm and market needs, being the expansion done accordingly with its market share growth and results (Bradley, 2002).

Also foreign investors may join with local investors to create a joint venture company in which ownership control, risk, and market experience is shared. The foreign firm might lack market knowledge, finance, and physical or management resources to operate alone, Kotler (2003), but other reason to form an international joint venture (IJV) can be to facilitate technology transfer, Bradley (2002).We can observe that IJVs rise especially in difficult markets, and they are motivated by the desire of the international partner to have access to local market knowledge and culture as well as to local distribution channels, since, accordingly with Bradley (2002, p. 291) "*General knowledge of the local economy is the key contribution a local partner can make to an international firm seeking entry to a new foreign market*".

In some countries, local conditions and/or government policies can orient the foreign investors to joint local companies and create joint ventures in which they share capital, technology, management and markets. Many countries are known to impose or to have imposed legal restrictions to foreign ownership of local enterprises, some relaxed these restrictions on the 80´s, others did so later or are doing it now or didn't make it yet (Barkema & Vermeulen, 1998). Government's regulations can require local equity participation for example; India, China, Brazil and Thailand were successful in imposing the joint venture option to foreigners (Govindarajan & Gupta, 2001). Chinese government has oriented foreign investors to joint venture with local firms and in some sectors still imposes it (Vanhonaker & Pan, 1997).

Firms may prefer joint ventures as a mode of entry in a culturally distant country, because joint ventures allow them to use the knowledge of the local partners, in local business practices, local institutions, distribution channels, customer preferences and so on (Kogut & Singh, 1988; Barkema & Vermeulen, 1998).

As per Terpstra and Sarathy (1994, p. 396) "*A joint venture is a foreign operation in which the international company has enough equity to have a voice in management but not enough to complete dominate the venture*", and this might be the problem with this entry mode since "*dominant managerial control exercised by foreign partners is significantly and positively related to performance*", (Bradley, 2002, p. 294).

Performance is affected by conflicts between partners, and these conflicts easily happen, when there are different incompatible business operational routines, management practices, strategic views and expectations, and dominant management controls don´t exist (Ding 1997) also the greater the cultural disparity the worst the joint venture performance, "*the cultural similarity between partners is a critical antecedent for IJV success. The greater the cultural similarity between IJV partners the better the IJV performance*" (Lin & Germain, 1998, p. 189).

But FDI research for itself developed its own theories which can be seen on light of an unified framework that attempt to incorporate the entry mode decision making of FDI, known as the Eclectic model or OLI (Ownership, Location, Internalization) model of Dunning as defended by Hill et al. (1990, p. 117) "*Dunning introduced a unified framework within which different factors can be placed and the relationship between them analyzed*". The model of Dunning (1980, 1988), brought together aspects of industrial organization and location economics, but was particularly dependent on internalization analysis to explain FDI. The model introduces explicit considerations of ownership factors which are very close to the ideas of resources, skills and capacities in the Resources Based View models of multinationals enterprises (MNEs) or FDI strategic choices, (Fladmoe-Lindquist & Tallman, 1997). The Eclectic model identifies ownership-specific advantages, location-specific advantages and internalization-specific advantages as relevant factors for entry mode.

Table 1. Some OLI Paradigm Advantages

Ownership of Specific Advantages (O)	Location Advantages (L)	Internalization Advantages (I)
Ownership rights, tangible or intangible; Innovation capacity; Accumulated experience; Exclusive access to certain inputs.	Inputs price, Quality and productivity; Transports costs; Infrastructures; Legislation.	Markets control; Externalities internalization; Avoid government barriers

Source: Adapted from Dunning and Lundan (2008, pp. 101-102)

Ekeledo and Sivakumar (1998, p. 276) concludes that "*Ownership advantages refer to firm specific assets and skills, such as firm size, multinational experience, or ability to develop and market a differentiated product. Location advantages refer to attractiveness of a foreign market. Internalization advantages refer to benefits of retaining assets and skills within the firm when market fails or there is a potential for opportunistic behavior by a partner*".

Dunning, 1980, concluded that the ability of enterprises to acquire ownership advantages is related to specific endowments of their country of origin which is the base of their ownership advantages, and of the countries they operate, this is particularly important because "*the possession of ownership advantages determines which firms will supply a particular foreign - market, whereas the pattern of location endowments explains whether the firm will supply that market by exports (trade) or by local production (non trade)*" p. 11.

Ownership of specific advantages is the ownership of unique factors, meaning resources, skill capabilities developed in country of origin. Location advantages are linked to specific factors of a foreign country that could include cheap labor, superior production processes, local image, governmental trade barriers or others, which justify to undertake production in that country in order to obtain new competitive advantages, that can be internally exploited, through complementary resources of the firm like an international distribution network, or finance capability (Table 1).

Dunning (1977, 1993) suggests three basic motivations to justify FDI, that are going to influence also its location; foreign market-seeking, efficiency (cost reduction) seeking and resource-seeking (including strategic-asset-seeking). Market-seeking FDI is frequently used by firms to have access to distribution networks and to enhance exports, efficiency-seeking FDI normally occur when firms seek lower-cost locations for operations or seek high technologic operation centers and resources-seeking FDI is practice by firms who need to acquire or guarantee the supply of raw-materials, energy and other kind of assets like knowledge and experience in a market, patents or famous brands.

This firm´s oriented analysis mostly done at the light of BI and marketing theories, can be complemented with the other side of the some coin, the macroeconomic theory, which looks for answers to the question; what is the relation between economic development of one country and its inward and outward foreign investment?. A widely used framework for researching the relationship between inward FDI from foreign firms, outward foreign investment of domestic firms and economic development of the host country is the Investment Development Path (IDP) proposed by Dunning (1981) and further develop mainly by Dunning and Narula (1996) and Durán and Úbeda (2001, 2005). According with these researchers, IDP is a dynamic sequential model, with 5 stages of economic development, analyzed in terms of GDP or GDP per capita that relates a country net outward direct investment (NOI) with its economic development stage. This progressive stages starts with a 1st stage where the country is a net receiver of inward FDI and goes to the 5th stage where the country is a net outward investor (see Figure 1).

Figure 1. Model - Investment Development Path Stages Source:Dunning and Narula (1996)

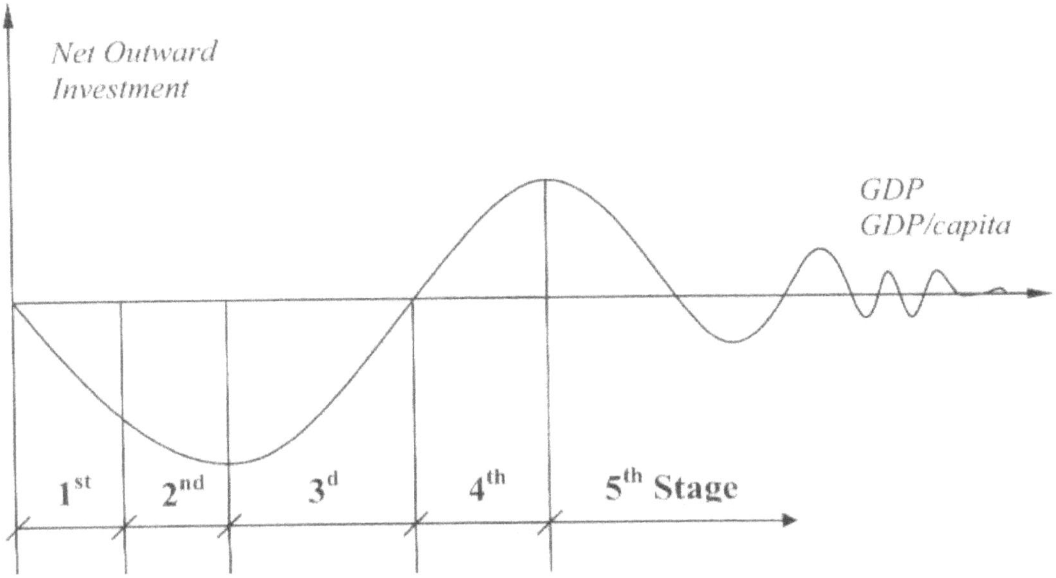

The empirical studies of Mendonça et al. (2009), confirmed this model showing a relationship between GDP per capita and NOI, the results were obtained in a sample comprising USA, Japan and 23 EU countries. Durán and Úbeda (2005) concluded, that the main variables that defines countries in the 1st, 2nd and 3d stages (less developed economies) is inward FDI, although outward FDI and NOI can be verified in a group of these countries, they demonstrated that 4th stage countries are focusing outward FDI, while the 5th stage countries are characterized by a positive NOI.

Dunning et al. (2001) developed the IDP model, considering countries with different types of products and industries, but belonging to the same development stages (based mostly in the GDP per capita, in terms of 1994 US$ values), being these stages the main determinants of the verified inward and outward investment flows. In the Stage 1 of IDP – pre-industrial stage (GDP per capita below US$1000 at 1994 prices) the country receives a small amount of inward FDI, due to insufficient market size, poor infrastructures and has insignificant outward FDI, the weak L-advantages explain it. The Stage 2 is characterized by a GDP per capita between US$1000- $3000, in which economic growth, development of infrastructures and enlarged market size, attracts market-seeking and resource-seeking FDI. Some OFDI can also begin as a result of the internationalization of domestic firms, but in this stage the country is a net recipient of FDI. In the Stage 3 with GDP per capita from $3000 to $10000, improvements in L-advantages go on being registered and the country continues to attract significant amounts of FDI. Domestic companies begin to develop O-advantages, and their internationalization process explain the exploitation of their own specific assets in foreign markets, increasing the amount of OFDI flows, but the country remains a net recipient of FDI. At the 4th Stage GDP per capita exceeds $10000 and the country is normally characterized by being a net outward investor, with OFDI increasing faster than FDI. Finally in 5th Stage the country is characterized by having high living standards and advanced technology, becoming major importer and exporter of FDI, and the FDI and OFDI are balanced.

Statistical observation confirms the relationship between the evolution of China economic

Figure 2. Chinese GDP Growth Source: IMF (2010)

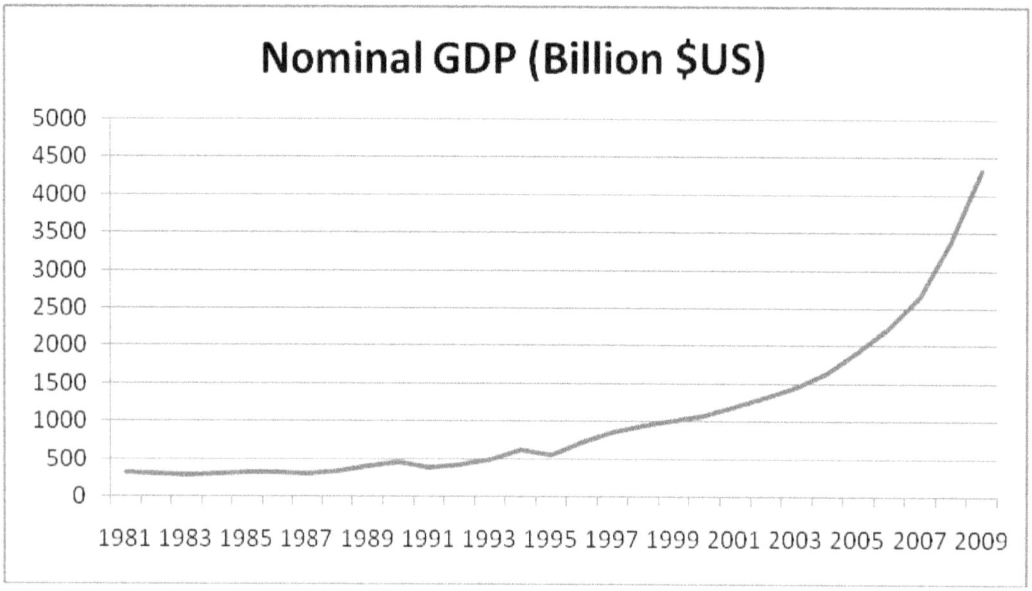

growth and inward and outward FDI; see Figure 2, Figure 3, and Figure 4. Per the classification of Dunning et al. (2001), China is now on stage 2 on the bridge to stage 3, remaining an important recipient of FDI but increasing substantively and at a very fast pace the amount of OFDI, although still being a net recipient of FDI.

IDP pattern vary significantly across developing countries but the case of China is so different that Dunning et al. (2001) and Liu et al. (2005) among others, consider GDP per capita insufficient to explain Chinese OFDI.

Liu et al. (2005) researched 3 supplementary variables, the local investment in human resources, the exports and the inward FDI. Their conclusions postulate, that the increase in GDP per capita and the investment in human resources are positively correlated with Chinese outward investment, but

Figure 3. Evolution of Chinese GDP per capita (nominal and PPP) Source: Source: IMF (2010)

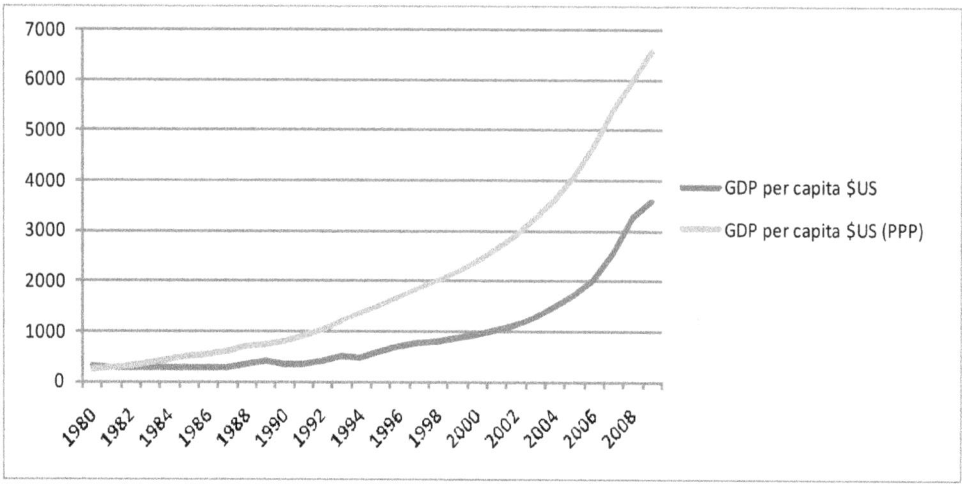

Figure 4. Evolution of Chinese Inward and Outward Investment Source: China Statistical Yearbook (1982-2010; Ministry of Commerce (MOFCOM) (2010)

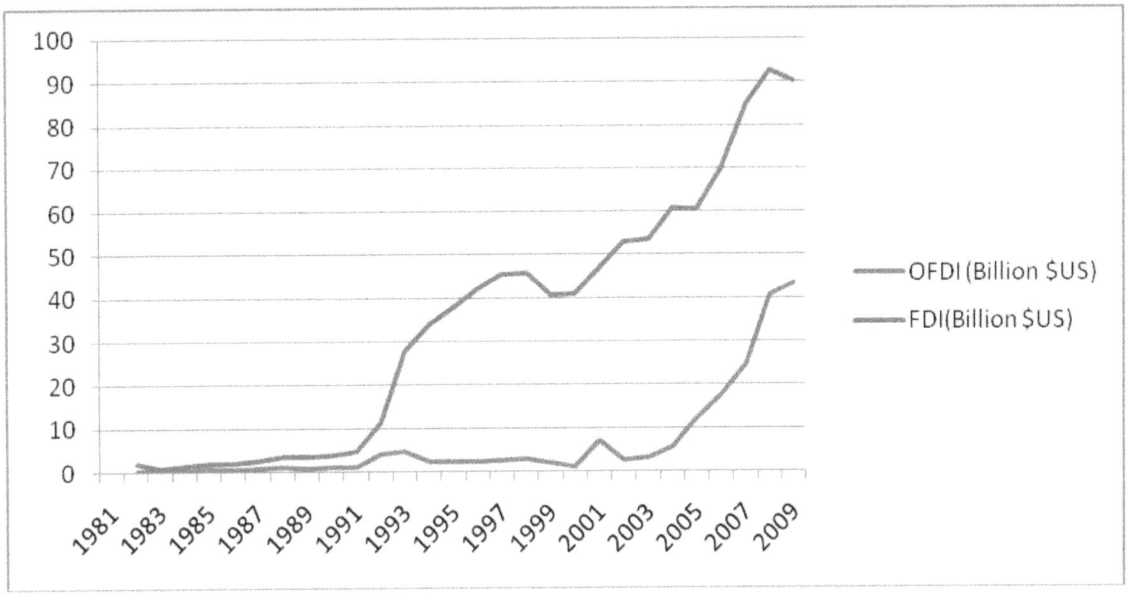

the exports effects in OFDI are ambiguous, however they found a significant interaction between inward FDI and OFDI. Other important conclusion is that China´s OFDI seems to be consistent with an economy at China´s level of economic development; the tested model suggests that policies specially directed to OFDI may be unnecessary, since OFDI seems to follow economic development automatically. The conclusions of Gao (2008) don´t differ most from the ones of Liu *et al.* (2005) he found that GDP per capita indirectly influence China´s OFDI and proved that the outward investment of China is directly influenced by the overseas network, knowledge building (investment in R&D) and market openness (inward FDI).

Buckley et al. (2007) researched if OFDI from China requires a special theory within Dunning general theory, calling our attention for three potential arguments; capital market imperfections, special ownership advantages of Chinese MNEs and institutional factors. Capital imperfection could be found in China, as SOEs can access capital below market rates, inefficient banking system may make soft loans available to OFDI as a policy, conglomerate firms may be funded with subsidies (state-sponsored soft loans) family owned firms can obtain cheap capital from family members. Chinese low cost capital for Chinese firms mostly SOEs made available by market imperfections can explain that M&As is often chose as the mode of entering a new market, and the perception of risk is not as rigorous as it is for firms of industrialized economies. This turns on to be an ownership advantage of Chinese MNEs, being others the flexibility and the networking skills of the Chinese Diaspora. State direct or indirect influence over these firms, is going to drive the decision of outward investment and its location pattern, with criterions that are not profit maximization as in the general theory.

Although policies specially directed to OFDI might not be necessary as per conclusions referred above, the fact is that governments in emerging economies such as India, China and Brazil actively encourage their domestic enterprises to go global (WIR, 2009). A good example is China, that in the last 30 years emerge from a marginal to an important source of OFDI (Luo et al., 2010), its domestic government policy was the main determinant of Chinese OFDI on the beginning of the Chinese internationalization, and after 2001 with the implementation of the "*Go Global*" policy, when the controls on outward investments start being steadily relaxed and at the same time that the Chinese companies begin to be encouraged to invest outward following its own commercial strategies. Through the implementation process of this policy the Chinese government passed from regulator to supporter of OFDI and we can even say to promoter within certain drivers (Figure 5).

Liu (2009) considered the following motivations for Chinese companies to invest abroad; seek resources, markets, technology or strategic

Figure 5. Key Stages of the Chinese Outward Foreign Direct Investment Source: Buckley et al. (2007);Liu (2009);Luo et al. (2010).

Stage 1 (1979-1985) - Cautious Internationalization
With the "Open Door Policy" Chinese OFDI was identified by government as one mean of opening up China to the world economy. Focus in FDI inflows rather than outflows (going global). OFDI was very residual, and only allowed to SOES (State Owned Enterprises), under MOFTEC (later MOFCOM, now Ministry of Commerce).
Stage 2 (1986-1991) - Government Encouragement
The government liberalized restrictive policies and allowed more enterprises to establish foreign affiliates, provided they have sufficient capital, technical and operational know-how and a suitable joint venture partner. Only companies who export could retain a share of its foreign exchange positive balance and invest outside China. With the objective of accumulate foreign exchange Chinese application process for OFDI was very strict. Approval of all the FDI projects was done by the National Planning Commission (NPC) or the State Council, after the evaluation of SAFE- State Administration of Foreign Exchange. The maximum limit was US$10 million and all the profits abroad had to be remitted to China
Stage 3 (1992-1998) - Expansion and Regulation
Chinese governmental actively promoted the international business activities of enterprises under their supervision especially in Hong Kong. Regulations indicate that Chinese OFDI should focus on using overseas technologies, resources and markets, applications should include a feasibility report, Though approval procedures for projects over US$ 1 million were to be done by NPC and over US$ 30 million by the State Council. Projects concerning SOEs should be approved by the State Council.
Stage 4 (1999-2001) – Implementation of the "Go Global" Policy
In June 1999 the "*Guidelines on Granting Credits for Overseas Processing and Assembling*" was published and in October 2000 the "*Measures of Capital Support for Small and Medium Enterprises to Develop International Markets*" were approved. These guidelines and measures had as objective encourage exports, provide loans to companies with mature technology and equipment to invest abroad, and encourage small medium enterprises (SMEs) to join international market. Mostly national Chinese "champions" in domestic market were encouraged to go global, through privileged loans, tax advantages, and "diplomatic" markets. OFDI in specific industries was actively encouraged with export tax rebates, foreign exchange assistance and direct financial support. In 2001 this encouragement policy was formalized in the 10th five-year plan. To achieve the "*Go Global*" policy the Chinese government created incentives for OFDI, lightened the administrative procedures and capital controls, decentralized to local authorities the approval to projects under a certain amount and provided information and guidance on international opportunities and investment risks.
Stage 5 (since 2002 till now) - Post- WTO Period
OFDI become part of China´s national strategy. Regulations were approved to implement this strategy, like the; "*Notice on Simplified Foreign Exchange Administration Related to OFDI*" (March 2003), "*Notice on Providing Credit Support to Key ODFI Projects Encouraged by State*" (May 2003), "*Notice on Using and Managing Special Funds for Foreign Economic Cooperation*" (December 2005), "*Encouraging and Supporting "Go Global" of Private Owned Enterprises*" (February 2006), "*Supplement Measures of Foreign Exchange Usage for OFDI*". The quota system to purchase foreign exchange for OFDI was released, the funds for domestic investors to invest abroad, can be constituted by self-owned foreign exchange fund and by domestic or overseas foreign exchange loans. A government system supporting OFDI has been approved in the 11th five-year plan in 2006, including a simplified approval system where only projects over US$ 100 million require provincial approval, provision of financial support, expansion of the business areas allowed for OFDI, taxation and foreign exchange management, overseas insurance. In September 2007, the Chinese Wealth Fund, (CIC-Chinese Investment Cooperation) a semi-independent quasi-governmental investment firm, was created with the objective of invest a portion of China foreign exchange huge reserves and funded with $US 200 billion initial capital. Chinese government provides financial support to enterprises making OFDI and China Export and Import Bank EXIM bank offer them credit funds and credit insurance.

assets and capabilities. M&As mode of entry, was identified, as a major form of doing it, since 2002. In fact in 2004 the State Administration Reform Commission and the EXIM bank published a guideline to promote OFDI, stating witch kind of OFDI was welcomed by Chinese government, namely" *(1) resources exploitation projects to mitigate the domestic shortage of natural resources, (2) projects that encourage the export of domestic technologies, products, equipment and labor, (3) overseas R&D centers to utilize internationally advanced technologies, managerial skills and professional contacts, and (4) mergers and acquisitions that could enhance the international competitiveness of Chinese enterprises"*, Luo et al. (2010).

OECD (2008), categorize Chinese OFDI into five types of projects; resource-seeking, market secking, strategic assets seeking, diversification seeking and efficiency seeking.

China needs an enormous amount of natural resources to fuel its economic growth, government consider this gap a national strategic issue, backing its SOEs, with preferential loans, soft loans or grant programs to get them and developing intense diplomatic relations with countries rich in natural resources even offering to some of them official development assistance (ODA) programs. Very often Chinese government offer to these governments infrastructure projects in exchange of deals of natural resources acquisions or exploration. Davies (2009) refer that one of the most well known motivations for Chinese OFDI is the resource-seeking investment to secure natural resources needed to fuel Chinese rapid economic growth. Although this is presently not the most important sector of OFDI, see Fig. 6, it received a lot of government backing including ODA programs. This sector of activity represented 48% of the total Chinese OFDI in 2003,

Figure 6. Chinese OFDI by Sectors of Activity in 2008 Source: Statistical Bulletin of China's Outward Foreign Direct Investment, Ministry of Commerce (2008)

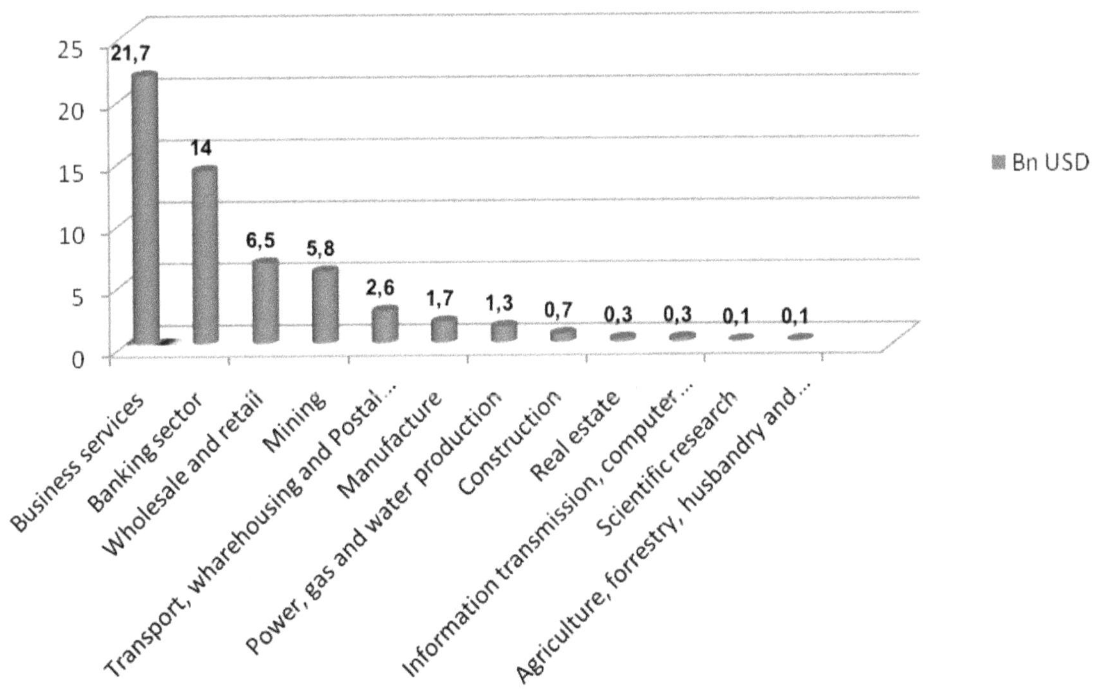

but accounted in 2008 for 10.4% of the Chinese outward investment, mainly in projects of oil and gas exploitation and non-ferrous metal extraction. Resource-seeking OFDI is dominated by few large SOEs, like; CNPC - China National Petroleum Corporation, CNOOC- China National Offshore Oil Corporation, Sinopec- China Petroleum & Chemical Corporation, Sinochem, Baosteel and China Minmetals among others who have access to huge loans of state-owned banks. This type of projects has been progressively changing from North America and Oceania in the 90s to Africa, Asia and Latin America.

Other changing trends can be seen in business services now the most important sector of Chinese OFDI (38.8), banking (25%), transport (4.8%) these sectors accounted with less of 3% of Chinese investment outward in 2003, wholesaling and retailing continue to be an important sector of investment for Chinese enterprises in foreign markets with 11.7% of the total. These trends can be explained by the importance of market-seeking projects and Chinese firm´s needs of distribution channels in foreign countries and services like shipping and insurance to support their exports. Lately many Chinese enterprises have been investing in production units and supply chains to enhance their competitiveness in foreign markets and sometimes as an entry mode in certain markets. This kind of projects driven by commercial reasons, are not considered crucial by the Chinese government and as a consequence they are normally small private firm's projects without much official support.

In certain circumstances these market-seeking projects turns on to be strategic-asset seeking projects, when what is in stake is the need by Chinese companies of strategic assets like state-of-art technology, global brands, customers networks and international marketing know- how. China multinational companies consider the acquisition of foreign international enterprises a quick way to obtain these strategic assets and are practicing M&A to acquire them; good examples of this strategy can be found in companies like the Nanjing Auto purchase of MG Rover or Lenovo buying of IBM personal computers unit. Preferential loans of Chinese state-owned banks, like EXIM bank and China Development Bank (CDB), are offered to this type of Chinese companies for M&A transactions, like for instance the acquisition of Ssangyong Motors by Shanghai Auto, which got 66% of the buying fund in this way (Lunding, 2006).

Also Chinese fierce domestic market competition obliges big Chinese SOEs to diversify to other business areas into foreign markets arena, if they want to survive and growth. Examples of Chinese SOEs pursuing diversification-seeking OFDI include Sinochem and China Resources Corporation; both were large trading companies under the control of MOFCOM, both are now listed in Fortune Global 500, with multinational businesses comprising agriculture, energy, chemicals, finance and real estate (OCDE,2008).

As Chinese enterprises moves up in the value chain, the most internationalized ones look for efficiency-seeking projects, either investing in labor intensive markets, either investing in capital and knowledge intensive markets, depending on their international marketing strategies. For instance the rise of labor costs in some regions in China, like the regions of Guangdong and Shanghai, Zhejiang among others, justifies the production investment of Chinese companies in locations like some ASEAN countries and Africa, but this represents only 3.2% of Chinese OFDI. We can find research centers of Chinese companies like Glanz, Konka, ZTE, Huawei, Haier, in developed countries like USA (Silicom Bay and Dallas) in Sweden (Stockholm) Germany, or in some specific technologic geographic clusters like Bangalore in India.

According with OECD (2008) the three former types of projects have been highly influenced by Chinese government intervention, while the latter two are more commercial oriented, for instance state-owned oil companies accounted for nearly

Figure 7. Evolution of Chinese OFDI by Regions (2003-2008)Source: Statistical Bulletin of China's Outward Foreign Direct Investment, Ministry of Commerce (2008)

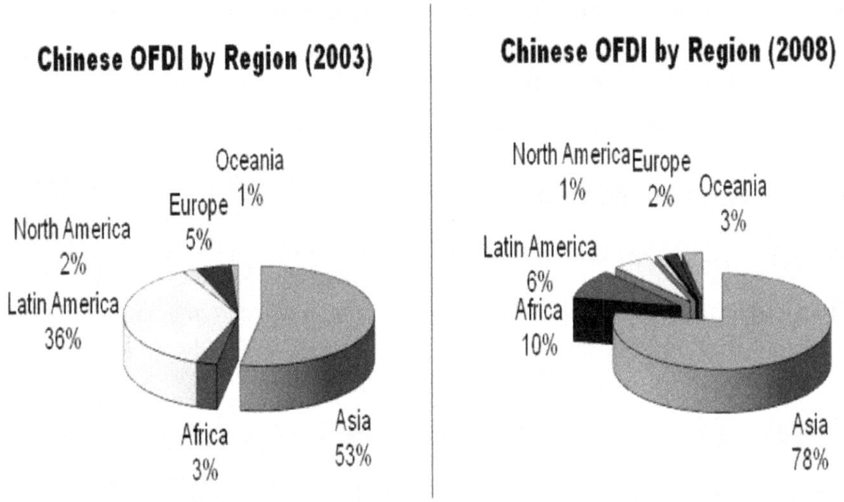

50% of the outward Chinese M&A in 2005, implementing Chinese government energetic policy (Lunding, 2006).

The evolution of the geographical destinations of Chinese OFDI reflects the evolution of the referred investment drivers, reinforcing the weight of Asia, where in 2008, 78% of Chinese OFDI was made, Figure 7. More than 50% of this investment is made in Hong Kong (being a part of it reinvested in China or in other Asian countries) and the rest is mainly applied in Chinese production facilities in Southeast Asia countries, with the objective of increasing its market share in countries like Japan, South Korea, Singapore, Thailand, Malaysia among others, and benefit from lower production costs in ASEAN countries. China is each time more the final stage of the production value chains of Chinese or foreigner MNEs. The importance of Southeast Asia is recognized by President of the World Bank, Zoellick (2010) in this affirmation "*Southeast Asia has become a middle income region of almost 600 million people, with growing ties to India and China, deepening ties with Japan, Korea, and Australia, and continuing links through global sourcing to North America and Europe.*

Africa was in 2008, the second destination of Chinese OFDI, with 10% of all these flows, replacing the importance of Latin America in last five years. According with OECD (2008), China increasingly consider Africa as an important source of energy and strategic natural resources, and a good expanding market for its manufactured products. China trade volume with Africa has increased more than China trade with the world and China is now the seconds' largest trading partner of Africa, after the USA and surpassing France and England.

CONCLUSION

Chinese OFDI don't conflict with Dunning general theory but cannot be explained only by it even after considering supplementary variables, such as the local investment in human resources, the exports and the inward FDI, the overseas network and knowledge building.

Buckley et al. (2007) researched if OFDI from China requires a special theory within Dunning theoretical framework, calling our attention for three potential arguments; capital market imperfections, special ownership advantages of Chinese MNEs and institutional factors

Capital imperfection could be found in China, as SOEs can access capital below market rates. Chinese low cost capital for Chinese firms mostly SOEs made available by market imperfections can explain that M&As is often chose as the mode of entering a new market, and the perception of risk is not as rigorous as it is for firms of industrialized economies. This turns on to be an ownership advantage of Chinese MNEs, being others the flexibility and the networking skills of the Chinese Diaspora. State direct or indirect influence over these firms is going to drive their decision of outward investment and its location pattern with criterions that are not profit maximization as in the general theory.

The conclusion is that Chinese government actively encourages their domestic enterprises to go global with the implementation of the "*Go Global*" policy, being Chinese companies encouraged to invest outward following its own commercial strategies. In last 30 years Chinese economic policy shaped Chinese OFDI.

REFERENCES

Barkema, H. G., & Vermeulen, F. (1998). International Expansion through Start-Up or Acquisition: A Learning Perspective. *Academy of Management Journal*, *41*(1), 7–26. doi:10.2307/256894

Bradley, F. (2002). *International Marketing Strategy* (4th ed.). Upper Saddle River, NJ: Prentice Hall.

Buckley, P. J., Clegg, J., Cross, A. R., Liu, X., Voss, H., & Zheng, P. (2007). The Determinants of Chinese Outward Foreign Direct Investment. *Journal of International Business Studies*, *38*(4), 499–518. doi:10.1057/palgrave.jibs.8400277

China Statistical Yearbook. (1982-2010).

Davis, K. (2009). While global FDI falls, China's outward FDI doubles. In *Proceedings of the Columbia FDI in Perspectives, Perspectives on topical foreign direct investment issues, Vale Columbia Center on Sustainable Investment.*

Ding, D. Z. (1997). Control, Conflict, and Performance: A Study of U.S.-Chinese Joint-ventures. *Journal of International Marketing*, *5*(3), 31–45.

Dunning, J., & Lundan, S. M. (2008). *Multinational Enterprises and the Global Economy* (2nd ed.). Cornwall, UK: Edward Elgar.

Dunning, J. H. (1977). Trade, Location of Economic Activity and the MNE: A Search for an Eclectic Approach. In B. Ohin, P. O. Hessalbom, & P. M. Wijkman (Eds.), *Proceedings of the International Allocation for Economic Activity*. London: Macmillan.

Dunning, J. H. (1980). Toward an Eclectic Theory of International Production: Some Empirical Tests. *Journal of International Business Studies*, *11*(1), 9–31. doi:10.1057/palgrave.jibs.8490593

Dunning, J. H. (1981). Explaining the International Direct Investment Position od Countries: Toward a Dynamic and Development Approach. *Weltwirtschaftliches Archiv*, *117*(5), 30–64. doi:10.1007/BF02696577

Dunning, J. H. (1988). The Eclectic Paradigm of International Production, a Restatement and Some Possible Extensions. *Journal of International Business Studies*, 1–31. doi:10.1057/palgrave.jibs.8490372

Dunning, J. H. (1993). *Multinational Enterprises and the Global Economy*. Reading, MA: Addison-Wesley.

Dunning, J. H., Kim, C., & Lin, J. (2001). Incorporating trade into the investment development path: A case study of Korea and Taiwan. *Oxford Development Studies*, *29*(2), 145–154.

Dunning, J. H., & Narula, R. (1996). The Investment Development Path Revisited: Some Emerging Issues. In Narula, J. H. (Ed.), *Foreign Direct Investment and Governance*. London: Routledge.

Durán, J. J., & Úbeda, F. (2001). The investment path: A new methodological approach and some theoretical issues. *Transnational Corporations*, *10*(2), 1–34.

Durán, J. J., & Úbeda, F. (2005). The Investment Development Path of Newly Developed Countries. *International Journal of the Economics of Business*, *12*(1), 123–137. doi:10.1080/1357151042000323076

Ekeledo, I., & Sivakumar, K. (1998). Foreign Market Entry Mode, Choice of Service Firms: A Contingency Perspective. *Journal of the Academy of Marketing Science*, *26*(4), 274–292. doi:10.1177/0092070398264002

Fladmoe-Lindquist, K., & Tallman, S. (1997). In Wortzel, H. V., & Wortzel, L. H. (Eds.), *Strategic Management in a Global Economy* (pp. 149–167).

Gao, L. (2008). *Determinants of China's Outward Foreign Direct Investment*. Paper presented in CEA Conference: China's Three Decades of Economic Reform (1978-2008), Cambridge, University, UK.

Govindarajan, V., & Gupta, A. (2001). *The Quest for Global Dominance*. San Francisco: Jossey Bass.

Hill, C. W. L., Hwang, P., & Kim, W. C. (1990). An Eclectic Theory of the Choice of International Entry Mode. *Strategic Management Journal*, *2*, 117–128. doi:10.1002/smj.4250110204

IMF-International Monetary Fund. (2010). *World Economic Outlook*. Retrieved from www.imf.org

Jornal Tribuna de macau. (2009). Retrieved from www.jtm.com.mo

Kogut, B., & Singh, H. (1988). The Effect of National Culture on the Choice of Entry Mode. *Journal of International Business Studies*, *19*, 411–432. doi:10.1057/palgrave.jibs.8490394

Kolter, P. (2003). *Marketing Management* (11th ed.). Upper Saddle River, NJ: Prentice-Hall International Inc.

Lin, X., & Germain, R. (1998). Sustaining Satisfactory Joint-Venture Relationship: The Role of Conflict Resolution Strategy. *Journal of International Business Studies*, *29*, 179–196. doi:10.1057/palgrave.jibs.8490031

Liu, H. (2009). *Chinese Business; Landscapes and Strategies*. New York: Routledge.

Liu, X., Buck, T., & Shu, C. (2005). Chinese Economic Development, the Next Stage: Outward FDI? *International Business Review*, *14*(1), 97–115. doi:10.1016/j.ibusrev.2004.12.003

Luo, Y., Xue, Q., & Han, B. (2010). How emerging market governments promote outward FDI: Experience from China. *Journal of World Business*, *45*, 68–79. doi:10.1016/j.jwb.2009.04.003

Mendonça, A., Passos, J., & Fonseca, M. (2009). *The Investment Path Hypothesis: A panel data Approach for Portugal and the Cohesion Countries, 1990-2007*. Paper presented at the 3rd National Congress of Portuguese Economists in Funchal, Madeira.

Meyer, K. (2004). Perspectives on Multinational Enterprises in emerging economies. *Journal of International Business Studies*, *35*, 259–276. doi:10.1057/palgrave.jibs.8400084

MOFCOM. (2010). *Ministry of Commerce the People's Republic of China -Foreign Trade and Investment Statistics*. Retrieved from www.english.mofcom.gov.cn

OECD. (2008). Investment Policy Reviews China. *China Outward Direct Investment*. Retrieved from www.oecd.org

OECD. (2008). *Benchmark Definition of Foreign Direct Investment* (4th ed.). Retrieved from www.oecd.org

Statistical Bulletin of China's Outward Foreign Direct Investment. (2008). *In*. MOFCOM.

Statistical Yearbook (1982-2010). In MOFCOM.

Terpstra, V., & Sarathy, R. (1994). *International Marketing* (6th ed.). Orlando, FL: The Dryden Press.

UNCTAD. (2009). *World Investment Report.*

Vanhonaker, W., & Pan, Y. (1997). The Impact of National Culture, Business Scope, and Geographic Location on Joint -Venture Operations in China. *Journal of International Marketing*, *5*(3), 11–30.

WIR–World Investment Report (2009). Transnational Corporations, Agriculture Production and Development. In UNCTAD.

Woodcock, C. P., Beamish, P. W., & Makino, S. (1994). Ownership-Based Entry Mode Strategies and International Performance. *Journal of International Business Studies*, (2): 253–273. doi:10.1057/palgrave.jibs.8490200

Yan, Y. (2005). *Foreign Investment and Cooperate Governance in China*. New York: Palgrave Macmillan.

This work was previously published in International Journal of Asian Business and Information Management, Volume 1, Issue 4, edited by Dasho Karma Ura and Patricia Ordóñez de Pablos, pp. 43-56, copyright 2010 by IGI Publishing (an imprint of IGI Global).

Chapter 13
No Strategy is an Island to Itself:
China First-Mover and Other FDI Strategies' Interaction Effects

Bradley J. Koch
Grand Valley State University, USA

ABSTRACT

The first-mover strategy for foreign investment is examined to determine under what conditions a significant effect exists when it is combined with other foreign investment strategies like partner selection, geographical market focus, joint-venture control, and resource commitment strategies. Using official audited data and survey data from Sichuan, the results reveal that there are significant interaction effects. The interaction effects can eliminate first-mover advantage, create a first-mover effect that previously didn't exist, or change the direction of the effect. Consequently, the author argues that it is better to analyze strategies as a set that is formed by a series of strategic decisions made by managers as they establish foreign joint ventures and wholly owned subsidiaries.

INTRODUCTION

First-mover advantages have become a foundational principle in strategic management by asserting that those firms that act earlier than their competitors obtain a competitive advantage that results in greater profits (Lieberman & Montgomery, 1988, 1998). This concept has been applied not only to the introduction of new products, but also in establishing joint ventures and WFO (Wholly Foreign Owned) in foreign markets. The first-mover strategy has been frequently invoked by firms entering post-socialist nations where first-mover advantages have been asserted to be beneficial in obtaining the best local partner, which leads to increased market share

DOI: 10.4018/978-1-4666-0276-2.ch012

and profits. This enthusiasm for the first-mover strategy has led researchers to begin to question whether merely being the first to enter a market is in of itself sufficient to produce the desired advantages, or whether it must be combined with specific firm attributes (e.g. Schoenecker & Cooper, 1998), competitors' responses (Ketchen, 2004; Lambkin, 1988), market environments (e.g. Luo & Park, 2001; Gaba, Pan, & Ungson, 2002, Sarkar, Cavusgil, Tamer, & Aulakh, 1999), or other strategies (e.g. Mitchell, 1989; Pan, Li, & Tse, 1999; Zhao & Luo, 2002) to be beneficial.

Despite an increased interest on how the first-mover strategy interacts with other strategies, a systematic examination is needed to determine first-mover effects as it is combined with other foreign investment strategies. This study intends to fill this gap by analyzing how the first-mover strategy interacts with partner selection strategies, geographical market strategies, joint-venture control strategies, and resource commitment strategies across three performance measures. The goal of this analysis is to identify the conditions under which there is a first-mover advantage in foreign investment by examining how the first-mover strategy interacts with other foreign investment strategies.

THEORY AND HYPOTHESES

First-Mover Strategy

First-mover advantages occur when an asymmetry develops between a firm and its competitors due to the firm possessing unique resources, insights, or luck. This asymmetry provides a competitive advantage by which the firm can extract economic profits by a variety of mechanisms such as technology leadership, preemptive control of resources, and buyer switching costs (Boulding & Christen, 2001, Lieberman & Montgomery, 1988). Empirical evidence for first-mover advantages has been found in a number of areas. Ramamurti (2000), for example, finds that first-movers in the Latin America telecommunications market obtained first-mover advantage because the initial foreign investment in these high risk countries forced the governments to discount state assets at the beginning of privatization. Vanderwerf and Mahon (1997) perform a meta-analysis of studies related to first-mover advantages and find significant positive correlation between first-movers and market share. Additional studies link first-mover advantages to corporate acquisitions (Carow, Heron, & Saxton, 2004) oligopolistic industries (Sarkar, Cavusgil, Tamer, & Aulakh, 1999), the money market mutual fund industry (Makadok, 1998), foreign investment (Isobe, Makino, & Montgomery, 2000), and to investing in developing nations' infrastructure (Doh & Ramamurti, 2003).

The high uncertainty environments of post-socialist and transitional economies such as China are quite different from developed economies, and thus require a re-examination of the first-mover strategy to determine its effects (Ketchen, 2004). In these environments, it has been argued that the weak institutions, poor infrastructure, and underdeveloped markets create barriers to entry for late-movers, and thus result in first-mover advantages that more than offset the additional uncertainty in these environments (Bhattacharyya & Rahman, 2003). First-mover research conducted in China supports this view by finding significant evidence for first-mover advantages. For example, Isobe, Makino, and Montgomery's (2000) study of 220 Sino-Japanese manufacturing subsidiaries found that first-mover firms had a high level of performance where performance was measured by employee turnover, perceived economic performance, and overall satisfaction with the firm. Similarly, Pan, Li, and Tse's (1999) sample of over 14,000 foreign firms also revealed significantly higher market shares and profitability for early entrants, and Luo's (1998) sample of 168 foreign invested light manufacturing firms in Jiangsu province revealed that first-movers had a significantly larger market share than the late-movers. In addition, other studies have also found first-mover advantages in China's transitional environment (e.g. Pan & Chi, 1999; Luo & Peng 1998). Based on the first-mover research conducted in China the following prediction can be made:

Hypothesis 1. There will be a positive relationship between foreign firms that establish joint ventures or WFOs in China before their competitors and the performance of these firms.

Interactive Strategy Effects

The interaction effects of the first-mover strategy have been more thoroughly researched in the area of marketing than in foreign investment. Szymanski, Troy, and Bharadwaj (1995) conducted a meta-analysis to determine how the first-mover strategy interacted with other marketing strategies such as product line breadth, product customization, relative product price, end customer purchase frequency, market growth rate, and eighteen other marketing variables. They find evidence that first-mover interaction effects can exceed main effects. In contrast, the interaction of the first-mover strategies with other strategies in the context of the foreign investment has not been systematically examined. This study intends to fill this gap by analyzing how the first-mover strategy interacts with partner selection strategies, geographical market strategies, joint venture control strategies, and resource commitment strategies across three performance measures.

Partner Selection Strategies. Two partner selection strategies are examined for their interaction with the first-mover strategy. First, unrelated diversification occurs when the foreign firm and the Chinese partner don't both have prior experience with the product or service. Although the unrelated diversification strategy is often considered a risk reduction strategy by the parent firm as it diversifies its product portfolio, with regards to the China entity it increases the likelihood of failure for two reasons. First, host governments prefer related product investments that are more likely to involve a greater degree of technology transfer (Brouthers & Bamossy, 1997; Doh & Ramamurti, 2003). Second, more synergy can be created with the local partner by focusing on the investing firm's core competencies (Bartlett & Ghoshal, 1989; Hennart, 1988). In support of the core competence perspective, Luo (2002b) found that foreign joint ventures in China where both partners had experience with similar products performed better than joint ventures with unrelated products. He attributed this to inter-partner learning, resource sharing, innovation improvement, and market power integration. Therefore, one important aspect of partner selection strategy is to decide whether to choose a firm that produces similar products and services.

The second partner selection strategy concerning foreign investment in China is to determine whether to choose partners from existing relationships or whether to search for a partner from all potential firms. Relationships are extremely important within the Chinese business cultural context. The Chinese word for relationship is "*guanxi*" and is by far the most researched Chinese business characteristic. It has been applied to inter-organizational ties among Chinese firms (Chen & Chen, 1998, Park & Luo, 2001), labor markets (Bian, 1997, 1999; Bian & Ang, 1997), relations among regions (Tan & Yeung, 2000; Yeung, 1997), marketing (Wong & Chan, 1999), market efficiency (Lovett et al., 1999), and to a firm's competitive advantage (Luo, 1997, Peng & Luo, 2000; Tsang, 1998). With regards to international joint ventures in China, Luo (2002a) found that previous cooperation bolsters contractual adaptability, which in turn nurtures current cooperation and results in better performance. Therefore, a partner selection strategy is to form joint ventures with Chinese firms where a good relationship has already been established.

The first-mover strategy is often classified as a high risk strategy because of the liabilities of newness that must be surmounted to make it a viable strategy (Fowler, Lawrence, & Morse, 2004). Therefore, firms that combine the first-mover strategy with lower risk partner selection strategies, such as focusing on core competencies and using existing trade partners as joint venture partners are more likely to succeed. This is indicated in the Hypotheses 2a and 2b below.

Hypothesis 2a. The combination of the first-mover and unrelated diversification strategies will have a negative relationship with the joint venture's performance.

Hypothesis 2b. The combination of the first-mover and having no prior relationship with the Chinese partner strategies will have a negative relationship with the joint venture's performance.

Market Geography Strategies. China does not currently consist of a single fluid market, nor did it historically have integrated markets. It has been asserted that 19th century China was not a single unified economic system, but rather nine "macroregions" defined by geographical barriers (Skinner & Baker, 1977), and that the early economic reforms implemented in the 1980s served to consolidate economic power at the provincial and local government levels (Walder, 1995). This decentralization resulted in local and provincial governments adopting protectionist policies to promote the local firms (Pigott, 2002, p. 38). Consequently, a group of researchers have conducted quantitative studies to estimate inter-province trade and the overall integration of China's markets. Naughton's (2000a) study is the only study to assert that China's markets were becoming more integrated during the 1987-1992 period. A second study, Young (2000) used data from 1978 to 1997 to model the degree of integration. Young concluded that twenty years of economic reforms resulted in "a fragmented internal market with fiefdoms controlled by local officials whose economic and political ties to protected industries resemble those of Latin American economies of past decades" (2000, p. 1128). Yet a third study, (Poncet, 2002) revisited the 1987-1992 time period used by Naughton calculating that consumers in 1987 purchased 12 times more local goods than goods from the rest of the country, and that this figure increased to 16 times more local goods in 1992. By 1997, this ratio had increased to 27:1. Based on this ratio, China is less integrated than EU. In a separate study, even Naughton (2000b) concluded that disintegration occurred during the 1992-1997 period. Thus, there is strong empirical evidence to support the view that China consists of fragmented markets.

The lack of a single fluid market makes the geographical market strategy an important consideration for foreign firms entering China. A common strategy is for foreign firms to begin in a single province and use it as a base to spread to other provinces (Luo, 1998). The interaction of this strategy with the first-mover strategy should have a strong positive relationship with performance because it is focused on obtaining market share in the domestic market. The following hypotheses can be derived from this reasoning.

Hypothesis 3. The combination of the first-mover and the local provincial market focus strategies will have a positive relationship with the joint venture's performance.

Control Strategies. Control is an important and complex matter in foreign invested firms that embodies multiple issues such as the relationship between informal and formal control structures (Yan & Gray, 1994), the relationship among resources committed, bargaining power, and equity ownership (Mjoen & Tallman, 1998), and the relationship between overall control and control over specific functions (Luo, Shenkar, & Nyaw, 2001). With regards to the first-mover strategy that is largely based on being able to find the best partner and establishing early market leadership, however, the emphasis must be on leveraging the local partner's expertise of the local environment. Yan and Child (2004) found a strong correlation in the investment of resources (e.g. management systems and services) and control. Thus, the greater control given to the Chinese partner the greater the value placed on its expertise to operate within the local environment. Foreign managers of late-moving firms can study first-movers and enter with more local market knowledge, but foreign managers of first-movers that attempt to control the joint venture are likely to be less successful then foreign managers that allow the Chinese managers to have control. Hypotheses 4a and 4b are derived based on this reasoning.

Hypothesis 4a. The combination of the first-mover and Chinese partner control strategies will have a positive relationship with the joint venture's performance.

Hypothesis 4b. The combination of the first-mover and foreign partner control strategies will have a negative relationship with the joint venture's performance.

Resource Commitment Strategies. The two most important aspects of resource commitment with regards to the first-mover strategy are financial/technological and human resource commitments. The financial and technological resources are difficult to separate. Leading edge technology is more costly than old technology. It is also more risky to invest leading edge technology because it may result in the creation of a formidable competitor. In spite of these obstacles, Isobe, Makino, and Montgomery (2000) in their study of 220 Sino-Japanese joint ventures found that there was a positive association between a high level of resource commitment to technology transfer and perceived economic performance. They assert that firms that are willing to commit resources are signally their overall commitment to establishing a successful joint venture.

Human resources are the second aspect of resource commitment. Technology transfer and management skills are often the primary reasons for Chinese firms to enter international joint ventures (Shenkar, 1990). In order for Chinese firms to obtain these goals there must be a significant amount of personal interaction between the Chinese and foreign partners in the joint venture (Luo, Shenkar, & Nyaw, 2001). Consequently, foreign firms must be willing to commit human resources to engage the Chinese partner. Luo (2002c) found that personal and structural attachments between joint venture partners have a positive relationship with joint venture performance. Thus, in order for the foreign partner to achieve their profit, sales growth, and market share goals, they must be as committed to placing personnel in the foreign venture.

The first-mover strategy is often classified as a high risk strategy that deters firms from entering the market, and that late-movers prefer to observe the first-mover's mistakes to ensure that they don't replicate the same mistakes (Lieberman & Montgomery, 1988). In this high risk environment, the first-mover is prone to make mistakes, but must be willing to persist to obtain the advantages of being a first-mover. Consequently, foreign investors must be willing to commit financial/technological resources and the human resources to overcome the obstacles associated with being the first-mover. Therefore, firms that combine the first-mover strategy with a resource commitment strategy are more likely to succeed. This is stated in the Hypotheses 5a and 5b below.

Hypothesis 5a. The combination of the first-mover and financial/technological resource commitment strategies will have a positive relationship with the joint venture's performance.

Hypothesis 5b. The combination of the first-mover and human resources commitment strategies will have a positive relationship with the joint venture's performance.

METHODS

Sample

The primary data used in this research is based on audited financial data and surveys. The official 2001 foreign investment data was provided by the Sichuan branch of the Ministry of Foreign Trade and Economic Cooperation (MOFTEC). This is the official audited financial statements that foreign invested firms are required to file annually and also includes data on the firms' characteristics and contact information for all 1613 foreign invested firms in Sichuan. The survey was originally written in English then translated and then back translated before being pre-tested. With the support of the Sichuan International Chamber of Commerce, the survey was faxed to 800 firms and 335 were completed and returned which is a 42 percent response rate.

Interviews with managers were used to assist in interpreting the results of the survey data. After the manager returned the survey, a subset of them were contacted to be interviewed. The focus of the interviews was to develop a deeper understanding of the manager's survey responses and probe into exactly what they perceived to be the advantages and disadvantages for their first-mover or late-mover firm. A total of 43 on-site interviews were conducted.

Dependent Variables

The dependent variables for this research are firm performance variables. Prior research indicates that firms and joint ventures in particular can have multiple performance goals (Beamish, 1993; Yan & Gray, 1994). In addition, when comparing first-movers to late-movers Luo (1998) found that first-movers performed better by some performance measures while late-movers perform better by other measures. Therefore, it is important to have a variety of performance measures.

This research uses annual sales growth, market share, and sales on investment ratio as its three performance measures. The annual sales growth and market share variables are self-reported from the survey data. Mangers were asked "What has been the average sales growth since the founding of the firm?" and "What is the current market share of the firm's main product?" In contrast, the sales on investment ratio is derived from audited sales data and the official foreign investment amount registered with Sichuan MOFTEC. Sales on investment ratio is calculated by dividing sales by investment amount, and from this point forward will be referred to sales/investment.

Independent Variables

The survey data was the source of all the independent variables. Since all of the independent variables are defined as being strategies, they are dichotomous variables that are coded with a 1 to indicate the presence of a strategy. The main independent variable in this research is the first-mover variable. This variable is coded 1 if the managers indicated that they were the first foreign invested firm in Sichuan for their product or service. Although this may be considered subjective, it is the best way of determining which firms were first-movers because other failed firms may have been first but no longer present in the data. The remaining independent variables that interact with the first-mover strategy are defined below.

Partner Selection Strategies. The two partner selection strategies are unrelated diversification and using established relationships. The diversification variable is coded a 1 when both parties do not have previous experience in the product or service of the joint venture, or when the WFO China's subsidiary doesn't have experience in the product or service being produced. The "no prior relationship" variable is coded 1 when the joint venture partners did not previously have a trade, licensing, supplier, personal, or other form of relationship prior to the establishment of the joint venture.

Market Geography Strategies. The first aspect of the market geography strategy is whether the foreign investment firm will produce for the domestic or foreign market. The survey asked the managers what percent of the firm's products were exported. If there were not any exports, the "no export" variable was coded with a one implying a purely domestic market focus. In addition, the managers were asked to list the top three sales provinces and the percentage of the firm's sales for each of these provinces. If Sichuan was not listed among the top three provinces, the "sales outside Sichuan" is coded with a 1 indicating a broader national market strategy. If Sichuan was listed as accounting for 100 percent of the firm's sales, then the "all Sichuan sales" variable is coded with a 1 indicating a very exclusive Sichuan focus.

Control Strategies. Control strategies focus on decision making which is crucial when it interacts with the first-mover strategy where moving decisively is of the utmost importance. The easiest way to measure control is by equity ownership. When the Chinese partner (CP) is the majority owner the "CP Majority Owner" variable is coded with a 1, and when the foreign partner (FP) is the majority owner the "FP Majority Owner" variable is coded with a 1. Since majority ownership cannot be assumed to always be the equivalent of being the decision-maker, additional variables are coded based on who the respondents identify as the main decision maker. Thus, the "CP decision maker" variable is coded a 1 when the Chinese partner is indicated to be the main decision maker, and the "FP decision maker" variable is coded a 1 when the foreign partner is indicated to be the main decision maker.

Resource Commitment Strategies. Since the first-mover is operating in a high-risk environment where the firm is likely to encounter setbacks, the foreign investor must be willing to commit additional resources to overcome unforeseen obstacles. If managers indicated that the foreign investor was willing to commit additional resources, the "more resources" variable is coded with a 1. An additional measure of the foreign investor's resource commitment is their willingness to assign foreign expatriates to the firm. If the manager indicated that one or more expatriates were working at the firm, the "expatriate on-site" is coded with a 1. If the firm's general manager is an expatriate, the "expatriate GM" is coded with a 1.

Control Variables

There are several important control variables that need to be included in the model. First, the mode of entry is often considered as a strategy that is chosen by the foreign investor. For Example, Pan and Chi (1999) found that joint ventures performed better than WFOs. In China, however, foreign investors have not been permitted to establish WFO subsidiaries in all industries. Since all firms in the sample didn't have the freedom to choose between joint ventures and WFO entry modes, the WFO variable is used as a control variable.

Second, size of the firm is important in how it relates to first-movers and the performance measures. Gaba, Pan, and Ungson (2002) found that larger U.S. firms were more likely to be first-movers in China. The size of the foreign firm is also often correlated with resource endowments, which have been found to be important to the success of early movers (Mitchell, 1989, 1991). In addition, large firms are more likely to achieve larger market shares. Therefore, the amount of the foreign investment and the number of employees in the firm are used as control variables.

Third, the year the firm began production is used to indicate the length of time that a firm has been operating, which is likely to have an effect on performance. Pan and Chi (1999) found that MNCs that entered China in an earlier year had a higher level of profit than those that entered in a later year. In addition, early entrants will have had more time to capture a larger market share. Since the list of firms is from 2001 financial records and the surveys were distributed in 2003 all firms would have been operating for at least two years.

Finally, particular industry characteristics may affect first-mover opportunities and performance. Song, Di Benedetto, and Zhao (1999) found that managers from manufacturing firms generally perceived first-mover advantages and risks to be greater than managers from the service sector. In addition, industry specific characteristics can determine the extent to which first-mover advantages exist (Schoenecker & Cooper, 1998). Therefore, each firm's product description was used to assign a two digit SIC code. Based on this SIC variable, the manufacturing, service, and real estate sectors accounted for 90 percent of the total sample, and 72, 11, and 7 percent respectively. Therefore, control variables were constructed for each of these three sectors. Furthermore, the overall large proportion of manufacturing firms in the sample necessitated additional control variables to be created for the food processing, chemical, and electronics sectors each of which comprised at least 10% of the sample.

RESULTS

Table 1 presents descriptive statistics for the correlations for the dependent and independent variables. The N column is the number of valid responses for each variable. This column highlights the difficulty in getting performance data from firms. Approximately half the firms either did not know their market share or were unwilling to provide it. In addition it should be noted that there were 103 WFOs in the sample for which the decision making and prior relationship variables were not applicable. These variables were coded as missing for WFOs. Finally, it is important to note that the three performance measures are not significantly correlated.

First-Mover Strategy

Table 2 presents the results of the GLS models that control for heteroscedasticity. The dependent variables for models 1, 2, and 3 are sales growth, market share, and sales/investment respectively. Based on these results, first-movers have significantly higher annual sales growth than late-movers. In contrast, there is no correlation between first-movers and obtaining market shares, and there is a relatively strong negative correlation between first-movers and the sales/investment measure although it is not quite significant at the .05 level. Therefore, Hypothesis 1 is only true for the for the sales growth performance measure.

Interactive Strategy Effects

Table 3 presents the results for the interaction effects between the first-mover strategy and other foreign investment strategies. The models used in Table 3 are the same models used in Table 2 except an additional strategy variable is added and

Table 1.Means, standard deviations, and correlations for independent and dependent variables

Variables	N	Mean	s.d.	1	2	3	4	5	6	7	8	9	10	11	12	13	14	15
1. Market Share	157	26.3	24.9	1														
2. Sales Growth	186	25.5	27.6	.06	1													
3. Sales/ Investment	335	25.3	99.9	.08	.12	1												
4. First Mover	323	.63	.49	-.02	.13	-.09	1											
5. Diversifi-cation	316	.16	.37	-.16*	-.03	.01	-.08	1										
6. No Prior Relation	219	.40	.49	-.16	-.03	-.10	.10	.07	1									
7. No Exports	325	.51	.50	-.16	-.07	.03	-.18 **	.09	.22 **	1								
8. Sales Outside Sichuan	255	.15	.36	.00	.10	.08	.10	.01	-.06	-.13 *	1							
9. All Sich-uan Sales	255	.10	.31	-.12	-.13	.00	-.24 **	.00	.05	.21 **	-.23 *	1						
10. CP Majority Owner	308	.63	.49	.05	.03	.10	.04	-.07	-.12	-.01	-.02	-.07	1					
11. FP Majority Owner	308	.28	.45	.02	-.01	-.07	-.11 *	.12 *	.03	-.01	.02	.08	-.85 **	1				
12. CP Decision Maker	230	.56	.50	.17	-.07	-.04	.02	.00	-.14 *	-.15 *	-.13	-.02	.46 **	-.33 **	1			
13. FP Deci-sion Maker	230	.16	.37	-.06	-.05	-.02	-.08	-.03	.03	-.01	.03	.04	-.38 **	.40 **	-.44 **	1		
14. More Resources	294	.49	.50	.05	.23 **	-.05	.13 *	-.15 *	.03	-.08	.02	-.11	-.23 **	.26 **	-.13 *	.09	1	
15. Expatri-ate On-Site	329	.43	.50	.00	.06	-.02	.08	.00	.10	-.07	.02	-.07	-.23 **	.21 **	-.24 **	.27 **	.20 **	1
16. Expatri-ate GM	334	.16	.37	-.10	.02	-.03	.00	-.09	.15 **	-.04	.01	-.03	-.32 **	.26 **	-.34 **	.34 **	.19 **	.51

$*p < .05$
$**p < .01$

Table 2. Results of GLS Regression Model for First Mover Effects Sales Growth Market Share Sales/ Investment Variables Model 1 Model 2 Model 3

Variables	Sales Growth Model 1		Market Share Model 2		Sales/Investment Model 3	
First Mover	10.27 *	(4.84)	-2.38	(4.54)	-10.81 +	**(5.99)**
WFO	-3.97	(4.96)	-2.62	(4.64)	-4.95	**(6.88)**
Investment (in millions)	-.05	(.06)	-.06 +	(.03)	-.27 *	**(.11)**
Number of Employees (100s)	-.21	(.59)	.39	(.39)	1.52	**(1.07)**
Year Production Began	2.06 *	(.98)	.24	(.66)	.421	**(.89)**
Manufacturing Sector	13.80 **	(5.19)	-7.72	(8.98)	-5.76	**(5.36)**
Food Processing Industry	3.58	(10.23)	-5.82	(6.01)	1.89	**(5.65)**
Chemical Industry	-1.94	(7.32)	-3.96	(4.74)	2.98	**(21.66)**
Electronics Industry	-11.28	(9.18)	-4.38	(7.83)	-2.59	**(4.42)**
Service Sector	-6.55	(6.23)	-4.84	(11.40)	-4.10	**(7.61)**
Real Estate Sector	15.31	(10.86)	-17.30	(10.90)	4.09	**(11.46)**
Intercept	-4090 *	(1947)	-432	(1316)	-814	**(1769)**
R^2	.10		.04		.03	
Overall F	1.5		.08		1.37	
df	11, 170		11, 143		11, 308	

[a] Standard errors are in parentheses.
$+p < .10$
$*p < .05$
$**p < .01$

the associated interaction variable. The interaction variables are derived by multiplying the dichotomous first-mover variable with other dichotomous strategy variables. The results are in sequence by hypothesis numbers where H1 is a reproduction of the Table 2 results for Hypothesis 1 and H2a are the results for Hypothesis 2a and so forth. All control variable results have been omitted.

The interpretation of models that include interaction effects in Table 3 is not straightforward and needs clarification. The results for the first-mover variable in Table 2 should be interpreted as the average first-mover effect. In contrast, the first-mover variable for H2a in Table 3 should be interpreted as the first-mover effect for non-diversifying firms. Similarly, the diversification variable should be interpreted as the effect of the diversification strategy for firms who are not first-movers. The interaction variable (First Mover × Diversification) indicates whether the diversification effect is the same for first-movers and late-movers, but not whether diversification significantly affects the expected level of sales growth for first-movers or whether the first-mover strategy significantly affects the expected level of sales growth for diversified firms. In order to determine these effects, the interaction effects must be added to the first-mover and diversification effects and the standard errors must be recalculated. These results are presented in the "First Mover + Interaction Effect" and "Diversification + Interaction Effect" rows. The addition of the interaction and the first-mover effects is interpreted as the difference in predicted sales growth between first-movers and late-movers among diversified firms. The addition of the interaction and diversification effects is interpreted as the difference in predicted sales growth between

Table 3. Results of GLS regression model with strategy interaction effects sales growth market share sales/investment

Hypothesis # and Variables		Sales Growth Model 1		Market Share Model 2		Sales/Investment Model 3	
H1.	First Mover	10.27 *	(4.84)	-2.38	(4.54)	**-10.81** +	(5.99)
H2a.	First Mover	9.73 *	(5.59)	-2.54	(5.58)	**-8.68**	(8.09)
	First Mover × Diversification	1.02	(14.32)	-3.34	(9.03)	**-4.25**	(10.76)
	Diversification	-2.56	(6.69)	-9.31	(7.68)	**-.11**	(10.77)
	First Mover + Interaction Effect	10.76	(13.03)	-5.88	(6.83)	**-12.92** *	(6.56)
	Diversification + Interaction Effect	-1.54	(13.74)	-12.65 **	(4.74)	**-4.36**	(3.73)
H2b.	First Mover	11.59	(7.43)	-6.55	(8.62)	**-12.71**	(11.71)
	First Mover × No Prior Relation	-5.53	(11.67)	13.94	(11.74)	**8.14**	(11.08)
	No Prior Relation	1.38	(8.11)	-18.26 +	(9.78)	**-14.82**	(11.77)
	First Mover + Interaction Effect	6.05	(9.09)	7.39	(7.01)	**-4.57**	(5.10)
	No Prior Relation + Interaction Effect	-4.16	(7.99)	-4.32	(6.11)	**-6.88**	(4.54)
H3.	First Mover	5.44	(5.34)	3.41	(5.36)	**-5.53**	(3.85)
	First Mover × Sales Outside Sichuan	16.98	(16.68)	-28.60	(17.36)	**-5.83**	(49.66)
	Sales Outside Sichuan	-1.99	(9.68)	20.33	(16.18)	**45.71**	(49.81)
	First Mover + Interaction Effect	22.42	(16.17)	-25.19	(16.33)	**-56.36**	(49.04)
	Sales Outside Sichuan + Interaction Effect	14.99	(13.90)	-8.27	(6.27)	**-5.13** +	(2.95)
H3.	First Mover	7.83	(6.12)	-3.81	(5.90)	**-16.19**	(11.32)
	First Mover × All Sichuan Sales	-10.85	(9.13)	14.27	(11.97)	**14.18**	(20.17)
	All Sichuan Sales	-8.84	(5.67)	-12.94 +	(7.53)	**-9.68**	(16.79)
	First Mover + Interaction Effect	-3.02	(7.16)	10.46	(10.12)	**-2.01**	(11.17)
	All Sichuan Sales + Interaction Effect	-19.69 **	(6.85)	1.33	(8.92)	**4.50**	(5.83)
H3.	First Mover	13.87 *	(6.78)	-16.62 *	(8.00)	**-5.60**	(5.21)
	First Mover × No Exports	-7.52	(9.59)	20.81 *	(9.53)	**-8.13**	(10.64)
	No Exports	-.18	(5.69)	-22.31 **	(8.22)	**5.84**	(10.16)
	First Mover + Interaction Effect	6.35	(6.68)	4.19	(5.33)	**-13.72**	(8.88)
	No Exports + Interaction Effect	-7.69	(7.23)	-1.51	(5.53)	**-2.29**	(3.79)
H4a.	First Mover	14.40	(7.65)	-2.13	(6.04)	**-10.69** **	(4.37)
	First Mover × CP Majority Owner	-7.55	(9.92)	-2.70	(9.27)	**-2.43**	(13.36)
	Chinese Partner (CP) is Majority Owner	10.58	(7.94)	1.49	(8.30)	**11.63**	(14.77)
	First Mover + Interaction Effect	6.85	(6.28)	-4.82	(7.00)	**-13.11**	(13.87)
	CP Majority Owner + Interaction Effect	3.04	(7.47)	-1.21	(5.67)	**9.20** *	(4.66)
H4a.	First Mover	15.03	(10.31)	-6.17	(8.04)	**-17.72**	(14.16)
	First Mover × CP Decision Maker	-13.14	(12.39)	10.27	(10.81)	**19.00**	(16.49)
	Chinese Partner (CP) is Decision Maker	3.29	(7.79)	2.46	(9.26)	**-14.11**	(14.25)
	First Mover + Interaction Effect	1.89	(6.31)	4.10	(7.72)	**1.28**	(4.90)
	CP Decision Maker + Interaction Effect	-9.85	(9.26)	12.73 *	(5.83)	**4.89**	(4.91)

[a] Standard errors are in parentheses.

+$p < .10$ *$p < .05$ **$p < .01$

Continued on following page

Table 3 Continued

Hypothesis # and Variables		Sales Growth Model 1		Market Share Model 1		Sales/Investment Model 3	
H4b.	First Mover	4.55	(5.63)	-7.64	(6.83)	**-10.48**	(12.06)
	First Mover × FP Majority Owner	11.65	(9.86)	8.31	(9.33)	**-3.25**	(11.34)
	Foreign Partner (FP) is Majority Owner	-9.35	(8.44)	.39	(8.38)	**-5.20**	(12.25)
	First Mover + Interaction Effect	16.20 *	(8.24)	.67	(6.27)	**-13.73 ***	(4.99)
	FP Majority Owner + Interaction Effect	2.30	(9.30)	8.71	(6.06)	**-8.45**	(4.43)
H4b.	First Mover	8.26	(5.87)	.61	(6.17)	**-10.93**	(9.88)
	First Mover × FP Decision Maker	-.18	(19.63)	-11.40	(14.36)	**6.17**	(14.48)
	Foreign Partner (FP) is Decision Maker	-2.78	(10.84)	5.51	(13.33)	**-8.13**	(15.83)
	First Mover + Interaction Effect	8.08	(18.91)	-10.79	(13.27)	**-4.76**	(9.98)
	FP Decision Maker + Interaction Effect	-2.96	(16.88)	-5.89	(6.69)	**-1.95**	(7.27)
H5a.	First Mover	5.38	(6.56)	4.53	(6.55)	**-16.17**	(10.97)
	First Mover × FP Willing to Invest More Resources	2.38	(9.39)	-13.53	(9.33)	**10.92**	(12.21)
	FP Willing to Invest More Resources	14.07 *	(6.24)	10.77	(7.45)	**-8.10**	(14.44)
	First Mover + Interaction Effect	7.76	(6.98)	-9.01	(6.64)	**-5.25**	(4.78)
	FP Willing to Invest More + Interaction Effect	16.45 *	(7.22)	-2.76	(5.60)	**2.82**	(5.85)
H5b.	First Mover	7.60	(5.78)	5.77	(5.69)	**-20.40 ***	(10.08)
	First Mover × Expatriate On-Site	4.13	(9.99)	-18.32 *	(8.87)	**20.95 +**	(11.33)
	Expatriate On-Site	1.73	(6.41)	11.34	(7.47)	**-11.74**	(10.12)
	First Mover + Interaction Effect	11.73	(8.07)	-12.55 +	(7.13)	**.54**	(4.86)
	Expatriate On-Site + Interaction Effect	5.86	(8.18)	-6.99	(5.16)	**9.21 ***	(4.07)
H5b.	First Mover	6.18	(4.87)	3.23	(5.14)	**-14.58 +**	(7.75)
	First Mover × Expatriate General Manager	16.69	(12.02)	-26.89 **	(10.01)	**17.85 +**	(10.41)
	Expatriate General Manager	-8.17	(8.65)	9.75	(8.74)	**-14.01**	(8.74)
	First Mover + Interaction Effect	22.87 *	(11.72)	-23.66 **	(8.76)	**3.27**	(5.34)
	Expatriate General Manager + Interaction Effect	8.52	(10.10)	-17.14 **	(4.83)	**3.84**	(3.98)

[a] Standard errors are in parentheses.
$*p < .05$
$**p < .01$

diversified and non-diversified firms among first-movers. For a complete discussion on the interpretation of interaction effects among dichotomous variables see Hardy (1993).

Partner Selection Strategies. The interaction effects between the first-mover strategy and the partner selection strategies are shown in Table 3, in rows H2a and H2b. There is no support for Hypothesis 2b, which asserts that not having a prior relationship with the Chinese partner would have a negative impact on performance. In addition, the inclusion of the no-prior-relationship and the interaction variables removes the significant first-mover effects that existed in Table 2 for the sales growth model.

There is some broad support for Hypothesis 2a. First, the first-mover effect for non-diversifying firms has a positive relationship with sales growth where first-movers who are not diversifying are predicted to have a sales growth rate 13 points higher than other firms. Second, among first-movers the difference in predicted market share between diversified and non-diversified is significant and predicts that the market share for first-movers will be nearly 13 percent less for diversifying firms. Third, the predicted sales/investment ratio for diversified firms is significantly lower for first-movers than for late-movers. In general, the combination of diversifying and being a first-mover has a negative impact on performance.

Market Geography Strategies. The interaction effects between the first-mover strategy and the geographic market strategies are shown in Table 3, in rows H3. In general there is no support for the domestic market strategy. With regard to first-movers that just focus on the domestic market and do not export, there is a positive and significant interaction effect in obtaining market share, which indicates that first-movers do better than late-movers in obtaining market share. This difference in market share, however, is not significantly different when the interaction effect is added to either the first-mover or no-export coefficients. Although the first-mover effect for exporting firms is negative with regards to market share, it is positive with regards to sales growth. Thus, overall there appears to be no benefit for first-movers to exclusively focus on the domestic market.

Although there may be no advantage for first-movers to focus on the broad national market, it is still possible that the segmented market structure could make it beneficial to just focus on a more local market. The results, however, show that for first-movers the difference in predicted sales growth rate is 20 points lower for those firms who only have sales in Sichuan. In addition, there is no evidence that using Sichuan as a production base to sell to other provinces significantly improves the first-movers performance. Therefore, there is no support for Hypothesis 2. Instead it appears that sales growth is optimized by having sales outside of the province and having some exports.

Control Strategies. The interaction effects between the first-mover strategy and the control strategies are shown in Table 3, in rows H4a and H4b. With regards to Hypothesis 4a, there does appear to be some positive effect between first-movers and Chinese partner control. For first-movers, the predicted sales/investment is 9 points higher for firms where the Chinese partner is the majority owner. In addition, the predicted market share for first-movers is 12 points higher for firms where the Chinese partner is the primary decision maker. Thus there is some support for the assertion that first-movers controlled by the Chinese partner perform better.

There are mixed results for Hypothesis 4b which asserts that first-movers controlled by the foreign partners will perform poorly. For joint ventures with the foreign partner as the majority owner the predicted sales/investment ratio for first-movers is 13 points lower than for late-movers. When performance is measured by sales growth, however, there is an opposite effect where the predicted sales growth for first-movers is 16 points higher than for late-movers. In addition, there are no significant effects when the first-mover strategy is combined with the Chinese partner being the primary decision maker. Therefore, there is no conclusive evidence to support Hypothesis 4b.

Resource Commitment Strategies. The interaction effects between the first-mover strategy and the resource commitment strategies are shown in Table 3, in rows H5a and H5b. Hypothesis 5a asserted that first-movers who were willing to commit resources would perform better. The results reveal that the predicted market share for first-movers is 16 points higher for firms willing to commit additional resources. There is no significant effect, however, on market share and sales/investment. Therefore, there is marginal support for Hypothesis 5a.

The results for Hypothesis 5b are mixed. In support of the Hypothesis 5b, the predicted sales/investment ratio for first-movers is 9 points higher when expatriate workers are working at the China location, and the predicted sales growth for firms with expatriate workers on site is 22 points higher for first-movers than late-movers. The results for the market share models, however, contradict this finding. The predicted market share for firms with expatriate workers on site is 23 points lower for first-movers than late-movers, and the predicted market share for first-movers is 17 points lower when expatriate workers are working at the China location. Therefore, support for this hypothesis varies depending on the performance used.

Robustness of First-Mover Advantages

Table 3 provides a means to test the robustness of the results for Hypothesis 1. Table 2 shows that there is a positive correlation between the first-mover strategy and annual sales growth. When additional strategies and their interaction effects are added to the sale growth models in Table 3, however, a first-mover effect only remains significant at the .05 level in 6 out of the 12 models, and in the case of firms having all Sichuan sales the effect switched from being positive to being negative. Therefore, when the first-mover variable is combined with another strategy variable and the associated interaction variable, the first-mover effect is no longer positive and significant at the .05 level with 58 percent of the strategies. When this same analysis is conducted on the market share models, 4 out of the 12 of the models that include other strategies and their interaction effects have statistically significant first-mover effects even though there was no first-mover effect for the market share model in Table 2. Likewise, in the case of the sales/investment models, 4 out of the 12 models have significant first-mover effects with two of the effects being negative and two being positive. This shows that drawing conclusions regarding first-mover effects on performance measures without considering the effects of other strategic decisions and their interaction effects may be tenuous at best.

DISCUSSION

The purpose of this study was to investigate the relationship between the first-mover investment strategy and performance, and to probe more deeply to determine under what conditions first-mover advantages and disadvantages exist. Since managers typically make a set of strategic decisions that ultimately combine and interact to create the firm's strategy, I combine different strategy decisions with the first-mover strategy decision to determine how each set of two strategies interact and influence performance. This method illuminates more clearly under what conditions the first-mover effects exist and whether they persist at all. There are three main findings and implications.

First, when other strategy effects are not controlled for there is an overall significant first-mover advantage for the sales growth performance measure, but no significant advantage or disadvantage for the market share and sales/investment performance measures (see Table 2). These findings are largely consistent with previous findings that control for firm characteristics and found positive effects for first-movers.

Second, the inclusion of other foreign investment strategies with the first-mover strategy and the associated interaction is overall quite influential on firm performance. The pairing of the first-mover strategy with other strategies and the associated interaction effects eliminated the first-mover advantage in many cases, and in other cases greatly enhanced the first-mover advantage. Therefore, it is important to conceptualize firm strategy as a broader set of strategies that interact. It is the set of strategies and the interaction among the strategies within the set that ultimately determine the firm's performance.

Third, the frequency in which the inclusion of an additional strategies and their associated interaction variables changed the first-mover effects of the models in Table 2 that only include the first-mover strategy calls to question the robustness of the first-mover effect. If analysts are to take strategy sets seriously and view managers as making a series of strategic decisions to maximize performance goals, then claims concerning first-mover effects (and any other particular strategic decision) should only be asserted after controlling for all other strategies. Consequently, claims concerning the advantages of the first-mover strategy and all other strategies must be made after controlling for the effects of all other strategies. If the first-mover is more likely to be acting strategically with asymmetric information whereas the late-mover is more likely to be merely copying the strategy of the first-mover, then first-movers are more likely to make the best set of strategic decisions whereas late-movers are more likely to copy a set of strategies that are inappropriate for its firm's particular situation (Carow, Heron, & Saxton, 2004; Lieberman & Asaba, 2006). Consequently, the first-mover advantage is not a result of being a first-mover, but rather from first-movers acting strategically to select the best set of strategies as they enter new environments full of uncertainty. Therefore, the first-mover advantage may be a proxy for acting strategically rather than blindly copying other firms, and thus will disappear when all other strategies are controlled for.

Fourth, the inclusion of multiple performance measures as dependent variables illustrate how the same strategy set can have different effects on different performance measures. Thus, among firms where the majority owner is the foreign partner there was a positive first-mover effect on annual sales growth, but a negative effect on the sales/investment ratio. This is significant because it is often difficult to determine what performance measures managers are actually trying to meet with their strategy set.

Theoretical and Methodological Implications

The primary theoretical contribution of this study is the conceptualization of the first-mover strategy as being one piece of a broader puzzle of strategic decisions that managers put together when making foreign investments. Analysis of the first-mover strategy within context of foreign investment strategies has lagged behind marketing applications to new product entry in realizing that a market strategy is really a mix of strategies that interact (e.g. price, product characteristics, distribution, and promotion). I assert that the foreign investment strategy also is comprised of a mix of strategies that interact. Consequently, the first-mover strategy cannot be analyzed in isolation of other strategies no more than a pricing strategy can be analyzed independent of other components of the marketing mix. This approach was confirmed in my interviews with managers. The managers always explained their foreign investment strategies as sets of interacting strategies rather than as a single isolated strategy.

CONCLUSION

I use methods similar to previous research and find some support for first-mover advantages in foreign investment just like prior research. I then deviate from prior research by conceptualizing foreign investment strategies as a mix of investment strategies similar to the marketing mix. This conceptualization requires an analysis of sets of strategies to identify the interaction effects among strategies. The introduction of an additional strategy variable and its associated interaction effects greatly change the results in two ways. First, the results that only include the first-mover strategy indicate that there is a positive association between first-movers and sales growth. The introduction of an additional strategy variable and its associated interaction effects, however, removes this positive relationship in half of the cases. Second, the results that only include the first-mover strategy indicate that there is no relationship between first-movers and either market share or the sales/investment ratio. The inclusion of an additional strategy variable and its associated interaction effects, however, reveals that there are significant first-mover and late-mover effects when the first-mover strategy is combined with certain other foreign investment strategies. Consequently, these results call into question a host of prior first-mover research that doesn't control for the effects of other strategic decisions that are integral components of the broader foreign investment entry strategy.

Future research should attempt to more completely conceptualize foreign entry strategies as being comprised of multiple strategic decisions. Within this more comprehensive framework, the decision to be a first-mover is only one piece of the larger puzzle. These findings illuminate the need for more research on other strategic decisions that are important components of the broader strategy package that managers develop before undertaking a foreign venture.

REFERENCES

Bartlett, C. A., & Ghoshal, S. (1989). *Managing across borders: The transnational solution*. Boston: Harvard Business School Press.

Beamish, P. W. (1993). The Characteristics of Joint Ventures in the People's Republic of China. *Journal of International Marketing*, *1*, 29–48.

Bhattacharyya, S. K., & Rahman, Z. (2003). Sources of first mover advantages in emerging markets - an Indian perspective. *European Business Review*, *15*, 359–369. doi:10.1108/09555340310500541

Bian, Y. (1997). Bringing strong ties back in: Indirect ties, network bridges, and job searches in China. *American Sociological Review*, *62*, 366–385. doi:10.2307/2657311

Bian, Y. (1999). Getting a job through a web of guanxi in China. In Wellmam, B. (Ed.), *Networks in the global v illage: Life in contemporary communities* (pp. 275–296). Boulder, CO: Westview Press.

Bian, Y., & Ang, S. (1997). Guanxi networks and job mobility in China and Singapore. *Social Forces*, *75*, 981–1005. doi:10.2307/2580527

Boulding, W., & Christen, M. (2001). Idea - First-mover disadvantage. *Harvard Business Review*, *79*, 20–21.

Brouthers, K. D., & Bamossy, G. J. (1997). The role of key stakeholders in international joint venture negotiations: Case studies from Eastern Europe. *Journal of International Business Studies*, *28*, 285–308. doi:10.1057/palgrave.jibs.8490102

Carow, K., Heron, R., & Saxton, T. (2004). Do early birds get the returns? An empirical investigation of early-mover advantages in acquisitions. *Strategic Management Journal*, *25*, 563–585. doi:10.1002/smj.404

Chen, H., & Chen, T. J. (1998). Network linkages and location choice in foreign direct investment. *Journal of International Business Studies*, *29*, 455–478. doi:10.1057/palgrave.jibs.8490002

Doh, J., & Ramamurti, R. (2003). Reassessing risk in developing country infrastructure. *Long Range Planning*, *36*, 337–353. doi:10.1016/S0024-6301(03)00069-4

Fowler, S. W., Lawrence, T. B., & Morse, E. A. (2004). *New ventures in the information economy: Exploiting virtual embeddedness and knowledge-based assets.* Paper presented at the annual meeting of the Academy of Management, New Orleans, LA.

Gaba, V., Pan, Y., & Ungson, G. R. (2002). Timing of entry in international market: An empirical study of U.S. fortune 500 firms in China. *Journal of International Business Studies*, *33*, 39–55. doi:10.1057/palgrave.jibs.8491004

Hardy, M. A. (1993). *Regression with Dummy Variables*. Newbury Park, CA: Sage.

Hennart, J. F. (1988). A transaction costs theory of equity joint ventures. *Strategic Management Journal*, *9*, 361–374. doi:10.1002/smj.4250090406

Isobe, T., Makino, S., & Montgomery, D. B. (2000). Resource commitment, entry timing, and market performance of foreign direct investments in emerging economies: The case of Japanese international joint ventures in China. *Academy of Management Journal*, *43*, 468–484. doi:10.2307/1556405

Ketchen, D. J. Jr, Snow, C. C., & Hoover, V. L. (2004). Research on competitive dynamics: Recent accomplishments and future challenges. *Journal of Management*, *30*, 779–804. doi:10.1016/j.jm.2004.06.002

Lambkin, M. (1988). Order of entry and performance in new markets. *Strategic Management Journal*, *9*, 127–140. doi:10.1002/smj.4250090713

Lieberman, M. B., & Asaba, S. (2006). Why do firms imitate each Other? *Academy of Management Review*, *31*, 366–385.

Lieberman, M. B., & Montgomery, D. B. (1988). First-mover advantages. *Strategic Management Journal*, *9*, 41–58. doi:10.1002/smj.4250090706

Lieberman, M. B., & Montgomery, D. B. (1998). First-mover (dis)advantages: Retrospective and link with the resource-based view. *Strategic Management Journal*, *19*, 1111–1126. doi:10.1002/(SICI)1097-0266(1998120)19:12<1111::AID-SMJ21>3.0.CO;2-W

Lovett, S., Simmons, L. C., & Kali, R. (1999). Guanxi versus the market: Ethics and efficiency. *Journal of International Business Studies*, *30*, 231. doi:10.1057/palgrave.jibs.8490068

Luo, Y. (1997). Guanxi and performance of foreign-invested enterprises in China: An empirical inquiry. *Management International Review*, *37*, 51–70.

Luo, Y. (1998). Timing of Investment and International Expansion Performance in China. *Journal of International Business Studies*, *29*, 391–407. doi:10.1057/palgrave.jibs.8490042

Luo, Y. (2002a). Contract, cooperation, and performance in international joint ventures. *Strategic Management Journal*, *23*, 903–919. doi:10.1002/smj.261

Luo, Y. (2002b). Product Diversification in International Joint Ventures: Performance Implications in an Emerging Market. *Strategic Management Journal*, *23*, 1–20. doi:10.1002/smj.203

Luo, Y. (2002c). Stimulating exchange in international joint ventures: An attachment-based view. *Journal of International Business Studies*, *33*, 169–181. doi:10.1057/palgrave.jibs.8491010

Luo, Y., & Park, S. H. (2001). Strategic alignment and performance of market-seeking MNCs in China. *Strategic Management Journal, 22*, 141–155. doi:10.1002/1097-0266(200101)22:2<141::AID-SMJ151>3.0.CO;2-O

Luo, Y., & Peng, M. W. (1998). First mover advantages in investing in transitional economies. *Thunderbird International Business Review, 40*, 141–163. doi:10.1002/tie.4270400205

Luo, Y., Shenkar, O., & Nyaw, M.-K. (2001). A dual parent perspective on control and performance in international joint ventures: Lessons from a developing economy. *Journal of International Business Studies, 32*, 41–58. doi:10.1057/palgrave.jibs.8490937

Makadok, R. (1998). Can first-mover and early-mover advantages be sustained in an industry with low barriers to entry/imitation? *Strategic Management Journal, 19*, 683–696. doi:10.1002/(SICI)1097-0266(199807)19:7<683::AID-SMJ965>3.0.CO;2-T

Mitchell, W. (1989). Whether and when? Probability and timing of incumbents' entry into emerging industrial subfields. *Administrative Science Quarterly, 34*, 208–230. doi:10.2307/2989896

Mitchell, W. (1991). Dual clocks: Entry order influences on incumbent and newcomer market share and survival when specialized assets retain their value. *Strategic Management Journal, 12*, 85–100. doi:10.1002/smj.4250120202

Mjoen, H., & Tallman, S. (1997). Control and performance in international joint ventures. *Organization Science, 8*, 257–274. doi:10.1287/orsc.8.3.257

Naughton, B. (2000a). China's trade regime at the end of the 1990s: Achievements, limitations, and impact on the United States. In Carpenter, T. G., & Dorn, J. A. (Eds.), *China's Future: Constructive Partner or Emerging Threat?* Washington, DC: Cato Institute.

Naughton, B. (2000b). *How much can regional integration do to unify China's markets?* (Working Paper No. 58). Palo Alto, CA: Stanford University Center for Research on Economic Development and Policy Reform.

Pan, Y., & Chi, P. S. K. (1999). Financial performance and survival of multinational corporations in China. *Strategic Management Journal, 20*, 359–374. doi:10.1002/(SICI)1097-0266(199904)20:4<359::AID-SMJ31>3.0.CO;2-9

Pan, Y., Li, S., & Tse, D. K. (1999). The Impact of Order and Mode of Market Entry on Profitability and Market Share. *Journal of International Business Studies, 30*, 81–82. doi:10.1057/palgrave.jibs.8490061

Park, S. H., & Luo, Y. (2001). Guanxi and organizational dynamics: Organizational networking in Chinese firms. *Strategic Management Journal, 22*, 455–477. doi:10.1002/smj.167

Peng, M. W., & Luo, Y. (2000). Managerial ties and firm performance in a transition economy: The nature of a micro-macro link. *Academy of Management Journal, 43*, 486–501. doi:10.2307/1556406

Pigott, C. A. (2002). *China in the world economy: The domestic policy challenges.* Paris: Organisation for Economic Co-operation and Development.

Poncet, S. (2005). A fragmented China: Measure and determinants of Chinese domestic market disintegration. *Review of International Economics, 13*(3), 409–430. doi:10.1111/j.1467-9396.2005.00514.x

Ramamurti, R. (2000). Risks and rewards in the globalization of telecommunications in emerging economies. *Journal of World Business, 35*, 149–170. doi:10.1016/S1090-9516(00)00030-4

Sarkar, M. B., Cavusgil, S. T., & Aulakh, P. S. (1999). International expansion of telecommunication carriers: The influence of market structure, network characteristics, and entry imperfections. *Journal of International Business Studies*, *30*, 361–381. doi:10.1057/palgrave.jibs.8490074

Schoenecker, T. S., & Cooper, A. C. (1998). The role of firm resources and organizational attributes in determining entry timing: A cross-industry study. *Strategic Management Journal*, *19*, 1127–1243. doi:10.1002/(SICI)1097-0266(1998120)19:12<1127::AID-SMJ7>3.0.CO;2-4

Shenkar, O. (1990). International joint ventures' problems in China: Risks and remedies. *Long Range Planning*, *23*, 82–90. doi:10.1016/0024-6301(90)90056-A

Skinner, G. W., & Baker, H. D. R. (1977). *The city in late imperial China*. Palo Alto, CA: Stanford University Press.

Song, X. M., DiBenedetto, C. A., & Zhao, Y. Z. L. (1999). Pioneering advantages in manufacturing and service industries: Empirical evidence from nine countries. *Strategic Management Journal*, 20.

Szymanski, D. M., Troy, L. C., & Bharadwaj, S. G. (1995). Order of entry and business performance: An empirical synthesis and reexamination. *Journal of Marketing*, *59*, 17–33. doi:10.2307/1252325

Tan, C. Z., & Yeung, H. W. (2000). The regionalization of Chinese business networks: A study of Singaporean firms in Hainan, China. *The Professional Geographer*, *52*, 437–454. doi:10.1111/0033-0124.00237

Tsang, W. K. (1998). Can guanxi be a source of sustained competitive advantage for doing business in China? *The Academy of Management Executive*, *12*, 64.

VanderWerf, P. A., & Mahon, J. F. (1997). Meta-analysis of the impact of research methods on findings of first-mover advantage. *Management Science*, *43*, 1510–1520. doi:10.1287/mnsc.43.11.1510

Walder, A. G. (1995). Local governments as local firms: An organizational analysis of China's transitional economy. *American Journal of Sociology*, *101*, 263–301. doi:10.1086/230725

Wong, Y. H., & Chan, R. Y.-K. (1999). Relationship marketing in China: Guanxi, favouritism and adaptation. *Journal of Business Ethics*, *22*, 107–118. doi:10.1023/A:1006077210425

Yan, A., & Gray, B. (1994). Bargaining power, management control, and performance in United States-China joint ventures: A comparative case study. *Academy of Management Journal*, *37*, 1478–1517. doi:10.2307/256796

Yan, Y., & Child, J. (2004). Investors' resources and management participation in international joint ventures: a control perspective. *Asia Pacific Journal of Management*, *21*, 287–304. doi:10.1023/B:APJM.0000036464.38236.34

Yeung, H. W. (1997). Cooperative strategies and Chinese business networks: A study of Hong Kong transnational corporations in the ASEAN region. In Beamish, P. W., & Killing, J. P. (Eds.), *Cooperative Strategies: Asian Pacific Perspectives* (pp. 22–56). San Francisco: The New Lexington Press.

Young, A. (2000). The razor's edge: Distortions and incremental reform in the People's Republic of China. *The Quarterly Journal of Economics*, *154*, 1091–1135. doi:10.1162/003355300555024

Zhao, H., & Luo, Y. (2002). Product diversification, ownership structure, and subsidiary performance in China's dynamic market. *Management International Review*, *42*, 27–48.

This work was previously published in International Journal of Asian Business and Information Management, Volume 1, Issue 4, edited by Dasho Karma Ura and Patricia Ordóñez de Pablos, pp. 35-53, copyright 2010 by IGI Publishing (an imprint of IGI Global).

Chapter 14
Study on the New Tendency of the International Industry Shifts and Chinese Enterprise Implementation Tactics

Jing Sun
Shenyang University, China

Mingyou Wang
Shenyang University, China

ABSTRACT

At the beginning of 21st century, the Multinational Corporation take globalization strategy as the foundation and the starting point, accelerate the industry restructure greatly, the industrial shift presented the depth and breadth tendency day by day, how does China hold this opportunity, attracted the foreign capital vigorously, promotion industrial structure, will be the next economic development strategic mission. This article analysis multinational corporation international industry shift new tendency and analysis China uses the international industrial shift's present situation and the superiority, proposed the development countermeasure.

1. INTRODUCTION

The international industrial shift is refers to the process of some industries through the international trade and the international investment and so on many kinds of ways shifts to another national and the local from a country and the area (Wang, 2003); it is comparison superiority dynamic development change result which because between the country's factor and so on resources, talent, market, position, industrial necessary environment, system environment differences form, it is not only representing the material capital movement and economic variable change, but also

DOI: 10.4018/978-1-4666-0276-2.ch013

is between developed country and developing country economic relationships substantive movement (Yang, 2005), and causes this international economy relations establishment in one kind of new international division of labor foundation, forms one kind of new international pattern of the trade and the investment benefit assignment: Knowledge intensity industry country, capital intensity industry country and between labor concentrated industry country relations. In the product manufacturing technological development's different period, is different to element of production's demand, the compared superiority along with product manufacturing essential factor crowded change, will urge the industry which occupy the international leading position to shift unceasingly between in the country which have modern scientific knowledge, have the abundant capital and skilled worker, have the rich labor resources.

1. MULTINATIONAL CORPORATION INTERNATIONAL INDUSTRY SHIFT NEW TENDENCY

1.1 The Whole World Industry Adjustment and Promotion Step Speeds Up, The International Industrial Shift Level Enhances Unceasingly

Along with economic globalization tendency expand rapidly and international competition intense day by day, the developed country in order to maintain the competitive advantage, speed up the step which the industry restructure promotes, prioritize innovative technology-intensive industry which has the higher added value, for example the high technology and new technology industry which take information technology as forerunner and the modern service industry which take finance, insurance, professional service and so on as core, but take the ordinary labor, capital and technology-intensive industry which added value low to shift to other countries and local, thus forms the new shift tide unceasingly (Guo, 2000). The developing country speeds up the industrial upgrading and the economic development through the industry shift, develops the traditional manufacturing industry vigorously, and leads related necessary industrial and service industry development by this, the industrialization advancement speeds up obviously. While the global industry adjustment and promotion step speeds up, the international industrial shift's level unceasingly also enhancing unceasingly, presents the tendency which promotes unceasingly from the labor concentrated industry to the capital intensity industry, the traditional industry to the emergent industry, the manufacturing industry to service industry, the low added value industry to the high added value industry. Moreover, along with knowledge economy further development, the international industrial shift's level will further also enhance

1.2 The Multinational Corporation Investment Growth Explosively, the International Industrial Shift Scale Expands Unceasingly

First, the world foreign trade total amount may reflect the international industrial shift scale approximately. Since the 1980s, the global foreign direct investment scale has expanded unceasingly, from 1980 to 1985, the international industrial shift average scale is about 140 billion US dollars, until 2005, the global foreign direct investment accumulation surpasses above 3,500billion US dollars. The globalization also leads the international investment growth explosively; the whole world foreign direct investment inflow storage quantity has occupied GDP proportion 9.3% in 1993 to enhance 22.3% in 2005. The world direct investment storage quantity increases suddenly from 560 billion US dollars in 1980 to 8,200 billion US dollars in 2005, grew 14 times, the yearly average rate of increment has reached as high as 33%. This time, the world foreign trade total amount's rate of rise also speeds up unceasingly (Council Development Research Center Industry Economical Research department, 2006). From 1980 to 1985, the world foreign trade total amount annual average growth rate was 28.8%; from 1985 to 1989 was 33.1%; from 1990 to 1995 was 35.7%; from 1995 to 2000 was 38%; achieves 45% in 2005. This also from a side indicated that the international industrial shift speed is speeding up.

Next, the main developed country productivity shift's proportion rises unceasingly (Bi, 2006). According to the preliminary statistics, the global sharing scale of production scale which between the developed country and the developing country forms already surpasses 800 billion US dollars, this is equal on 30% of the whole world manufacturing industry trade. According to the American Department of commercial affairs statistics, approximately some 2500 American Corporation take the produce link to shift to the overseas, the result is that the American manufacturing industry output value drops unceasingly, the employment proportion from 36% to drop to present 10%.

1.3 The International Industry Shift's Key Point Will be Knowledge Service Industry and High Technology and New Technology Manufacturing Industry in the Future

In the future 5 years, estimated that will have the new round international industrial shift upsurge, will change the knowledge service industry and the high technology and new technology manufacturing industry with emphasis, like financial, information; energy, medicine, bio-engineering global dispositions. Chinese economy continues and fast growth and fund need strong, the transnational capital and service industry investment constitution in China have the obvious change, the market actuation transnational investment which sells internally primarily will become the mainstream tendency. At present, the world 500 strong enterprises already started to set up the organizations and so on research and development, design, operation, physical distribution allocation, consultation and training in China, and sped up the localization advancement (Wang & Chen, 2006). Like Intel Corporation has set up the Chinese laboratory at the end of 2000, develops research and development which face the global; Microsoft Corporation set up Chinese Research Institute in Beijing, is its foreign biggest scientific research institution (National Development and Planning Commission Macro Economic Research Institute Foreign Economics Department, 2007).

1.4 The International Industrial Shift Presents the Group Type, and Industrial Chain Whole Shift Tendency

The Multinational corporation society cooperation degree is high, crosswise relates broadly. Along with the competition intensifies (Hu, 2007), the multinational corporation no longer follows the traditional industrial shift gradient stage to carry on the investment, but leads on own initiative and guides related investment, encourages its overseas supplier to investment to the host country, enlarges implementation dynamics which the spare part supplies localization strategy, development necessary industry and establishes the industrial group, put the strip whole industry chain relocation and shift to the developing country. Multinational Corporation besides shift tradition manufacturing industry also starts to shift to other production operation link to other local, like research development, design and Base Company and so on, also starts to other local shift. This new industrial shift's tendency follows the enterprise size to expand unceasingly as well as the position condition change appears, it is advantageous in raises enterprise's resources deployment efficiency, and promotion overall enterprise's competitive power.

1.5 Multinational Corporation Investment Way Diversifies Day by Day, Leads the Global Industry Adjustment and Shift

The direct investment and each stockholder's rights arrangement have been the fundamental modes which the Multinational Corporation invests, but the new round international industrial shift investment way diversifies day by day. Besides the original direct investment and each stockholder's rights arrangement, the indirect investment and the non-stockholder's rights investment are getting more and more. The international indirect investment rate of rise in the international direct investment quickly, the international indirect investment scale already surpassed the international direct investment scale (Song, 2004). Statistics indicated that the developed country carries on the direct investment to the developing country annually rate of rise was 8% from 1995 to 2005, but the corresponding indirect investment rate of rise was 15%. The securities investment and the transnational enterprise merger and acquisition become the international investment and the industrial shift fundamental mode day by day. While developed country enterprise purchase and annexation fast development, developing country also more and more has selected this method in international direct investment and industrial shift.

2. CHINA FACES THE NEW OPPORTUNITY WITH CHALLENGES

By the end of 2005, China accumulates authorization Foreign invested company 508941, uses foreign merchant investment amount 562.1billion US dollars actually, already continuously 11 years became the foreign direct investment inflow biggest developing country, and have become foreign merchant direct investment storage quantity scale biggest country in the Asian and Pacific area.

Looked from the international industrial shift trend of development, the international industrial shift has had the major impact to Chinese industrial growing, the technology advancement as well as the structural adjustment (Dai, 2007). At the same time, the Multinational corporation industry shift promoted Chinese manufacturing industry productivity enhancement, has led Chinese industries upgrading, sped up Chinese technology advancement step, in the certain extent raised Chinese manufacturing industry overall technical level, and promoted Chinese export competitive power promotion and the export structure promotion optimization from this. For example, the electronic information industry has realized the leap frog development depending on the foreign capital, in the information industries increase in value and the export, the Foreign-funded enterprise occupies 2/3 and above separately 4/5, Chinese electronic information industry sales revenue amounted to 1,880 thousand billion Yuan in 2003, surpassed Japan to leap to the world second. Moreover, Multinational corporation large-scale industry entered, promoted the Chinese profession competition main body multiplication, intensified the enterprise competition, has had the positive competition demonstration effect to the domestic enterprise, some Domestic enterprises have had the study opportunity from Multinational corporation's industry and the technology transfer process, strengthened the participation international competition ability (Lai, 2007). While the international industrial shift gives Chinese economic development to bring the numerous opportunities, also has brought many challenges, mainly displays in the following four aspects:

2.1 The International Industrial Shift Creates the Region Flows to not Balanced

The international industrial shift concentrates in the eastern area of China, eastern area new batch of enterprise quantity, the contract foreign capital, actual accounts for the total quantity using the foreign capital the proportion to surpass 80%, has played the major role to the eastern area economic development; But is small to the mid-west investment's absolute quantity and the proportion, its function extremely is also limited, like this intensified the inter-zone development not to be balanced.

2.2 China Accept the International Industrial Shift is that Labor-Intensive Form, the Technique Content Low Manufacturing Industry Primarily

The industrial shift does not mean that the technical synchronization shift; China accept the international industrial shift and introduction international vanguard technology not synchronization (Sun, 2006). The developed country considered stemming from the national interest that takes the restrictive measure to the high-tech exports, the US to China high-tech exports limit policy is one of American toward China policy important components. In the actual industrial shift, the developed country emigration's industry is not own country most advanced industry, but is the industry of technology enters the standardized stage in the product life cycle. While large-scale development international industry shift, Developed country enterprise control patent and technical know-how in industrial shift strictly, shifts the newest technology and the core technologies not easily, and prevents that the technical proliferation and overflow as far as possible, controls firmly the core technologies in own hand. Although many Chinese manufacturing industry products occupies the market share is very big in the international market, but because the key technologies and the market were still controlled by the Multinational corporation, Chinese enterprise can only earn the small part processing assembly profit, the profit is meager.

Because Chinese Enterprise oneself lacks the core technologies, lacks the proprietary intellectual property rights, Chinese manufacturing industry superiority lies in the market to be broad and the factor cost is low, moreover this kind of low cost, the low price superiority mainly exist in the product of the low technical level industry and in the processing assembly link (Lu, 2004). Looking from the international industrial division of labor pattern, China accepted the manufacturing industry major part is in the industrial chain's low end, the technical level is not high. Under this kind of vertical division of labor pattern, developed country offset industry take labor concentrated industry and low added value industrial link primarily, After developed country enterprise industrial offset, they stresses in makes "the brand", and more disposes the resources in the product research and development, the market development and the brand mold.

2.3 Chinese Financial System Whole Opening Degree is Not High, Accepts the Industry Shift Which by Ways and So On Securities Investment, Stockholder's Rights Transaction With Difficulty

The securities investment is the international capital movement mainstream, the Multinational corporation to speed up the industry to shift and to enter the local market step, favors more and more draws support from the money market, carries on the transnational annexation and the purchase to the domestic enterprise through ways (Qu, 2005) and so on securities investment and stockholder's rights transaction, with the aim of realizing the industry to shift and to form the productivity as soon as possible, simultaneously is also advantageous when the need withdraws the investment. But because Chinese financial market opening degree is not high, still had the very strict management and the control to the circulation of capital, did not favor China expansion foreign capital intake scale.

2.4 Foreign Capital Monopolizing Merger and Acquisition and the Multinational Corporation Restrictive Competition Behavior in China Have Harmed the Market Effective Competition, Does Not Favor Chinese Enterprise Competitive Power Enhancement

In the last few years, the foreign capital direct merger and acquisition Chinese enterprise case increases gradually particularly. According to the UNCTAD issued that 2005 world investment report, China became biggest goal country which the transnational merger and acquisition in the developing country in 2004, the whole year transnational merger and acquisition resultant is 6.8 billion US dollars. After multinational corporation merger and acquisition Chinese enterprise, often have two procedures: First, terminates by the merger and acquisition enterprise's operation, achieves to eliminate competitor, gain domestic market share goal. Second, takes the enterprise through the merger and acquisition to transform into its subordinate processing industry, takes one link on a Multinational corporation whole world production chain, is already not had the intellectual property rights by the merger and acquisition enterprise, also does not have the core technologies (Shang, 2006). The Multinational corporation abused the market superiority status behavior and the monopolizing merger and acquisition weakened the market competition degree, has harmed the enterprise and consumer's benefit, did not favor to create fair market competition environment and promotion enterprise competitive power.

3. CHINA ACCEPTS THE INTERNATIONAL INDUSTRIAL SHIFT PROSPECT AND SUPERIORITY

3.1 China accepts the international industry shift prospect

At present the international industrial shift also has the very big space, especially service industry and the high technology and new technology industry shift is on the rise, is containing the huge opportunity (Ji, 2007). It is estimated that to 2010, the global outsourcing market will amount to 20 thousand billion US dollars, was 4 times to 2004, and was current Multinational corporation output value 5 times. China has advantageous obviously on the market, the cost, the prospects for development, the necessary ability and so on, and will maintain the quite long time, this is advantageous in China expands and attracts international industry shift. China will become "the Industrial workshop of the world" in the future (Hu, 2006). China accepts the software outsourcing service to start to be late, but the speed-up is very high, yearly average growths surpassed 70% from 2000 to 2003,2003 years to amount to 2 billion US dollars (Diao & Zhang, 2004). The American Gartner Company predicted that China will become third biggest international project outsourcing base besides India, and the Philippines after 2010, it is recognized as the emerging outsourcing center.

3.2 China Accepts the International Industrial Shift Superiority

China has the obvious space and the superiority in the international industry shift aspect, the main performance is:

3.2.1.

The time in the global economic growth low speed and paced back, Chinese GDP still maintained the yearly average 7.7% swift growths, this has created the good economic environment for the industrial shift (Rong, 2006). Meanwhile, Chinese government formulated many preferential policies which have encouraged the foreign investment, these policies may attract more the foreign direct investment, and has provided the broad space to international industry shift.

3.2.2.

Chinese economy was still in the industrialization intermediate stage at present, introduces the high technology and new technology industry and the partial blank industry urgently needed, by enriches and fills and promotes Chinese industry ability.

3.2.3.

Chinese market potential is huge, the labor force cost is low, has the enormous attraction to the international property and the industry.

3.2.4.

Chinese eastern area particularly Zhujiang Delta and the Changjiang Delta, has formed enterprise group and the industrial group with the product necessary degree very high, has laid the foundation for the international industry shift (Dai, 2006).

4. CHINA SHOULD ADOPT COUNTERMEASURE DEALS WITH THE INTERNATIONAL INDUSTRY TO SHIFT NEW TENDENCY

4.1 Development the Transnational Operations, Realizes Bidirectional Shifts

A perfect opening economy, impossible is the pure introduction, but should establish the bidirectional circulation mechanism; both must introduce, and must walk out. From looked the industrial gradient shift theory, a country, if wants to realize the industrial upgrading, must during admission international industry shift, to shift partial industries to the foreign, realize the industrial promotion pattern inevitably (Hu, 2004). Facing Chinese industrial structure unbalanced, the secondary production capacity of production excess, situations and so on high technology and new technology industry rising momentum insufficiency, shift excess production ability, development transnational operations are one brilliant plan. Looking from the situation of China participates in the international industrial shift, the international industrial shift is also a bidirectional choice to China, two aspects from the industrial introduction and the industry offsets to have the influence to Chinese industrial development, China may also play the role from these two aspects (Tian, 2005).

4.2 Unify Accepting the Industrial Shift with China Region Overall Plan Development

China should formulate the pointed measure according to the different area level of development to accept different type industry, and also considered to cultivate the leading industry colony and atonement industrial structure gap (Gong, 2003). Therefore, needs to continue promote industry level of the coastal economy developed area, mainly through the technology acquisition promotion, the special area for processing export products introduction promotion and the industrial colony introduces and so on promotion; the not developed mid-west area mainly through raises the project service level to Accept the international industrial shift.

4.3 The New Investment Project Must Be Advanced Technological Project and Strengthens Digestion to Absorb to the Introduces Technical

The developing country use foreign capital goal is mainly atonement two gaps in the atonement economic development process, namely capital gap and technological gap (Song & Niu, 2005). According to Chinese actual situation at present stage, capital gap already was not prominent contradictory. Domestic accumulation of capital ability and the foreign exchange reserve are the majority developing countries hope to attain with difficulty, the contradictory principal aspect is the high technology and new technology industry lacks the core technologies, therefore China uses the foreign capital to change with emphasis improves the quality and the level. How does obtain the vanguard technology, enhancement ability of digestion absorbance to the introduce technical and independent innovation ability is the most prominent duty (Rong, 2006). Regarding technology-intensive industry and the significant technical equip which relates Chinese industrial upgrading, and has the broad market prospect, should insist that the Chinese side controls stock, cooperative production, transfer technology primarily, the spare part which the Domestic enterprise cannot make temporarily may buy from outside, but must strengthen with digestion absorption to the introduction vanguard technology, but should not depend to purchase the long-term from the overseas.

4.4 Speeds Up the Acceptance Service Industry Shift Dynamics to Enhance Chinese Modern Service Industry Level

Service industry will be international industry shift key field in current and the future, involve the profession class will be numerous; to will lead the economic development and the employee enlargement has the widespread influence (Song & Niu, 2005). Chinese service industry development long-term lag, the marketability, the industrial production, the socialized degree are not high, accounts for the proportion to be somewhat low in the economy, technical knowledge content high modern service industry develops the insufficiency especially, is restricting Chinese economy coordinated development and the overall competitive power enhancement seriously. Along with China join the World Trade Organization transition-period soon conclusion, China service industry opening to the outside world degree will enhance further, especially service domains and so on bank, insurance, negotiable securities, telecommunication, law, consultation, traveling will start and continue to expand to the foreign capital opening level, but these domains were also precisely the present transnational investment and the service industry international shift key field. Therefore holds the opportunity and initiative contract service industry international shift, it is advantageous in speeds up Chinese service industry particularly knowledge service industry development, enhances the service industry quality and level comprehensively.

REFERENCES

Bi, J-Y. (2006). The economic globalization time international industry shifts new tendency with our country face the opportunity and challenges. *China Price*, 9-18.

Council Development Research Center Industry Economical Research Department. (2006, November 9). Continues the international industrial shift: Our country must deal with six aspects the challenge. *China Economical Times Daily*, 11-09.

Dai, H. (2006). *The international industrial shift with the Chinese manufacturing industry develops*. Beijing, China: People's Publishing Agency.

Dai, H. (2007). International industrial shift new tendency and to our country's enlightenment. *International trade, 2*, 13-14.

Diao, D., & Zhang, L. (2004). Implementation the science and technology trade strategy and development high-tech product to export. *Commercial Research*, *3*, 12–13.

Gong, Y. (2003). The special area of export products is for processing export-orient economy development another position. *Modern management science, 6*, 21-22.

Guo, K. (2000). The foreign direct investment to Chinese industrial structure influence studies. *Management World*, *2*, 37–38.

Hu, J. (2007). International industrial shift basic rule and tendency. *International Trade Question research, 8*, 46-47.

Hu, S. (2006). The international industrial shift optimizes with our country industrial structure. *Exploration*, *1*, 27–28.

Hu, X. (2004). The Thought of China Continues international industrial shift to promote regional economies development. *Economical latitude and longitude, 5*, 33-34.

Ji, J. (2007). Multinational Corporation direct investment new characteristic and countermeasure. *The Economic Study*, *2*, 9–11.

Lai, Y. (2007). Our country continues the international industrial shift basic trend of development. *Foreign economy and trade practice, 2*, 9-10.

Lu, B. (2004). Brief analysis our country continues new round international industrial shift strategy. *Yunnan Finance and Trade Institute Journal*, *3*, 38–40.

National Development and Planning Commission Macro Economic Research Institute Foreign Economics Department. (2007). The international industrial shift and the Chinese countermeasure. *China Economical Times Daily*, 06-08.

Qu, R. (2005). China becomes the strong trade nation strategic way. *Economic Theory and Economy Management*, *9*, 28–19.

Rong, Y. (2006). Our country's countermeasure in new round international industrial shift. *Shanxi finance and economics college journal, 1*, 17-18.

Shang, Y. (2006). International industrial shift to our country industrial development influence and countermeasure. *Economic Performance*, *10*, 15–16.

Song, H. (2004). International industrial shift new tendency. *Economics Daily*, 06-16.

Song, Y., & Niu, C. (2005). International industrial shift tendency and our country's essential countermeasure. *Industrial Technology Economy*, *8*, 27–27.

Sun, Y. (2006). The international industrial shift's new tendency and our country's strategy chooses. *Contemporary management of economy, 5*, 35-36.

Tian, Z. (2005). New tendency of International industrial adjustment and shift - international outsourcing. *Commercial Research*, *17*, 17–19.

Wang, R., & Chen, W. (2006). The international industrial shift's new tendency with deals with the mentality. *Scientific and Technical Information Development and Economic development, 3*, 21-22.

Wang, Z. (2003). *The Multinational Corporation develops the new tendency in China*. Beijing, China: New China Press.

Yang, F. (2007). Foreign direct investment with our country advance in technology. *Fujian Forum*, *5*, 24–25.

Yang, Z. (2005). International industrial shift with China's countermeasure. *Asking Journal*, *3*, 15–18.

This work was previously published in International Journal of Asian Business and Information Management, Volume 1, Issue 1, edited by Dasho Karma Ura and Patricia Ordóñez de Pablos, pp. 1-22, copyright 2010 by IGI Publishing (an imprint of IGI Global).

Chapter 15
Blue Ocean Strategy:
A Vehicle for Entrepreneurship Development and Economic Growth in Rural Areas

Vinay Sharma
Indian Institute of Technology, India

Piyush Seth
Sahara Arts and Management Academy, India

Shujauddin Niyazi
Sahara Arts and Management Academy, India

ABSTRACT

This paper emphasizes on the role of entrepreneurship in economic development of a society with the perspective of Blue Ocean strategy used by organizations across the globe and especially in the context of Indian rural markets. The paper also utilizes live examples of organizations going towards uncontested rural markets and is suggestive of the crux and the core definition of entrepreneurship in correlation to the core philosophy of Blue Ocean strategy. In addition, the paper emphasizes that the soul of entrepreneurship lies within the development of rural markets, which can be done through the effective utilization of Blue Ocean strategy.

INTRODUCTION

This paper is proposing the role of entrepreneurship in economic development of the society with the perspective of blue ocean strategy, with special emphasis on the usage of the said strategy at the levels of the Rural Markets (Figure 1).

DOI: 10.4018/978-1-4666-0276-2.ch015

Business strategies and marketing skills have been concentrating on fundamentals, especially on buying capacity. This paper argues on four major aspects:

a) We should be looking at the 'Size' of the market and not at the buying capacity as a fundamental aspect in the case of highly

Figure 1. Entrepreneurship, rural markets, the blue ocean strategy and the economic development

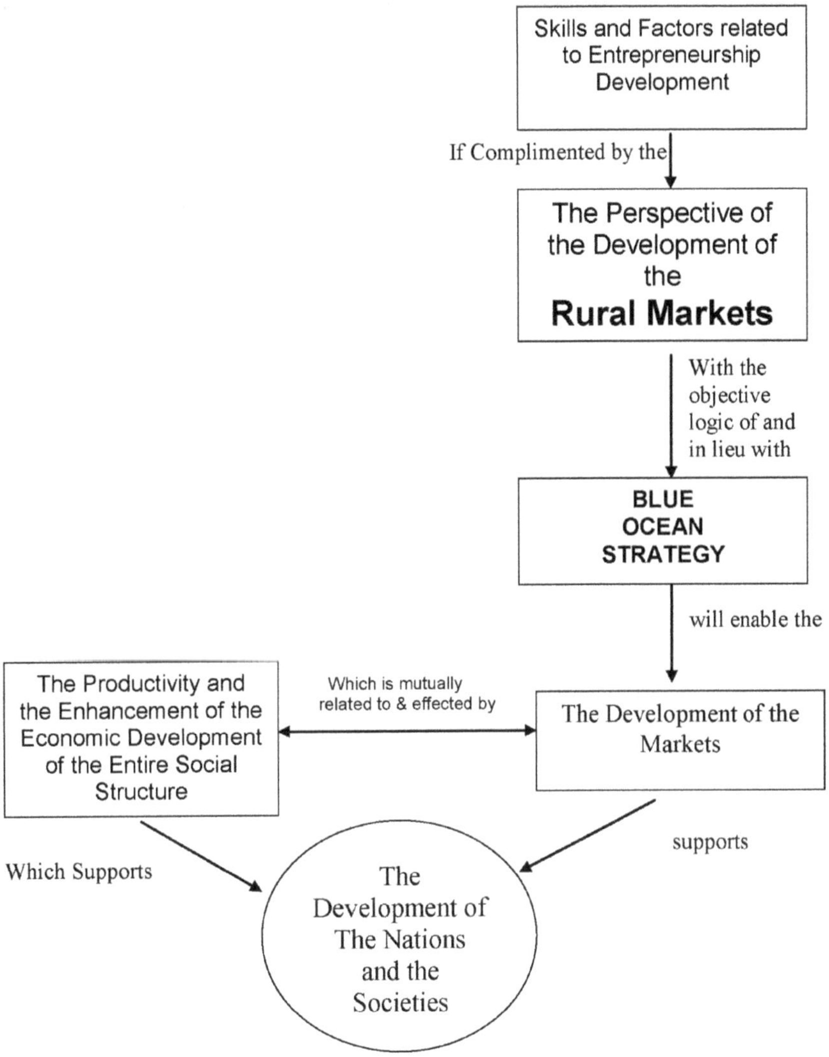

populated developing and the underdeveloped nations.

b) We should think of improving market conditions and should try to bring in 'Freedom of Market Transaction' (Sen, 2000) through providing better services to the rural and the poor masses.

c) Better services and the market conditions are the human right of the masses and should also be seen as a social responsibility by the marketer along with being seen as a business opportunity.

d) This paper also wishes to suggest that looking at the size of the market as an asset, we can complement the same with strategic generation of buying power i.e. by 'creating buying power'.Aptly suggested by Prahalad (2002):

"Global firms have demonstrated their ability to create wealth around the world. But the benefits of the capabilities of these firms and of the global market system do not yet reach most of the 4

billion people who live in relative poverty at the bottom of the economic pyramid. What if it were possible to expand the global market system to include those who now have no stake in it-to grow the market at the bottom, providing direct benefits and expanded opportunity to poor communities? (Prahalad & Hammond, 2002)."

ENTREPRENEURSHIP A PERSPECTIVE

The understanding of entrepreneurship owes much to the work of economist Joseph Schumpeter and the Austrian School of economics. According to Schumpeter (1950), an entrepreneur is a person who is willing and able to convert a new idea or invention into a successful innovation. Entrepreneurship forces "creative destruction" across markets and industries, simultaneously creating new products and business models. In this way, creative destruction is largely responsible for the dynamism of industries and long-run economic growth. The traditional microeconomic theory of economics has had little room for entrepreneurs in its theoretical frameworks (instead assuming that resources would find each other through a price system) (*The Economist*, 2006, p. 67; see http://en.wikipedia.org/wiki/Entrepreneurship).

THE PERCEIVED CORRELATION OF CO-CREATION OF VALUE AND ENTREPRENEURSHIP

Many efforts have been made in relation to the generation and the development of the buying power and the financial independence of the rural and the poor masses, especially women. The unprecedented reference of Amul known to be as white revolution of India (Dugdh Kranti as they say, http://www.amul.com), Sewa, The self employed women's association (http://www.sewa.org), Lijjat papad (http://www.lijjat.com/new.htm), and others suggest "that Co-Creation of Value and Entrepreneurship Practices are intensely interrelated".

CAPABILITY DEVELOPMENT AND ENTREPRENEURSHIP PRACTICES

The development of the 'capabilities' (Sen, 2000) on the part of the rural population if looked upon with the perspective of Entrepreneurship Practices, will bring to fore every sociological aspect which has been studied and stated with the motive of Social as well as the Market

Development. If an organization lets the development of the avenues for the development of the 'capabilities' (Sen, 2000) of its customers it implies that they want its customers to experience 'freedom' (Sen, 2000) and that is what is suggested by our Age Old Scriptures and this what is embedded and imbibed in the cultural ethos of the Indian Society.

BLUE OCEAN STRATEGY AND ENTREPRENEURSHIP PRACTICES

Kim and Mauborgne (2005) in their path breaking expression called Blue Ocean Strategy state that "Our research confirms that there are no permanently excellent companies, just as there are no permanently excellent industries. As we have found on our own tumbling road, we all, like corporations do smart things and less than smart things. *To improve the quality of our success we need to study what we did that made a positive difference and understand how to replicate it systematically*". Blue Ocean Strategy is suggestive of going apart from conventions or breaking away from conventions. It gives a new dimension to INNOVATION and simultaneously Entrepreneurship, it puts together the aspect of VALUE with innovation and terms it as VALUE-INNOVATION and suggests that value innovation occurs only when companies align innovation with utility, price and cost positions (Kim & Renee, 2005), which all together are the aspects of Entrepreneurial Innovation.

EXAMPLES FROM DAILY LIFE AS WELL AS FROM THE MNCS, USING BLUE OCEAN STRATEGY WITH THE PERSPECTIVE OF ENTREPRENEURSHIP PRACTICES

Grameen Bank

Grameen Bank (GB) has reversed conventional banking practice by removing the need for collateral and created a banking system based on mutual trust, accountability, participation and creativity.

GB provides credit to the poorest of the poor in rural Bangladesh, without any collateral. At GB, credit is a cost effective weapon to fight poverty and it serves as a catalyst in the overall development of socio-economic conditions of the poor who have been kept outside the banking orbit on the ground that they are poor and hence not bankable. Professor Muhammad Yunus, the founder of "Grameen Bank" reasoned that, if financial resources can be made available to the poor people on terms and conditions that are appropriate and reasonable, "these millions of small people with their millions of small pursuits can add up to create the biggest development wonder."

As of May, 2007, it has 7.21 million borrowers, 97 percent of whom are women. With 2431 branches, GB provides services in 78,659 villages, covering more than 94 percent of the total villages in Bangladesh.

The success of this approach shows that a number of objections to lending to the poor can be overcome if careful supervision and management are provided. For example, it had earlier been thought that the poor would not be able to find remunerative occupations. In fact, Grameen borrowers have successfully done so. It was thought that the poor would not be able to repay; in fact, repayment rates reached 97 percent. It was thought that poor rural women in particular were not bankable; in fact, they accounted for 94 percent of borrowers in early 1992. It was also thought that the poor cannot save; in fact, group savings have proven as successful as group lending. It was thought that rural power structures would make sure that such a bank failed; but the Grameen Bank has been able to expand rapidly. Indeed, from fewer than 15,000 borrowers in 1980, the membership had grown to nearly 100,000 by mid-1984. By the end of 1998, the number of branches in operation was 1128, with 2.34 million members (2.24 million of them women) in 38,957 villages. There are 66,581 centres of groups, of which 33,126 are women.

Group savings have reached 7,853 million taka (approximately USD 162 million), out of which 7300 million taka (approximately USD 152 million) are saved by women.

It is estimated that the average household income of Grameen Bank members is about 50 percent higher than the target group in the control village, and 25 percent higher than the target group non-members in Grameen Bank villages. The landless have benefited most, followed by marginal landowners. This has resulted in a sharp reduction in the number of Grameen Bank members living below the poverty line, 20 percent compared to 56 percent for comparable non-Grameen Bank members. There has also been a shift from agricultural wage labour (considered to be socially inferior) to self-employment in petty trading. Such a shift in occupational patterns has an indirect positive effect on the employment and wages of other agricultural waged labourers. What started as an innovative local initiative, "a small bubble of hope", has thus grown to the point where it has made an impact on poverty alleviation at the national level (http://www.grameen-nfo.org/bank/index.html).

A BUSINESS SCHOOL FOR RURAL WOMEN

Satara: Thirty-four-year old Lakshmi Kikade could never have dreamed of going to a Business Management School to learn the nuances of how to run a successful enterprise - for she had never received formal education. But today she is a budding entrepreneur in her own right with a business management diploma, thanks to the grooming she received from **Mann Deshi Udyogini - a Business School for Rural Woman**. "I make and sell ladies bags in markets like Mumbai," she says proudly. What makes Mann Deshi Business School unique is that it is probably among those institutions that do not ask for any educational degree. All one needs to get admission is a burning entrepreneurial spirit coupled with an unquenchable thirst to make it big in life.

This business school was started by a Non-Governmental Organisation called the Mann Deshi Mahila Sahakari Bank in collaboration with the Indian arm of HSBC, a leading global bank. According to Manndeshi trustees, it was the first rural bank to receive a cooperative license from the Reserve Bank of India (RBI). It provides micro-credit to poor women in the Satara district of Maharashtra. "Such training centres help to equip women in rural areas with requisite tools and nurse them to become efficient and successful entrepreneurs in their chosen fields," Naina Lal Kidwai, country head for HSBC India, said.

This B-school has just three rooms where classes take place on shared basis and its fee structure starts from Rs 150 for the entire course that may go up to Rs 600. The school offers a wide variety of courses on how to run enterprises in areas such as purse and bag making, photography, screen-printing and mobile telephony kiosks, apart from teaching women how to manage books of accounts and finance, the school's volunteers said.

When it started in December 2006, almost 150 women enrolled themselves in various courses and in the next five years it plans to admit over 350 students per session, which lasts between three-six months. "This is a path breaking institution as it recognises that women need relevant and appropriate training to enable them become good entrepreneurs, managers of their businesses and financial affairs," said Malini Thadani, head of public affairs and corporate responsibility with HSBC India.

The B-school also offers counseling sessions every Friday, free of cost, for women who are not able to decide which course to pursue and which would be the potential area best suited for their aptitude. Among its other services, it also offers a fully equipped gym where the women can attend exercise classes - also free of cost. Already, the school has opened four branches in Maharashtra and plans to spread out to Karnataka by tying up with the local non-government organisations there. "This school has now become the lifeline for the poor rural women here. It has not only given them a better life but also a chance to stand up and take charge of their lives," said Chetna Gala Sinha, founder of the institution.

"Initially when we started, we got a lukewarm response, but today it is teeming with students. We are now focusing on young girls, who, often due to family pressure, are married off without consent," Sinha, a Yale alumna, said. "Most of them want to work, and we want to give them the right platform to face the world," said Sinha, adding that she was planning to organise a student exchange programme with the Universities of Yale and Michigan. As a part of this programme, students from these Ivy League American colleges would hold classes for women and also to have a better understanding of rural India (http://www.indiaedunews.net/Management/A_business_school,_for_rural_women_987/).

YESHASVINI HEALTH INSURANCE SCHEME

The scheme originated in the mind of Dr. Devi Shetty, a cardiac surgeon who has pioneered the spread of telemedicine as well as low cost cardiac operations in India. Dr. Shetty has been acutely concerned with problems of access to sophisticated health care of the rural population. He attempted to solve this problem through the tele-medicine, using local providers and doctors in large urban hospitals connected via the internet. While this is still in the experimental stages in several Indian hospitals, what was clear is that the poor were still unable to afford medical care. Hence he turned to insurance.

A pilot study commissioned by his heart Hospital in Bangalore regarding the potential of introducing health insurance revealed some interesting results. First, and as expected, it revealed critical health issues. It was for instance common for aged men to suffer from kidney failure for want of a simple prostrate operation, or from premature blindness that is easily rectifiable by a cataract extraction procedure. Even more important, many middle aged women suffered from excessive bleeding because of a diseased uterus that could be removed by an inexpensive operation. Many children were dying of appendicitis, another easily curable condition. The study also raised several other issues that has been also uncovered in prior research, i.e., the fact that poor households spend a larger percentage of their budget on healthcare, and the fact that the major chunk of the healthcare budget of rural poor go to earning/male member of the family, followed by children and only then women. The study further revealed that the average poor farmer who perhaps owns two acres of land and maybe a cow would exhaust all of his savings just to get a serious problem diagnosed i.e. including running all diagnostic tests and traveling to a small town. By the time he/she has to go to a larger city hospital for the procedure he / she would be already in debt. This informal survey thus made it clear that it was the capacity of poor people in rural areas to pay for simple operations that was the problem that needed attention. The implications were clear.... any self-financed health insurance scheme for rural poor would have to be based on low premiums but at the same time provide benefits for surgical procedures and hospitalization expenses, as well as cover all of the costs associated with that procedure. In Dr. Shetty's mind, the only solution to this problem was to create a really large health insurance scheme, where the law of large numbers would overcome the basic financing problem associated with the small schemes of the past. This was the basis for the design of the scheme. Interestingly this pilot study also revealed that there was no (relative) lack of healthcare infrastructure in terms of hospitals and dispensaries. The average occupancy rates in Karnataka's hospitals, was, on average only 35%. The utilization of operation theatres was even lower. At this point, we do not have comparative data for other Indian states, so we do not know whether Karnataka is an exception or the rule. Thus, supply of doctors and hospitals was not the major issue (http://www.isec.ac.in/Karnataka_Kuruvilla16.5.05_aligned.pdf).

SKS MICROFINANCE

India has nearly 400 million poor, more than any other country in the world. These poor live on the brink of subsistence. SKS Microfinance was founded in 1998 to combat that poverty. It is guided by the principle that the poor need opportunity, not charity. SKS seeks to eradicate poverty by providing basic financial services (loans, savings, and insurance products) to the poor through a process of peer lending. In addition, SKS provides these services through a business model that enables SKS to be financially sustainable. This allows SKS to continue to expand its outreach to poor families throughout India. SKS launched its microfinance initiative in one of the poorest parts of India—the drought-prone Deccan region of southern India. Since its inception, SKS has made great strides in reaching out to the poor and achieving one of the fastest growth rates of any microfinance institution in the world. To manage this rapid growth, in

2003, SKS converted from a non-profit society to a for-profit company and received its Non-Banking Finance Company (NBFC) license in 2004. This legal structure enables SKS to access sufficient commercial funding, to expand its microfinance operations and to achieve its vision of reaching out to 300,000 poor women by 2008 (http://www.sksindia.com/SKS_Brochure_2005.pdf).

PROJECT SHAKTI

For Rojamma, Project Shakti means being able to educate her daughters. Rojamma is a single parent living in Kurumurthy, a small rural village 150 kilometers south west of Hyderabad in the Indian state of Andhra Pradesh. From a very poor background, she was married at seventeen to a man with whom she had two daughters but who then left her to fend for herself. At first she earned a few rupees working in her mother's field but she found it difficult to live on. A few years ago she joined a women's self-help group that was formed in the village to help women like Rojamma. "It felt good to be part of a group but that's not the same as eating food", she remembers. But then a man from Hindustan Lever came to Kurumurthy and told the women about *Project Shakti*. "From that moment my life changed". Hindustan Lever is Unilever's business in India. Its brands include such well-known names as *Lifebouy*, *Lux*, *Surf*, *Vim*, *Pond's* and *Lipton*, as well as local brands, such as *Kissan*, *Annapurna*, *Lakme*, *Fair & Lovely* and *Wheel*. The company generates around half its business from India's towns and cities and half from rural areas, where its products are sold in some 100,000 villages with populations of 2,000 or more. By the end of the 1990s, however, the company realised that to increase its market share it had to expand the market. The challenge was how to reach the 500,000 villages with smaller populations in more remote parts of the country, where there are millions of potential consumers but no retail distribution network, no advertising coverage and poor roads and transport. Hindustan Lever's solution, called *Project Shakti* (which means 'strength' in Sanskrit), was both bold and innovative. The company decided to tap into the growing number of women's self-help groups that had been springing up around the country. These groups, about one million of which now exist across India, are usually formed to help women save money and borrow from each other to avoid the excessive demands of unscrupulous moneylenders. Hindustan Lever made presentations at rural self-help group meetings, initially in Andhra Pradesh, and invited women, including Rojamma, to become direct-to-consumer sales distributors. The company provides self-help group women with training in selling, commercial knowledge and book-keeping, teaching them to become fully-fledged micro-entrepreneurs.

The women who are trained can then choose to set up their own business or to become *Project Shakti* distributors – or *Shakti Ammas* ('mothers') as they have become known. Each woman who becomes a distributor invests 10,000 –15,000 rupees (US$220-330) in stock at the outset – usually borrowing from self-help groups or micro-finance banks facilitated by Hindustan Lever. Each aims to have around 500 customers, mainly drawn from her village's self-help groups and from nearby smaller villages. Most generate sales of 10,000-12,000 rupees a month, netting a monthly profit of 700-1,000 rupees (US$15-22). For those with husbands who work in the fields, this typically doubles the household income. For single mothers like Rojamma, it is a far cry from the handful of rupees she earned working in her mother's field.

Project Shakti has proved to be a great success for Hindustan Lever and for women in India. The project started in a few pilot villages in Andhra Pradesh in 2000. In 2002 it expanded to two states and by the end of 2004 had grown to over 13,000 *Shakti* women entrepreneurs covering 50,000 villages in 12 states, selling to 70 million consumers. This represents a 30% increase in rural population reached. Hindustan Lever has had strong support from over 300 partners, including NGOs, banks and both state and local government departments, who recognise the potential for economic growth by encouraging women to become entrepreneurs. Andhra Pradesh typically had a 3% success rate in creating entrepreneurs among women's self-help groups prior to *Project Shakti*. This initiative has a 90% success rate so, not surprisingly, Andhra Pradesh's Women's Empowerment Commissioner, Ms Ramalakshmi, requests monthly updates on *Project Shakti's* progress. In 2003 Hindustan Lever started to pilot an information technology initiative called *i-Shakti*. This is designed to meet rural villagers' information needs and provide organizations with communications access to those parts of the country not reached by TV, radio and newspapers. This involves creating village 'kiosks' containing internet-linked computers run by entrepreneurs.

i-Shakti was formally launched in partnership with the Government of Andhra Pradesh in November 2004, and aims to have 3,500 *i-Shakti* kiosks on stream by the end of 2005. Mostly housed in the homes of *Shakti* entrepreneurs, *i-Shakti* kiosks provide villagers with free information on a wide range of topics, including health and hygiene, agriculture and horticulture, child and adult education, finance, employment, and entertainment.

Content is in the local language and has been specially developed by institutions and NGOs with experts in these fields, including the Azim Premji Foundation for children's education, the Tata Consultancy Services' Adult Literacy Programme and ICRISAT (International Crops Research Institute for the Semi-Arid Tropics) for information on agriculture. *i-Shakti* also includes an interactive service in which villagers can email questions to a panel of experts and receive a response within 24 hours. Sharat Dhall, Hindustan Lever's Business Head for the *Shakti* initiative, is excited about *i-Shakti's* capabilities: "Farmers can find a quick solution to pest problems with their crops, villagers can email their symptoms to a doctor and get a diagnosis in hours rather than days, and computer programs with voiceovers will teach people who are illiterate". Hindustan Lever and its partners are funding the initiative in the initial stages, but in future it is planned to charge content providers and brand advertisers to make the initiative self-financing and to generate incomes for *i-Shakti* entrepreneurs. While Hindustan Lever is intent on building its rural sales and market share, it is equally committed to improving the lives and livelihoods of people in India. *Shakti Vani* ('Voice') takes *Project Shakti* a stage further. Hindustan Lever is now training rural women to give talks to villagers about basic health practices, such as good hygiene, disease prevention and pre- and post-natal care. *"Project*

Shakti's role in creating incomes for underprivileged rural women and helping to empower them economically is more important than sales alone", says Sharat Dhall. "*Creating incomes is more important than sales alone*". He believes *Project Shakti*,*i-Shakti* and *Shakti Vani* will have the potential to act as a catalyst for creating new markets and generating rural micro-economies. He recognises this is something that Hindustan Lever cannot achieve alone and is actively building links with non-competitive partners, such as ICICI Bank, which specialises in micro-loans. At the same time, competitors and companies from other sectors are watching *Project Shakti* very closely and are expected to develop similar distribution models. *Project Shakti's* goal is to recruit 100,000 *Shakti* entrepreneurs covering 400,000 villages and 400 million consumers by 2008. "I believe it can become the biggest rural operation in the history of Indian business and change the way companies look at reaching consumers living in the smallest of villages", says Sharat Dhall. For the thousands of women like Rojamma who have become *Shakti* entrepreneurs, this initiative has already changed their lives in ways that are much more profound than the income they earn selling soaps and shampoos. It has brought them self-esteem, a sense of empowerment and a place in society. As Rojamma says: "When my husband left me I had nothing except my daughters. Today everyone knows me. I am someone now". It has also meant she has been able to send her daughters to school, giving them the chance in life she didn't have, although

Rojamma's aspirations for them remain modest: "I hope they have happy marriages and they too become *Shakti Ammas*"(http://www.unilever.com/Images/es_Project_Shakti_tcm13-13297.pdf).

CONCLUSION

The core suggestion that this paper extends is that, basically Entrepreneurship is at the core of almost every Social Development effort as first it leads to the Economic Development which subsequently elevates and develops the societies. This fact though has to be and has been fathomed from the literature time and again but, also has been strengthened through the scientific researches and propositions made through path breaking studies like Amartya Sen's Capability Approach, C. K. Prahalad's Fortune at the Bottom of the Pyramid, Kim and Renee's Blue Ocean Strategy, Stuart Hart's Capitalism at the Cross Roads and many other similar works of extreme credibility. Practically, these value, innovation and Entrepreneurship based aspects are demonstrated by many organizations like Grameen Bank, SKS Microfinance, Mann Deshi Udyogini, HLL's Shakti, Devi Shetty's Narayan Hrudayalaya through approaches like Yeshaswini and several others whose efforts and approaches are not so well known. The paper is finally suggestive of and expects that such approaches are not only everlasting but also the only possible solution for the development of the Societies.

REFERENCES

Giri, A. K. (1998). *Values Ethics and Business Challenges for Education and Management*. India: Rawat Publications.

Huntington, S. (1996). *The Clash of Civilizations and the Remaking of World Order*. New York: Simon & Schuster.

Kim, W. C. & Mauborgne. (2005). *Blue Ocean Strategy: How to Create Uncontested Market space and Make Competition Irrelevant*. Boston, MA: Harvard Business School Press.

Michale, P. T. (1993). *Economic Development in 3rd world*. Hyderabad, India: Orient Longman Limited.

Prahalad, C. K., & Hammond, A. (2002, September). What Works: Serving the Poor Profitably. A private sector strategy for global digital opportunity. *Harvard business review*.

Ragner, N. (1973). *Problem of Capital Formation in Underdeveloped countries*. Delhi: Oxford Universities Press.

Sen, A. (2000). *Development as freedom*. Oxford, UK: Oxford University Press.

Sharma, V. (2002, February). Visionary Enterprise. In *Proceedings of the First International Conference on Vedic Values and Corporate Excellence,* Shantikunj, Haridwar, Uttarakhand (pp. 22-24).

This work was previously published in International Journal of Asian Business and Information Management, Volume 1, Issue 1, edited by Dasho Karma Ura and Patricia Ordóñez de Pablos, pp. 32-41, copyright 2010 by IGI Publishing (an imprint of IGI Global).

Chapter 16
Study on Environmental Tax: A Case of China

Huifeng Li
Beijing Union University, China

Xuanwei Zhang
Beijing Union University, China

ABSTRACT

Since the industrial age, along with the promotion of economic development, problems of ecological destruction and environmental pollution are becoming more serious. To solve this problem, the implementation of environmental tax is an effective measure. Based on the analysis of the theoretical foundation of environmental taxes, international practice and experience in environmental tax system, in this paper, the authors describe the status quo of China's environmental tax system, and frames the further step strategy in building China's environmental tax system.

1. INTRODUCTION

With the constant expansion both in area and in depth of human economic activities, human living environment continues to deteriorate with resources depleted and ecosystems destructed. Environmental problems facing the world today, which is restricting economic development and threatening the survival of humanity, have become the globally common issue. Environmental pollution is the typical phenomena of externality for the market mechanisms. In order to overcome the shortcomings of market mechanisms, countries are actively and cooperatively exploring effective approaches for a living and sustainable development. Theoretically, governmental policies and measures for environment protection are of two types, regulative measures and economic measures. Regulative measures refer to the ban, the regulations and other measures such as discharge standards and so on, especially the provisions of emission standards and technical standards, to prohibit or limit the emissions of certain pollutants, or to limit certain pollution activities to a certain extent, so that the ultimate control of pollution and the efficient use of resources

DOI: 10.4018/978-1-4666-0276-2.ch016

can be realized. Economic measures refer to policies that can influence the polluters' costs and benefits of alternative activities to guide them to carry out environmentally-friendly options, such economic policy measures are mainly environmental taxes.

In recent years, with the environmental problems intensified, the traditional "command - control" types of control measures (command and control instruments) cannot be persisted due to its potentially high cost and low efficiency and the increasing market-oriented and decentralized behavior by various kinds of economic agents and so its drawbacks are increasingly exposed. In this context, environmental taxes, as a market-based regulatory tool, by exerting a commitment of environmental responsibility to the polluters to internalize their externality, limiting their damages to the environment and thus correcting the effect of market failure, are paid more and more concerns by more and more countries. An increasing number of countries regulate the economy running through the establishment of an environmental tax system. In March 29, 1993, the publishing by OECD of a report entitled "Tax and the Environment", which comprehensively introduced and summarized the status and experience in the implementation of environmental taxes by its member states, marks the international progress on this matter. As environmental problems are plaguing the whole world, China is no exception. In consideration of China's weak legislation on environment tax and the prevailing wave of low-carbon economy, the study of the status of China's environmental tax system, aiming at the building of the strategy for the protection of China's fragile environment and ecology, and the promoting of China's socio-economic sustainability, is a must and of practical significance.

2. ENVIRONMENTAL TAX AND ITS THEORETICAL BASIS

Environmental taxes are the kind levied by the government, in order to protect the environment and resources, to all its units and individuals involved in environmental resources development and utilization in accordance with their extent of development and utilization of natural resources or the extent of their damage to the environment.

Environmental taxes can be broadly divided into three categories: First, direct taxation on the amount of pollutants discharged, such as emission taxes. Emission tax is levied on the emission of pollutants to the environment at its final stage. As a method of taxation at the source of pollutants emission, it has the advantages of low cost in taxation, targeted pollution control, etc., but the drawback is that it is focused only on certain pollutants and thus not conducive to addressing the related environmental problems; Second, indirect taxation on goods and services, i.e. by changing the rate of sales tax, VAT and other indirect taxes to prevent environmental pollution, which through increasing taxation on goods or services that are environmentally harmful or decreasing taxation on goods or services that are environmentally friendly to help promote environmentally friendly goods or services to achieve environmental objectives. At the same time, such taxes can also be a specific source of income for environment protection; Third, environmental tax exemptions. It is an exemption by the Government to certain acts conducive to environmental protection or to environmental facilities, which remission can be viewed as a kind of governmental subsidy. Western countries mainly use the former two kinds of environmental taxes. The main function of environmental taxes is to regulate the economic behavior of taxpayers in order to reduce pollution, promote the rational allocation of resources and the sustainable economic development, rather than to congregate national wealth.

The source of ideas for environmental tax is "Pigovian Tax". Modern British economist Arthur C. Pigou, through the development of the theory on externality, first proposed the idea that government tax revenue could be used to regulate polluting acts. Pigou, as early as 1920 in his "Welfare Economics", advocated that enterprise which produces external effects should be levied a "Pigou tax" to regulate their polluting behavior. According to Pigou, if polluters can be imposed a certain tax of emissions, which is equal to the difference between the marginal social cost and the marginal private cost, then the external social cost of the polluters can be internalized which can make polluters face the marginal social cost of their private activities. Li et al. (2007) indicated that stimulated by the Pigou tax, polluters can control their amount of pollution in two ways: First, to adopt pollution control technology to reduce emissions from their activities; Second, to reduce the total amount of pollution. Polluters will make their own best choices on the basis of their cost and profit function while also optimal to society. Pigou's theory of externality is the original theoretical source of environmental taxes.

Public goods theory is the theoretical foundation of environmental taxes. Comparing to private goods, public goods, according to Samuelson's definition, have two essential characteristics: non-exclusive to consumers and non-competitive of consumption. Because of their nature of indivisibility, the atmosphere, water and other environmental resources on which the property rights are difficult to define or the definition leads to high cost, possess the characteristics of public goods, namely, they are indivisible of utility, non-competitive of consumption and non-exclusive from benefit.

Modern economics theorized that free market transactions can achieve efficient production of private goods, but environment, national defense and other public goods should be delivered by the government. The establishment of an environmental tax system was originally designed for the market in which public goods cannot be produced. In the field of environmental protection, the "collective rationality" and "public choice" is necessary, however, the individual rationality does not naturally produce collective rationality and individual choice is usually opposite to the best interests of the society. Therefore, environmental protection, ecological compensation, etc. the kind of collective rationality, good for all of us, cannot be self-realized. In the field of environmental protection, only emphasizing the role of markets is not enough. The role of a pure market is difficult to ensure the effective protection of ecological environment, thus the Government is a must. By the imposition of an environmental tax to all people using the public goods, market failures can be corrected. By the use of tax revenue in pollution control and environmental protection, the harmonization between economy and environment and the sustainability of development can be achieved.

3. STATUS OF RESEARCH ON ENVIRONMENTAL TAX

Environmental tax, as an effective measure to solve the problem of environmental pollution, has attracted widespread concern form scholars both home and abroad. The related researches are enormous, especially since the 80s of last century, which constantly enriched the integrity of environmental taxes and charges system. The current research results on the issue of environmental taxes and charges can be summarized as the following three aspects:

First, the Design of Environmental Taxes

Joint research projects such as "Discussion on a number of issues of environmental taxes" by Natural Science Foundation of China and "An exploration on the design of environmental tax system of China" by Beijing Natural Science Foundation have carried out rather comprehensive analysis on the topics of environmental tax system design and proposed the scope of the corresponding tax, types of taxes, etc.. These researches are of practical significance.

Second, Studies on the Acceptability of Environmental Taxes

In order to ensure the public acceptance to the environmental tax policy issued by government, cautions should be particularly taken in the selection of tax base and tax rates. As for the determination on standards of environmental taxes and charges, Pigou derived a general equilibrium model to demonstrate, which showed that pollution taxes levied on a unit should be equal to the marginal social damage caused by it. The establishment of the theory has laid a corner stone for the reduction of tax distortions and the effective allocation of resources. However, in practice, due to the difficulties in the estimation of marginal social costs and benefits, time price, etc., Pigouvian tax rate is hard to quantify. In the 1990s, Bovenberg and Goulder (1996) also used the general equilibrium model to analyze the optimal environmental tax rate when tax distortions exist in the economic system. They pointed out that the optimal emission rate tended to be lower than the Pigouvian tax rate, and the greater the distortions caused by other taxes, the lower the optimal tax rate. Guo et al. (2008) indicated that: "This standard has broadened constraints on the determination to tax rate and is also more practical and closer to the foundations on which the economic system functions. Comparatively, this theory is more instructive to practice."

Third, Studies on the Impact of the Distribution of Environmental Tax

Environmental tax will have a greater impact on distribution, and will increase the possibility of an unfair tax burden. Such effects include the impact of environmental tax on the distribution among groups of different income, among different types of families, between firms and families as well as among different types of enterprises or industries and so on. Some economists, especially those from EU and the United States, not only recognized the role of environmental taxes and charges on environment protection but proposed that environmental tax revenue, by way of reducing tax distortion, can benefit people in non-environmental aspects. Tullock (1967), Pearce (1991) believed that the use of income derived from environmental taxes can reduce existing distorted taxes, therefore reducing the additional tax burden, and that is the so-called weak form win-win effect. In recent years, relevant scholars broadened the implications of win-win effect by the result of the raise in the level of social welfare, the increase in employment opportunities. Ballared and Medema (1993) also applied simulative mathematical models in analyzing the impact on welfare of pollution reduction approaches by taxes and by subsidies, and obtained the conclusion taxes are more effective than subsidies.

Chinese scholars have done a great deal of research on the environment taxes. Zhang (2003), Wang (2004) qualitatively analyzed the effects of environmental taxes and summarized that environmental taxes have such functions as increase revenue, change the pattern of human behavior, equalize social wealth and achieve self-regulation, promote technological innovations, and so on. China's domestic research related with environmental taxes can be summarized as the following aspects:

First, researches on the basic theory of environmental taxes focus on such topics as the relation between environmental tax and externality, the definition of the concept of environmental taxes, environmental taxes and China's sustainable development theory, etc.

Second, researches focused more on the introduction of foreign environmental taxes, particularly those of developed countries. However, rare if few research have a systematic research, sorting and summarization on the experience of successful implementation of national environmental taxes by typical countries, which leads to a lack of systematic experience that we can learn from.

Third, researches focused on the concrete policies on implementation of environmental tax in China, which mainly evaluated China's current tax system from the perspective of environment protection and tax reform and submitted policy recommendations on the establishment of the types of environmental taxes, on determination of tax base and on the choice of tax rate, etc., but there is a lack of in-depth analysis on main problems existed in China's current taxation system and the environment-related taxes and the direction of reform, etc., therefore, to sort out and to sum up the experience of international practice on environmental taxes, to carry out in-depth analysis on China's existing main problems related with environmental taxes and charges, and to improve the existing tax system to make it more compatible with sustainable development are the main tasks in building of China's environmental tax system.

4. INTERNATIONAL PRACTICE AND EXPERIENCE ON ENVIRONMENTAL TAXES

With the rapid development of global industrialization, the ecological environment is severely damaged. According to Deng (2007), the development of environmental tax system can be summarized into three phases: the first stage is the 70s to 80s of 20th Century, during which to impose a tax on the "explicit contamination" by waste water and solid wastes is the main practice. This kind of environmental tax is, in fact, fees for the compensation of the costs, i.e., in order to increase investment in pollution control, directly charge a certain fee to the polluters, mainly including user fees, charges for a particular purpose, etc. This is the embryonic form of environmental taxes. For example, the Netherlands since 1970 started levying taxes on living water sewage and waste disposal. The second stage is the 80s to 90s of 20th Century during which more and more countries in Europe and the United States began to use eco-taxes, green tax and other specific tax to maintain the ecological environment. These taxes are characterized by incentives to stimulate polluting producers to change their behavior to the environment. The third stage is the beginning of the 90s of 20th Century, during which the tax policies on environmental protection in developed countries got new development, moving from taxation on waste gas, waste water, garbage, noise, pollution, agriculture, etc., into a comprehensive "green" tax, from the scattered and individual environmental taxes to the formation of a green tax system. Apart from the introduction new green taxes, adjustments on the original taxes to make them eco-friendly are also in place, aiming at the fulfillment of a comprehensive "green" environmental taxation system.

The popular environmental taxes in current Western countries are:

Air Pollution Taxes

Air pollution taxes, including sulfur dioxide tax and carbon dioxide tax, is levied on the amount of polluting sulfur dioxide and carbon dioxide emitted into the air. In 1972, the US took the lead in the introduction of sulfur dioxide tax. The act provides that: in areas where sulfur dioxide concentration meeting primary and secondary standards, for each pound of sulfur emissions 15 cents and 10 cents were levied respectively, to induce producers to install pollution control facilities as well as to switch to low sulfur fuel. Sweden also began the introduction of a sulfur tax in January 1991, which was levied basing on sulfur content in oil and coal. Currently other countries that levy this tax are: the Netherlands, Norway, Japan, Germany, etc.

Carbon dioxide is the major cause for global warming. Sweden, in 1991, imposed a carbon tax which is levied on oil, coal, natural gas, liquefied petroleum, gas and domestic aviation fuels. The tax base is the average carbon content and calorific value of a variety of fuels. The rate is SEK 0.25

per kilogram of carbon dioxide emission. Different fuels, due to their difference in amount of emission, have different tax rates. For example, in 1992, the rate for per ton of coal and carbon was SEK 620. Its most prominent role is to curb the use of fuel producing carbon dioxide.

Lithuanian Government also promulgated, in 1991, "pollution tax", and levied tax on the emission of air pollutants according to certain standards.

Water Pollution Tax

Water pollution tax is levied on water pollution behavior. This tax system is most complete in Germany. Germany, since 1981, introduced pollution taxes to waste water and implemented a nationwide unified tax rate on "pollution units" of waste water. All tax income, as local revenue, is used to improve regional water quality, and good social benefits thus appeared. However, water pollution tax in the Netherlands is the kind levied by the Government on units and individuals who discharge waste, pollutants and toxic substances to the surface water and purification plants, either directly or indirectly. Tax rate is based on the oxygen consumption of emitted substances and heavy metal content. The rates are different in different water conservation areas. In Germany and the Netherlands the taxation is in accordance with a comprehensive water quality and quantity estimates, whereas in Singapore, New York and Moscow, taxation is calculated according to the quantity of sewage.

Solid Waste Tax

Standards for this tax are of two kinds: one is based on the actual volume of the solid waste collection, such as Canada, Sweden and the Netherlands; another is based on the quota for volume and type of the solid waste collection, such as Finland and France. Taxable objects are beverage packaging, paper and paper products, tires and so on. Tax on beverage containers is to encourage the production of light container, promote the re-use of containers. The tax on used tires is to encourage renovation of old tires and save resources and reduce pollution.

Noise Tax

The tax levied on special source of noise louder than a certain decibel. The main purpose is to raise funds for the municipal government to install noise-proof facilities in the vicinity of the airport or to accommodate the relocation of residents. The United States, Germany, Japan and the Netherlands have this tax. In Japan and the Netherlands, this tax is levied on the number of landing airplanes. In the United States, noise control tax is one U.S. dollar for each passenger and for per ton of cargo in Los Angeles airport.

Garbage Tax

Garbage tax, originated in the Netherlands, is a tax levied on the family and collected by the government for financing the collection and disposal of refuse. Each family is a collection unit. Families with fewer members can get some relief. The Netherlands has also introduced the Government's garbage collection tax (municipal governments can choose between both kinds of taxes), which is collected according to the amount of waste generated by each family. The amount can be counted by using a small trash bin. Each family is taxed according to the number of bins filled and the volume of the bin. Apart from the Netherlands, Finland and the United Kingdom also introduced the garbage tax in 1996.

Registration Tax

European Union countries popularly levy a registration tax. Registration tax is an effective measure to control environmental pollution at the source. For example, Sweden imposed a registration tax on new pesticides containing active substances by 1400-5600ECU and still on the registered products by 700ECU annually.

Main experiences learnt from the western countries in the practice of environmental taxes are as follows: types of tax showed the trend of diversity. For instance, in the United States where governments at the federal and the state level have introduced varying degrees of environmental taxes, environmental taxes are of many species, including energy, daily consumer goods and consumer behavior, etc.; Execute tax incentives and exemptions to environmentally-friendly behavior. For example, in the US, preferential policies on environmental tax are represented mainly by the direct tax relief, investment tax credits, accelerated depreciation, etc.; Tax burdens gradually shifted from the taxation on income to that on behavior harmful to the environment as represented by Denmark, Sweden and other Nordic countries, where environmental tax, by making adjustment on the overall structure of tax revenue, is gradually transferred from the focus on income to the focus on acts harmful to the environment, namely, redistribute tax revenue among labor, natural resources and pollution by gradually shifting from taxation on wage income to that on consumption and production having side effects on the environment.

According to Wang (2004), after years of practice and exploration, the implementation of environmental taxes in western countries has produced significant effects on environment improvement. Since its introduction in European market in 1985, unleaded gasoline has shown the trend of a rapid replacement to leaded gasoline. In the United States, since the implementation of environmental taxes in the 70s of 20th Century, carbon dioxide emissions have reduced by 20%, carbon monoxide emissions by 39%, sulfur dioxide emissions by 42%, suspended particles reduced by 70%. In Sweden, only one year after the implementation of environmental taxes, sulfur dioxide emissions solely reduced by a sharp 26%, not to say other emissions. In Finland where there is a good environment tax policy and good implementation, carbon dioxide emissions reduced from an annual 60,000 ton in the early 80s of 20th Century to today's annual tens of thousands. Today, its pulp and paper production is more than doubled compared to that in the early 70s of 20th century, but the discharge of pollutants has decreased by 95%.

5. ENVIRONMENT-RELATED TAXES IN CHINA AND THEIR PROBLEMS

5.1 Environment-Related Taxes in China's Current Tax System

China's legislation on environment tax is very weak. Very few tax measures are taken in environmental protection. However, since the tax reform in 1994, many of the taxes have had an account for environment protection. Through in-depth research on China's current taxation system, Li (2007) summarized the taxes having direct or indirect links to environmental protection mainly into the following categories:

The Resource Tax

The existing resource tax started in action on January 1, 1994. The main purpose of this tax is, by curbing the indiscriminative abuse, digging and plundering exploitation of mineral resources, to protect mineral resources and to promote its rational development and effective use.

Consumption Tax

It is a tax newly set in the turnover tax in the 1994 tax reform. The state introduced high energy-consumption, non-renewable and environmentally-unfriendly consumer goods such as firecrackers, fireworks, gasoline, diesel oil and other commodities into the scope of consumption tax, aimed at limiting production and consumption harmful to the environment.

The Urban Land Use Tax

A tax levied, since November 1 of 1988, on units and individuals using land resources within China by the actual land area occupied. The main purpose is to enhance land management and promote rational use of urban land.

Cultivated Land Occupation Tax

Beginning in 1987, this tax is levied on units and individuals according to the actual area of their occupation of arable land in China for use of building houses, engaging in other non-agricultural construction, etc. The purpose is to limit and regulate the behavior of taxpayers in use of farmland, to strengthen land management.

Urban Development and Maintenance Tax

The tax, as a "special tax." levied since 1985, is becoming the financing source for environmental protection,

Vehicle Use Tax

Vehicle use has large energy and environment significance. Taxation on the use of vehicles has a certain role in curbing energy consumption and reducing pollution.

VAT

For the ordinary taxpayer continuously producing bricks and lime using his/her own mined sand, soil, stone or other minerals, etc., or for products of building materials made of raw materials mixed with coal gangue, stone-coal, fly ash, bottom slag, etc., value-added tax is collected in a simple way. For the general taxpayer running business of waste materials, value-added tax is collected with a 70% of refund.

In addition, the business tax, income tax, property tax and other taxes, also provide policy incentives for environmental protection and comprehensive utilization of resources provides.

5.2 Problems of the Environment-Related Taxes in China

Though the tax system has played a certain role in environmental protection, facing with the deteriorating environment, the role of these measures is very limited.

Main problems in China's current taxation system are:

First, there is a lack of the type of tax specifically established on purpose of environment protection. Experience from countries around the world has shown that among the various measures for the protection of the environment, taxation has a decisive role and the environmental tax is one of the main taxes. At present, China has not a complete and effective environmental protection taxation system. Though there are many types of environment-related taxes within current taxation system, because their starting points for policy formulation is not merely based on the purpose of environmental protection, so their roles in environmental protection are very small, and thus limited the ability of taxation on pollution-control, weakened the role of taxes in environmental protection.

Second, in the existing taxes related with environmental protection, provisions for environment protection are not perfect and the scope of taxation is too narrow; the terms on environment are not detailed enough; the tax burden is unreasonable; the importance of environment protection is not highlighted; and so the protection on environment is far from enough. For example as for the resource tax, the tax rate is too low; the gap between two tax levels is too small; there is no obvious regulatory effect in reaching the reasonable utilization of natural resources. These taxes, with their narrow base, essentially belong only to the kind of mineral-resources-occupation tax. Many important natural resources such as forests, grasslands, oceans, famous and expensive herbs, fresh water resources and so on, are not included in the scope of taxation, which is not conducive to the comprehensive protection of natural resources. This is severely unsuitable to the situation of shortage of resources, low efficiency of utilization and serious waste in our country.

Third, tax preferences considering environmental factors are too simplistic, lack of relevance and flexibility, which affect the implementation effect of tax incentives. Such international practice as accelerated depreciation, tax rebate on reinvestment, tax defer, etc. can all be applied to environmental tax to increase the flexibility and effectiveness of tax policies, but, to our dismay, taxation in China almost has no such content.

Fourth, a number of tax incentives, in fact, generate the opposite consequences on environment and thus offset the environment functions of other taxes. For instance, provisions of VAT exempt for sales of gold and silver products, to a certain extent, supported the excessive exploitation of gold and silver resources, intensified resource destruction and environment pollution; provisions of value-added tax exempt on agricultural plastic films and pesticides, especially the highly toxic pesticides, adversely affected soil and water resources. Therefore, if environmental taxation system is to have a positive effect, the redevelopment of a coherent and unified environmental tax system is necessary.

The basic situation of China's environmental problems are: the degree of ecological damage is still increasing; environment pollution is still expending; land resources are undergoing serious damages; forest resources are facing with depletion, water resources are in crises; mineral resources are losing in large amount; air pollution is serious; water pollution is widespread; waste pollution is disastrous. In short, the problem of ecological destruction and environmental pollution has become the "bottleneck" seriously constrained the sustainable development of China's economy. Therefore, the introduction of environmental taxes to improve the existing tax system is in urgent need for the promotion of sustainable socio-economic development of China's.

6. STRATEGY FOR THE BUILDING OF CHINA'S ENVIRONMENTAL TAX SYSTEM

6.1 Turn Some of the Environmental Pollution and Sewage Charges into Environment Taxes

From 1979, when sewage charges were first introduced, to 2003, when the "Regulations on the Collection and Use of Sewage Charges" was issued, the introduction of sewage charges in China has undergone three phases, i.e., piloting, implementing and improving. It has become one of the most widely used economic means in environmental protection in China. Currently, China's sewage charges are of five categories as water, gas, solid waste, noise and radioactive waste, which can be cast into 113 items. Although the fees and taxes both could internalize the external costs by pollution, in view of practical operation, the existing sewage charges, in practice, has the problem of wide range and disorder, unreasonable charging criteria, disorder in types of fees, etc. In addition, from the point view of management, fees and taxes are essentially different. Tax is mandatory, while the fees are administrative in nature and optional. Therefore the role of pollution control for the existing charging system is greatly weakened and so it is necessary to change the sewage charges into pollution taxes.

With respect to China's actual situation and the experience in the introduction of environmental taxes in the Western developed countries, the introduction of China's pollution tax must be based on China's most important environmental issues and the real practice of sewage charges and to be implemented by different stages and batches.

First, the most serious environmental problems facing China is air and water pollution. Each year cost on human health damaged by air pollution and water pollution is huge in China. Second, solid waste pollution and municipal garbage pollution have become increasingly prominent. Third, practical experience has been accumulated in sewage charges on sewage, exhaust fumes (sulfur dioxide), solid waste and noise. Finally, in the western developed countries, when introducing environmental taxes, not all potentially polluting emissions are included in the scope of taxation, but the tax objects, resources advantages and industrial competitiveness are comprehensively accounted for. Those environmental taxes that affect the country's resources advantages and competitiveness should be deferred, relieved or exempted and those that have little effect on resources advantages and competitiveness should be collected earlier and larger in sum.

In response to the circumstances mentioned above, currently, environmental pollution taxes suitable for introduction in China are: water pollution tax, sulfur dioxide tax and garbage tax.

Water pollution tax should not only be collected on industrial wastewater, but also on domestic sewage and agricultural wastewater. Standards for water pollution tax rate should be determined basing on the marginal cost of governance and incentive mechanism should be used for sewage treatment and to encourage the whole society to save water. Sulfur dioxide tax rate should be determined according to its marginal control costs, i.e. equal to or slightly higher than the marginal control cost. Using incentive mechanisms to motivate and encourage the agents responsible for sulfur dioxide emissions to take active measures in the control and treatment of sulfur dioxide. Introducing garbage tax to make agents responsible for waste generation bear the financial responsibility for the dumping of waste and the tax rate should be the marginal treatment cost, so that externality of pollution caused by garbage producers internalized.

Further broaden the tax base after a period of time of collection. This way, on the one hand, can directly stimulate enterprises to improve pollution control, or to introduce advanced technology and equipment to reduce the emission of pollutants; and on the other hand, enable enterprises to choose their own pollution control methods suitable for their characteristics of production to achieve the optimum allocation of resources.

6.2 Adjust to the Existing Tax System

Adjusting the Existing Resource Tax

The main objective of the reform on current resource tax is to give its full play on the optimal allocation of resources to curb the over-exploited mining and waste of resources, which can be tackled from the following aspects:

First, expand the scope of resource tax collection. At present, in many countries, the scope of resource taxation covers mineral resources, land resources, water resources, animal and plant resources, forest resources, grassland resources, marine resources and geothermal resources. Whereas, China's resource tax basically only covers mineral resources. Therefore, the scope of resource tax collection should be timely expanded. Taking into account the current level of tax collection and tax policy, operational issues, etc., water resources and forest resources which are in serious shortage, damage and waste, should be first included in the scope of resource tax collection.

Second, increase collection standards of resource tax to correct the over low resource prices. Low in tax rate and small in gap between the tax levels are the main problems facing with current resource tax in China. An over low resource tax rate is insufficient to affect the economic behavior of taxpayers. Thus, while rationalizing the pricing system, appropriately increase the resource tax collection standard is necessary. Differential tax rates can be applied to resources with different levels of pollution, and heavy tax should be collected on non-renewable, non-alternative and scarce resources.

Third, improve the tax basis of resource tax. Changing resources taxation from sales-based or "amount of self use" based to output-based will enable resources mined but not sold or own resources also become taxable resources and make the company to bear the burden on their resource and products inventory and thus increase the cost of this part of the product. This can guide enterprises, for the purpose of profit-maximizing, to conserve national resources and avoid over-exploitation.

Certain Adjustment on Consumption Tax

Consumption tax is popular worldwide. Currently, in China, consumption tax is levied only on part of contaminable products, such as the firecrackers, fireworks, gasoline, diesel and motorcycle, cars and other consumer goods. The consumption tax rate on cars is different according their engine capacity. Whereas, coal, as the main body of energy consumption as well as the main source of air pollution, and some other easily polluting products such as batteries, throw-ways (cameras, razors, disposable supplies for hotels and restaurants, non-degradable plastic lunch boxes, plastic bags, etc.), Freon, phosphorus washing powder, nitrogen-containing paper products, CFC products, toxic pesticides and so on, are not put in the scope of taxation. So China's consumption taxation needs to be improved. For different products, according to their degree of friendliness to environment, design differential tax rates to encourage the use of cleaning products. From the perspective of environmental impact assessment, as long as the products have a direct or indirect damage on the environment, all should be levied a high consumption tax, so that the externality of consumption or production can be internalized and consumption tax can be a kind of highly eco-friendly tax through the adjustment to it.

Improve the Fuel Oil Tax

In China, reform on oil tax came into operation, since January 1, 2009. Abolish the original 6 charges collected outside the price of finished oil products, such as road maintenance fee, waterway maintenance fee, highway transportation management fee, road passenger and freight surcharges, water transportation management fee, waterway passenger and freight surcharges, etc. Gradually and orderly cancel the toll on second grade highways paid by the government on loans. At the same time, increase tax on price of gasoline by 0.8 Yuan per liter, i.e. from 0.2 Yuan to 1 Yuan per liter and increase diesel consumption tax by 0.7 Yuan per liter, i.e. from 0.1 Yuan to 0.8 Yuan per liter. Introduction of fuel tax, from the environmental point of view, on one hand, is beneficial to constraining the consumption of oil resources with economic measures and can also encourage people to use environmentally-friendly means of transport, or to use alternative sources of clean energy on the other hand, thereby reducing atmosphere pollution by fuel.

Through current implementation, fuel tax, in addition to its justification to the fees and charges collected by government, has a role in promoting energy-saving and emission reduction and also in enhancing awareness of fuel-saving. However, large amount of problems, related with the distribution of benefits, on policy-kit construction have not been properly resolved, such as distribution between transport sector and the taxation department, between central government and local government, etc. After the introduction of fuel tax, road maintenance fees and various traffic charges will be abolished. This involves the issue of interest-balance between the central government and local governments. In view of the reality that road maintenance belong, in the past, to local governments, but now fuel tax belong to the central government, unless the central government and local governments share the fuel tax, or else there will be more traffic-based charges in the local places, which will result in the formation of a co-existence of "fuel-Tax + transport fares" in China. This result is contrary to the original purpose of fuel tax policy which seeks to achieve a clean-up to taxes and charges and to promote energy-saving and emission reduction, therefore, reform on fuel tax needs to be further improved.

7. CONCLUSION

The improvement of environmental taxes will help the formation of the environment-friendly and resource-saving society. Environment tax policy, as an important economic and legal system, has accumulated rich experience through practice by western countries on environment-related taxation such as the trend of diversification in tax revenue categories, implementation of tax incentives and exemptions to the environmentally - friendly behavior, the shift of tax burden gradually from the taxation on income to the taxation on behavior harmful to the environment, etc., which is worthwhile for China to learn from and to emulate. China's current tax policies related to environmental protection can hardly protect the environment because there is a lack of the type of tax specifically designed for the purpose of environmental protection and the existing taxes related to environmental protection are not perfect, incentives considering environmental factors are too simplistic, a number of preferential tax policies generate opposite consequences and are not conducive to environmental protection and so on, so the strength is not enough to form a systematic protection of the environment. In addition to changing some of the sewage charges into pollution tax, taxing on waste water, sulfur dioxide and garbage and making it the main body of China's environmental tax system, adjusting the present resource tax and consumption tax, improving the fuel tax, borrowing the experience of national environmental tax from western developed countries to build China's environmental tax system and to promote sustainable economic development are the important topics in the construction of China's environmental tax system.

REFERENCES

Ballared, C. L., & Medema, S. G. (1993). The Marginal Efficiency Effects of Taxes and Subsidies in the Presence of Externalities. *Journal of Public Economics*, *52*(2), 199–216. doi:10.1016/0047-2727(93)90020-T

Bovenberg, A. L., & Goulder, L. H. (1996). Optimal environmental taxation in the presence of other taxations: general equilibrium analysis. *The American Economic Review*, *86*(4), 985–1000.

Deng, J. A. (2007). Comparative Study of environmental taxation and its impact on China. *Tax and Economics*, (3), 95-100.

Li, H., et al. (2007). Using of international practice on the environment tax system in the building of China's pollution tax system. *Circulation Economy*, (5), 34-36.

Pearce, D. W. (1991). The role of Carbon Taxes in Adjusting to Global Warming. *The Economic Journal*, *101*, 938–948. doi:10.2307/2233865

Tullk, G. (1967). Excess Benefit. *Water Resources Research*, (3): 643–644. doi:10.1029/WR003i002p00643

Wang, J. (2004). Environmental taxes in OECD countries and the Enlightenment to China. *Comprehensive Utilization of Resources in China*, (5), 33-36.

Wang, M. (2004). Discussion on Tax Reform and Environmental Protection. *Henan Social Science*, (3), 121-123.

Xia, G., et al. (2008). A review of researches on environmental taxes and charges. *Industrial Technology and Economy*, (5), 156-157.

Zhang, M., et al. (2003). Current analysis of the introduction of green taxes in China. *China Environmental Management*, (4), 4-5.

This work was previously published in International Journal of Asian Business and information Management, Volume 1, Issue 2, edited by Dasho Karma Ura and Patricia Ordóñez de Pablos, pp. 12-23, copyright 2010 by IGI Publishing (an imprint of IGI Global).

Chapter 17
Study on Chinese Low Carbon Economic Model

Huifeng Li
Beijing Union University, China

Xiaofang Wang
Beijing Union University, China

ABSTRACT

Global warming, which is caused by over consumption of fossil energy during economic development in human society, threatens global ecological balance, tampers social and economic development, imperils energy security, ecological safety, water and food safety, and endangers the entire human race. Low carbon economy is a new economic development model based on low energy consumption, low pollution, and low-emission-based. This paper analyzes the Status quo and Limits of the Chinese Low Carbon Economy, making deep research on Design and Implementation Strategy of Chinese Low Carbon Economic model.

INTRODUCTION

Global warming, which is caused by over consumption of fossil energy during economic development in human society, threatens global ecological balance, tampers social and economic development, imperils energy security, ecological safety, as well as water and food safety, and even endangers the entire human race. Hence, the international community pays great attention to global warming and starts to reflect on existing economic development mode. From the Kyoto Protocol to Bali Roadmap, many countries have been endeavoring to cope with global warming, attempting to reduce carbon dioxide emission incurred in economic growth, slow down climate change, and avoid irreversible influences on human race and natural ecological system. Therefore, the development mode based on low carbon economy is widely recognized in the world and becomes a goal of human development in the future.

Low carbon economy is a new economic development model based on low energy consumption,

DOI: 10.4018/978-1-4666-0276-2.ch017

low pollution and low-emission-based, is another significant progress after agricultural civilization and industrial civilization in human history. Low-carbon economy covers almost all industrial fields. It calls for reducing the influence brought by economic development to carbon recycling in the ecological system, that is, decreasing carbon flux in the ecological system caused by economic activities of human beings, improving self-adjustment of the ecological system through cutting down carbon dioxide emission, reducing carbon sources and increasing carbon sink, and maintaining carbon balance in the biosphere and even ecological balance on the earth. Low carbon economy, as a matter of fact, relates to energy exploitation efficiency and clean energy structure. Its core lies in technological innovation, institutional innovation and ultimate change of human outlook on survival and development. Low carbon energy is thc basic guarantee, clean production is the key point, recycling is the effective way, and sustainable development is the direction. On the one hand, China is short of energy, Chinese people have low per capita resource, and domestic environment is heavily polluted, so low carbon economy helps raise energy efficiency, improve energy structure and enhance energy security. On the other hand, low carbon economy is a necessary economic development mode in accordance with the China National Programs to Cope with Climate Change. China, as a big emission country of greenhouse gases, needs to lessen heavy realistic pressure and take international responsibility in ways of saving energy, reducing emission, transforming economic growth mode, creating an environment-friendly low carbon economic model, and gradually playing down dependence on high-carbon energies according to practical situations.

RESEARCH OVERVIEW

As the new economic development model, low-carbon economy attracts attention of domestic and foreign scholars. This article will present the main three research results:

RESEARCH ON CONNOTATIONS AND FEATURES

Low carbon economy was first found in Our Energy Future, Creating a Low Carbon Economy, British energy white book of 2003. The energy white book notes that through less natural consumption and less environmental pollution, carbon economy produces more economic outputs, creates new business opportunities and more jobs. The report In the Black: The Growth of the Low Carbon Economy issued by the Climate Group summarizes market development, analyzes benefits brought by low carbon mode, and makes the conclusion that the new mode turns out higher investment return by largely increasing output, shortening production period, enhancing production reliance, promoting product quality, improving working environment, and heating staff passion.

According to Zhuang (2005), one of the earliest Chinese scholars who study low carbon economy, the so-called low carbon economy is essentially to deal with energy efficiency and clean energy structure, focus on technological and institutional innovation, and aim to decelerate climate change and realize sustainable development. Bao (2007) points out that carbon emission amount has become a new index to evaluate economic growth mode, the new index will change modern industrial civilization based on fossil energy in all-round way, and the world will shift to ecological economy and civilization. Liu (2009) believes low carbon economy in modern times is a rational human choice to balance economic growth and welfare improvement, and economic development and environmental protection, and is a kind of reasonable cognition of harmonious relations between mankind and nature, mankind and society, and people and people.

RESEARCH ON NECESSITY AND POSSIBILITY

Johnston, Lowe, Bell (2005) and other scholars study technical feasibility of reducing carbon dioxide emission of housing in Britain, and their research result is that it's possible to reduce emission by 80% of the emission amount of 1990 at the end of 2050 at the premise of currently existing technologies. Treffers, Faaij, Sparkman, and Seebregts (2005) study whether it is possible for German to reduce GHG emission by 80% of the amount of 1990 at the end of 2050. Kawase, Matsuoka, and Fujino (2006) look back and describe long-term climate stability in the past, and divide emissions into three factors of carbon dioxide intensity, energy efficiency and economic activities. Finally, they draw the conclusion that both promotion rate of energy intensity and reduction rate of carbon dioxide intensity shall be two to three times the average rates over the past four decades to realize the objective of reducing 60-80% emission. Koji Shimada,Yoshitaka Tanaka, Kei Gomi, and Yuzuru Matsuoka (2007) create an approach to describe long-term low carbon economic development in urban areas and apply the method to Shiga, Japan. Chinese scholar Cui (2009) thinks China is in the midterm of industrialization, Chinese Environmental Kuznets Curve is going up, and environment pollution, resource consumption and carbon emission are overt. In the following 20-30 years, China's environment pressure will increase; the average per person carbon dioxide carbon withdrawal will be close to world average level. Jiang (2007), Fu (2008), and Xia (2008) elaborate the necessity of the development low-carbon economy separately.

RESEARCH ON CHINESE POLICIES ON LOW CARBON ECONOMY

Chinese scholars have carried positive exploration on how to develop low-carbon economy in China. Jin (2009) and other scholars put forward Chinese government shall formulate a national plan of low carbon economy and an implementation roadmap according to general national development plans and energy-saving and emission-reducing programs as soon as possible, establish a statistical and evaluation system for low carbon economy, incorporate such a plan into the Twelfth Five-Year Plan, and finally turn out a feasible long-term implementation plan of developing low carbon economy with the will of the state. Song (2009) and other scholars claim that effective policies will guarantee the implementation of low carbon economy, Chinese low carbon economy policies must follow up marketization progress, policy design shall be shifted from administration orientation to market orientation, and production and consumption of microeconomic parts shall be urged to turn from high carbon to low carbon.

There are two aspects characteristics on low carbon economy research:

Firstly, research on low-carbon economy connotation characteristics. Although they research perspective differently, they express the similar connotation: through the technological innovation, reduce the greenhouse gas emissions as far as possible, realizes the economical and society's clean development.

Secondly, research on the necessity of low carbon economy in China. But how? Therefore, it's urgent to clear up the Status quo and Limits of Chinese Low Carbon Economy, make deep research on Design and Implementation Strategy of Chinese Low Carbon Economic model.

STATUS QUO AND LIMITS OF CHINESE LOW CARBON ECONOMY

Current Situation

As the second largest energy producer and consumer, the second largest CO2 emitter, China highlights global climate change problem. Chinese government has pushed up low carbon economic development by bringing forward a significant strategy of building up a resource-conservation and environment-friendly society, strengthening measures to cope with climate change, and issuing a series of policies on saving energy and reducing emission. China has issued three important guidelines, including the National Program for Medium- to Long-term Scientific and Technological Development, the National Climate Change Assessment Report, and the Eleventh Five-Year Plan for Environmental Protection. These documents specify technologies concerning energies, water resource, and environmental protection as priorities, and emphasize R&D of energy-saving technologies, renewable resource technologies, clean and high-efficiency coal utilization technologies, and technologies of reducing GHG. Apart from guidelines, the State Council has also issued specific policies since 2003, including the Medium to Long-term Special Plan for Energy Saving, the Circular on Doing Well in Recent Key Jobs of Building a Conservation-minded Society, Opinions on Accelerating the Development of Recycling Economy, and the Decision on Reinforcing the Work of Energy Saving. In addition, Chinese government also set the China National Programs to Cope with Climate Change according to the regulations of the United Nations Framework Convention on Climate Change and the Kyoto Protocol, and officially promulgated it in 2003, indicating determination and courage of Chinese government to save energy, reduce emission, and develop low carbon economy. According to Guo (2009), in recent two years, a great deal of backward plants were washed out, including 21,570,000 kw of small-sized thermal power plants, 11,200 small-sized coal mines, 46,590,000 tons of iron production, 37,470,000 tons of steel production, and 87,000,000 tons of cement production. General investment in environment protection in China reaches RMB 550B, accounting for 1.24% of GDP during the same statistical period. Ten key energy-saving projects have been initiated, and desulphurization in thermal power plants has made remarkable results. The central government invested in 691 water pollution prevention and control projects in major basins. In 2007, energy consumption on each GDP unit decreased by 3.27% compared with 2006, and both COD and SO2 emission decreased by 3.14% and 4.66% respectively than that of 2006. This was the first time that both COD and SO2 emission reduced. But compared with other countries, Chinese carbon dioxide emission intensity per unit of GDP is 1.86 times that of India, 1.69 times of Japan, and 1.6 times of Western European countries. According to the prediction of the US Department of Energy, Chinese carbon dioxide emission gross amount has a tendency of rapid growth (average annual growth rate 2.6%) owing to domestic increasing economic scale and the energy structure with coal at the core, and will exceed the US in 2020, getting to 8,145 million tons and becoming the largest in the world. Accordingly, China is under the heavy pressure of reducing GHG emission, and it is urgent for China to develop low carbon economy.

LIMITS ON SHIFTING TO LOW CARBON ECONOMY

China is confronted with the following challenges under current circumstances:

Firstly, the energy structure that coal dominates inside is unreasonable. Within Chinese primary energy consumption structure, coal takes up over 2/3, and this structure is quietly different from the mainstream structure that oil and gas dominate in the world. In global primary energy consumption composition in 2007, coal takes up only 27.8%. Coal consumption proportion in developed countries is less than 20%, while it is as high as 69.5% in China. Chinese energy resource conditions and changes of supply and demand show Chinese energy structure won't change a lot by 2020, and coal will remain at a high rate of 60% or so in the structure. Carbon dioxide emission factor of oil is 80% of coal, and 60% of gas. If coal consumption is large, therefore, carbon dioxide emission intensity will be high, and economic development is characterized by high carbon. Obviously, China is facing serious challenges in deal with environment pollution and climate change in the following few years.

Secondly, energy consumption drastically grows. China is experiencing a rapid growth period of industrialization, urbanization and modernization. As social and economic development and constant increase of population, infrastructure construction in urban and rural areas and personal consumption structure upgrade demand much more and more energy. Energy consumption gross increased from the amount equivalent to 1,386 million tons of coal in 2000 to 2,656 million in 2007, with the average annual growth rate of 9.7%. According to the research by Jin in 2009, energy consumption will continue the growth trend in a short term, and Chinese energy demand will reach the amount equivalent to 5 billion tons of standard coal by 2020. High carbon emission limits Chinese sustainable development in a large extent.

Thirdly, the industrial structure is not proper at all. The ratio of Chinese primary, secondary and tertiary industries remain at the level of 1:5:4. The major part of Chinese economy is the secondary industry, which determines the main part of energy consumption is industry, particularly heavy chemical industry. The proportion of the tertiary industry and service industry that consume low energy is low. Statistic shows Chinese energy intensity increased from 4.21-ton standard coal/RMB 10,000 in 2001 to 4.77 in 2007. Most of the 31 provinces, autonomous regions and cities directly under the central government failed to achieve the goal of reducing per unit GDP energy consumption by 20% proposed in the Eleventh Five-Year Plan, and only a few regions fulfilled the objective of annual per unit GDP energy consumption.

Lastly, China stands at a low position in global industrial labor division system. In the current global industrial labor division system, Japan, the US and European countries have come into the phase of service economy and stand at the leading position in the system, but China is still in the low and is weak in technological element, added values and competitiveness. In the new international industrial restructuring, China takes a lot of labor and capital-intensive sectors of high consumption and serious pollution. When China becomes a world manufacturing base, the country suffers great losses in environment. According to the research done by Tyndall Centre for Climate Change sponsored by British government in 2007, Chinese carbon dioxide emission from net export products in 2004 was 1.1 billion tons, accounting for 23% of the total emission. IEA (International Energy Agency) estimated that the carbon dioxide emission in relation to energy during the production of domestic export commodities was 1.6 billion tons in 2004, accounting for 34% of the total emission.

DESIGN AND IMPLEMENTATION STRATEGY OF CHINESE LOW CARBON ECONOMIC MODEL

Design of Chinese Low Carbon Economic Model

There's no time to delay in transforming economic model of China, which emits a great deal of GHG. Due to Chinese low carbon economic development situation and restrictions, Chinese low carbon economic model (shown in Figure 1) is different from the free market economic model or the environment management model tightly controlled by the government. It's a development pattern with mutual participation, interaction and influences of three parties.

First, the government plays the instruction, the guidance and surveillance role in the low-carbon economic type. The Chinese government should formulate the Chinese low-carbon economic development strategy as soon as possible and carries on the legal supervision system. Then,

Figure 1. Chinese Low Carbon Economic Model

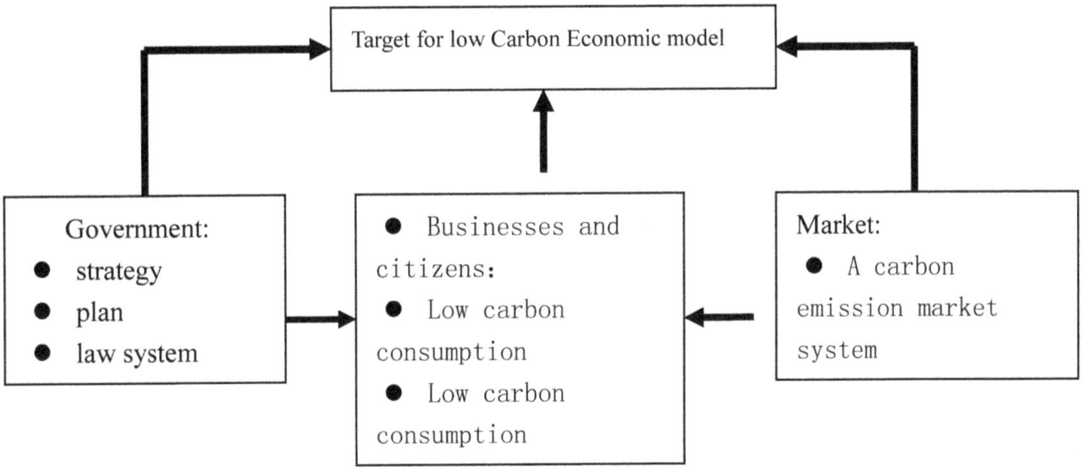

a carbon emission market system with Chinese characteristics shall be established, and the current market system shall be guided to a low carbon direction. At last, through the visible hand and the invisible hand, business production pattern and people's way of life shall also be led to the low carbon direction. Energy saving technologies shall be upgraded, and emission reduction capacity shall be further promoted to form a low carbon technology development atmosphere from the top down and actively develop low carbon products. Moreover, low carbon consumption guide, concept education and good habit publicizing shall also be carried through.

That model wipes off the deep impression that the government acts as the only responsible part to engage in economic model, stresses mutual effect brought by the government, markets, businesses and citizens, and expedites the transformation from high carbon to low carbon of microeconomic production and consumption.

Strategies to Realize Chinese Low Carbon Economic Model

Strategies to realize Chinese low carbon economic model consist of the following three aspects with reference to restrictions on developing low carbon economy and relations of the government, micro-economic entities and the market in the model:

Develop a National Strategic Framework and Establish a Legal System as Soon as Possible

Chinese long-term development strategy shall refer to, absorb and understand low carbon economy concepts. Based on existing national programs, operation plans, and guidelines and measures that respond to global climate change, the state shall rapidly deploy national strategies of low carbon economic development, include low carbon economy into the Twelfth Five-Year Plan, and make out a feasible long-term implementation plan of developing low carbon economy with the will of the state.

A legal system shall also be set up to guarantee low carbon economy and prepare a legal basis for impelling low carbon economy. Chinese relevant legislations consist of: a) the Law on Promoting Clean Production (put in force on January 1, 2003) and the Law on Promoting Recycling Economy (put in force on January 1, 2009), which is the basic legal guarantee of energy saving, emission reduction and low carbon economy; b) the Energy Law (under drafting), which is an important legal background to carry through China's energy development strategy; and c) attached documents to the Renewable Energy Law and the Energy Conservation Law are under drafting. Furthermore, China plans to revise laws in connection with environment and resource, including the Environmental Protection Law, the Law on Environmental Impact Assessment, the Law on Prevention and Control of Atmospheric Pollution, the Mineral Resources Law, the Law of Coal, and the Law of Electric Power. Generally speaking, China is creating a legal system frame that guarantees environmental protection and low carbon economy, and shall make progresses in a few fields. Legislation on environmental protection shall be in line with international practice, international cooperation on environmental protection shall be carried out, and foreign experience on environmental protection legislation shall be learned. Legislation shall be considered in an overall situation, local protectionism shall be fought against, and inconsistencies and even conflicts between local regulations on environmental protection shall be avoided. Legislation of procedural laws regarding environment shall be enhanced. Most environmental laws and regulations are substantive, and there are only a few procedural laws that are found in many places. This fails to realize environmental management according to law.

Optimize Energy Structure; Develop Alternative Energy and Renewable Energy

The forms of low-carbon economy is to optimize energy structure, improve energy efficiency, develop alternative energy and renewable energy, In the near future, China should pay more attention to energy conservation and clean use of coal. Energy saving technologies shall be upgraded, and emission reduction capacity shall be further promoted. In the medium term, increase the proportion of renewable energy; promote new energy technologies such as hydrogen fuel and carbon capture. In the long term, sustainable energy system should be established, and the main text is the renewable energy source, the cleaner coal, the advanced nuclear power and so on. Besides, speed up the development of alternative clean energy, especially enhances the renewable energy source, to "hydrogen supply with low-carbon energy resource". Particularly reduce the exploitation of underground carbon energy, make great efforts to develop solar, hydro, wind, nuclear, geothermal and other new energy sources.

Explore to Set Up Chinese Carbon Balance Trade Platform by Making Use of Market Mechanism

China shall be ready to learn from western countries, make out a reasonable carbon balance pattern, and pave the way for building up carbon fund system and ecological compensation system in China. China Environmental Culture Promotion Association and Chinese Association for Development Strategy have issued the project report of the Research on Chinese Carbon Balance Trade Framework on December 5, 2008, suggesting pushing carbon source-carbon sink trade system at the provincial level. The province-level carbon source-carbon sink balance account (specifying carbon sink trade rules and market tools of carbon budget, carbon credit and carbon fund) suggested in the project is preliminary for carbon source-carbon sink trade system. Carbon balance information, mainly carbon source amount and carbon sink amount, in 31 provinces, autonomous regions and cities directly under the central government shall be collected and analyzed. Ecologically benefited regions shall contribute economic surplus from external benefit to compensating ecological protection areas. In this way, carbon emission space is taken as a kind of rare resource, and carbon sink absorption capacity is deemed as a means of earning income. As different regions have diverse carbon source and sink amounts, effective swap forms and reasonable trade price turn ecological service from a free pattern to a paid one.

CONCLUSION

As a new economic development model, low-carbon economy is, and the only, way towards sustainable development. However, China is confronted with the following challenges under current circumstances: the energy structure that coal dominates inside is unreasonable, energy consumption drastically grows, the industrial structure is not proper at all; China stands at a low position in global industrial labor division system. There's no time to delay in transforming economic model of China. The Chinese Low Carbon Economic model stresses mutual effect brought by the government, markets, businesses and citizens, and expedites the transformation from high carbon to low carbon of microeconomic production and consumption. Strategies to Realize Chinese Low Carbon Economic Model are as follows: Develop a national strategic framework and establish a legal system as soon as possible, Optimize energy structure, develop alternative energy and renewable energy, Explore to set up Chinese Carbon Balance Trade Platform by Making Use of Market Mechanism.

ACKNOWLEDGMENT

This research is sponsored by Beijing Research Base under Grant BJXJD-KT2010-A04, and this support is gratefully acknowledged.

REFERENCES

Bao J. (2007). Reconstruction of the traditional concept of circular economy. *Research of Nature of Processes*, (4), 91-94

Cui, D. (2009). About Low-carbon Economy. *Environmental Education*, (7), 19-21.

Fu, Y. (2008). Low-carbon Economic Development Model, China's population. *Resources and Environment*, (3), 14-23.

Jiang, T. (2007). Promote Ecological Civilization Construction. *Environmental Protection*, (11), 46-48.

Jin, Q. (2009). China's Development Strategy for Low-carbon Economy. *China's National Conditions*, (10), 7-9.

Johnston, D., Lowe, R., & Bell, M. (2005). An Exploration of the Technical Feasibility of Achieving CO_2 Emission Reduction in Excess of 60%Within the UK Housing Stock by the Year 2005. *Energy Policy*, (33): 1643–1659. doi:10.1016/j.enpol.2004.02.003

Kawase, R., Matsuoka, Y., & Fujino, J. (2006). Decomposition Analysis of CO2 Emission in Long-term Climate Stabilization Scenarios. *Energy Policy*, *34*(15), 2113–2122. doi:10.1016/j.enpol.2005.02.005

Koji, S., Yoshitaka, T., Kei, G., & Yuzuru, M. (2007). Developing a long-term local society design methodology towards a low-carbon economy: An application to Shiga Prefecture in Japan. *Energy Policy*, (35): 4688–4703.

Lau, L. (2009). Low-carbon Economy and Development of Human Society. *Xinhua Digest*, (1), 26-27.

Treffers, T., Faaij, A. P. E., Sparkman, J., & Seebregts, A. (2005). Exploring the Possibilities for setting up Sustainable Energy Systems for the Long Term:Two Visions for the Dutch Energy System in 2050. *Energy Policy*, (33): 1723–1743. doi:10.1016/j.enpol.2004.02.011

Xia, L. (2008). Develop Low-carbon Economy And Achieve Sustainable Development of Urban. *Environmental Protection*, (2), 33-35.

Yin, G. (2009). Low-carbon Economic Development Experience and Implications for China. *Reform and Strategy*, (10), 176-179.

Zhuang, G. (2005). The Way of Chinese Low Carbon Economy. *International Technical and Economic Study*, (3), 79-87.

This work was previously published in International Journal of Asian Business and information Management, Volume 1, Issue 4, edited by Dasho Karma Ura and Patricia Ordóñez de Pablos, pp. 57-64, copyright 2010 by IGI Publishing (an imprint of IGI Global).

Section 4
Managerial Issues in China

Chapter 18
Mutual Guarantee Systems of Small and Medium-Sized Enterprises in China:
An Application of Self-Organization Theory

Yi Yang
Beijing Union University, China

Hong-yan Yan
Beijing Union University, China

Ze-yun Yang
Beijing Union University, China

ABSTRACT

Mutual guarantee systems (MGS) of small and medium enterprises (SMEs) are complex system mainly serving the SMEs, including the mutual guarantee institutions formed by SMEs, financial institutions, industry associations and credit re-guarantee institutions. From the perspective of system theory, this paper illuminates the self-organization mechanisms and the self-organization characteristics of openness, far-from-equilibrium, nonlinearity and fluctuation of mutual guarantee systems of SMEs. The implication of self-organization theory is that under some conditions, mutual guarantee systems have the internal tendencies of autopoietic, self-evolution and evolution from disorder to order and from low level to high level to solve SMEs' financing problems. Using self-organization theory, this paper proposes suggestions and measures to improve the operation efficiency of mutual guarantee systems in China.

DOI: 10.4018/978-1-4666-0276-2.ch018

1. INTRODUCTION

A large body of literature has shown that small firms experience difficulties in accessing the credit market due to informational asymmetries. Banks can overcome these asymmetries through relationship lending, or at least mitigate their effects by asking for collateral. Small firms, especially if they are young, have little collateral and short credit histories, and thus may find it difficult to raise funds from banks. Informational asymmetries between small firms and banks may be so pronounced that profitable investment opportunities are not financed (Petersen & Rajan, 1994; Berger, Udell, 1998; Coco, 2000; Berger & Udell, 2002; Berger & Udell, 2006). Small enterprises may mitigate this problem by posting collateral or building close relationships with lenders. Nevertheless, these solutions are of little help to firms which lack collateral or credit history.

In these cases, other contractual schemes may emerge to mitigate severe asymmetric information problems. In particular, borrowers may join together and share a joint responsibility for default. Mutual guarantee system (MGS) can be a viable solution to the problem of access to credit from banks for those small entrepreneurs who cannot offer sufficient collateral. In this case, instead of lending to each single borrower, banks lend to a group of borrowers linked by a joint responsibility for a single loan granted to all of them. empirical analysis shows that small firms affiliated with MGS pay less for credit compared with similar firms which are not MGI members (Francesco, Leonardo & Paolo, 2009).

MGS are quite widespread in Europe: on the basis of the latest available data provided by the European Mutual Guarantee Association, there are more than 1.4 million small and medium-sized enterprises (SMEs) affiliated with an MGI in the European Union. MGS are particularly widespread in Germany, France, Spain and Italy. Italian MGS represent the largest component of the European mutual guarantee sector, accounting for 37 per cent of the total outstanding volume of guarantees to SMEs.

Some researchers have taken a critical view on Chinese MGS. Mutual guarantee institutions have advantages of information gathering, monitoring and transaction costs, allowing the government to achieve optimal management of information asymmetry under the conditions of limited resources, comparing with other guarantee institutions (Ba, 2007).The friendly economical co-operations among SMEs provide convenient conditions for the co-operative credit guarantees that has become a well-applicable development pattern for the current financing system of SMEs, and it can be the main composing of diversified credit guarantee system of SMEs in the future in China (Peng, 2008).

In this paper we focus on Chinese MGS which can deal with banks on an equal level, helping SMEs gain access to credit that is normally unavailable. The characteristic of self-organizing system and present situation of mutual guarantee institutions in China are analyzed with the theory of self-organization.

2. THEORY OF SELF-ORGANIZATION

In general, the organization refers to the order within the system structure or the formation of this ordered structure. German theoretical physicist Hermann Haken (1983) found that the evolution of form from the organization point of view, it can be divided into two categories: he organized and self-organization. If a system formed by the external command organization, is his organization; if there is no external instruction, the system in accordance with the tacit agreement of some kind of rules, fulfill their responsibilities and who coordinated automatically to the formation of ordered structure, is self-organization.

One of the earliest references to the term 'self-organization' in management literature (Ashby, 1947) defines it as processes in which systems become more highly organized over time, and these are self-induced changes of organization, with no external order being imposed. Molleman (1998) who considers self-organization as the local autonomy to make decisions on both the transactions to be realized and the way transformation processes are organized to achieve these transactions.

Self-organization theory is the late 60s of the 20th century to establish and develop a system theory. Its main object of study is complex self-organizing systems (living systems, social systems) the formation and development mechanisms, which under certain conditions, how the system automatically from the disorderly to orderly. It has three main components: dissipative structure theory (Dissipative Structure), synergetic (Synergertios), catastrophe theory (Calastrophe Theory).

Dissipative structure theory led to pioneering research in self-organizing systems. A dissipative system is a thermodynamically open system which is operating out of, and often far from, thermodynamic equilibrium in an environment with which it exchanges energy and matter. A dissipative structure is a dissipative system that has a dynamical régime that is in some sense in a reproducible steady state. This reproducible steady state may be reached by natural evolution of the system, or by human artifice, or by a combination of these two. Energy exchange relations shall be based on the structure of dissipative structures, such as city, life, etc. Far from equilibrium, the system is open, between the different elements within the system is dissipative nonlinear system structure is the three conditions. Far from equilibrium, that the various regions within the system of matter and energy distribution is extremely uneven, the gap much (Willems, 1972).

Synergetic mainly focus on the internal co-operative mechanism of the system. The system always deviated from the average state value, this deviation is called fluctuation. Fluctuation is due to the independence movement and the cooperative movement of the system elements. When the system is in the transition from one steady state to another steady state, any small fluctuation will be amplified quickly to the giant fluctuation spread to the entire system, promoting the system into the order state.

Originated by the French mathematician Rene Thom in the 1960s, catastrophe theory is a special branch of dynamical systems theory. It studies and classifies phenomena characterized by sudden shifts in behavior arising from small changes in circumstances. Catastrophes are bifurcations between different equilibrium, or fixed point attractors. Due to their restricted nature, catastrophes can be classified based on how many control parameters are being simultaneously varied. For example, if there are two controls, then one finds the most common type, called a "cusp" catastrophe. If, however, there are more than five controls, there is no classification.

Over the last decades a variety of features have been identified as typical for self-organizing systems. Not all of these features are present in all systems able to self-organize. Self-organizing systems are open, far-from-equilibrium, dynamic, and sometimes employ autocatalytic amplification of fluctuations.

3. SELF-ORGANIZING CHARACTERISTIC OF SMES MUTUAL GUARANTEE SYSTEM

3.1 Mutual Guarantee System of SMEs

In a mutual guarantee system (MGS) a private society gives a guarantee in favor of an SME member of the group vis-a-vis a lender, usually a bank. MGSs are based on a mutual or self-help philosophy. There are variations in the schemes in China, where SMEs belong to a special professional association, and these bodies are partners of the guarantee scheme instead of individual firms' being members Mutual guarantee system of SMEs is a system set up by SMEs, local government and banks in specific region or industry chain, helping SMEs gain access to loan. SMEs mutual guarantee system consists of three parts: the core part, which is the mutual guarantee institution (MGI) formed directly by SMEs and are usually located in the headquarters of the business associations, aiming at improve their borrowing capacity; the supporting part, which is the financial institution such as rural credit cooperative, commercial bank and policy bank; auxiliary part, including the industry association and credit re-guarantee institution.

MGIs are formed directly by enterprises and are usually located in the headquarters of the business associations that promote them, or hosted by chambers of commerce. This helps MGIs to acquire information that firms share within the business association. MGIs' primary activity is to provide members with guarantees to be posted as collateral to back bank loans. Well-functioning mutual guarantee institution can have hundreds of members (Columba, Gambacorta, & Mistrulli, 2009). In countries where mutual guarantee associations are common, they have strong bargaining power with local banks and can negotiate flexible and affordable financial services. This is represented in Figure 1.

Figure 1. SMEs mutual guarantee system

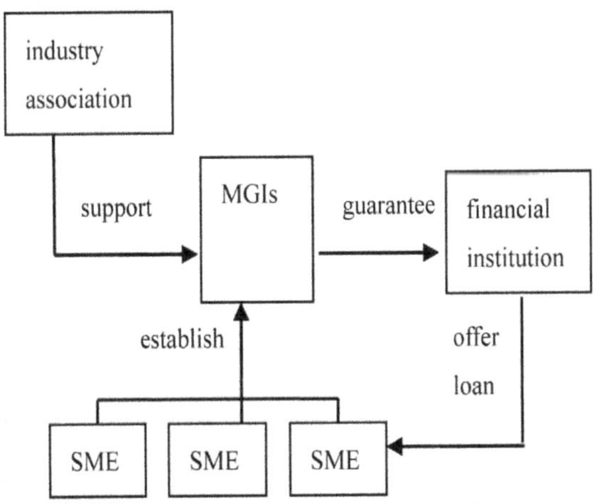

Mutual guarantee system can be considered as a result of a self-organized system. An originally random distribution of SMEs, connected by routes of randomly distributed capacity organizes itself by the forces of capacity extension and population growth into a highly structured network (Zhao & Xia, 2009). Capacity extension of a route is caused by the necessity to connect larger population conglomerates connected by this route and an extended route capacity causes further growth in the population along this route. As a result, initial fluctuations in carrying capacity and population distribution are amplified by this autocatalytic process until other restrictions like resource limitation constrain the growth.

3.2 Entropy Reduction in Open Systems

Entropy reduction is concept of local entropy decrease, referring to a hypothetical process wherein an activity or transformation is said to lower entropy and thus increase order. Entropy reduction was discussed early on by Erwin Schrödinger (1967). It relies on energy supply to create a metastable nonequilibrium state and electrical, pressure and chemical potential gradients across semipermeable membranes. The self-organizing systems are open to flow of energy, substance and information for route extension with the environment. An open system is a system that regularly exchanges feedback with its external environment. Open systems are systems, of course, so inputs, processes, outputs, goals, assessment and evaluation, and learning are all important. Healthy open systems continuously exchange feedback with their environments, analyze that feedback, adjust internal systems as needed to achieve the system's goals, and then transmit necessary information back out to the environment. The external environment includes a wide variety of needs and influences that can affect the organization, but which the organization cannot directly control. Influences can be political, economic, ecological, societal and technological in nature. A highly effective organization is regularly exchanging feedback with its external environment - it is an open system. Healthy organizations regularly try to understand their environments through use of environmental scanning, market research and evaluations (Ronald, 1971; Saadia, 2009). These organizations often try to influence their external environment, as well, for example, through use of public relations, advertising and promotions, lobbying and advocacy, and educating industry and local leaders. The environment of mutual guarantee system of SME means the factors collected with the system including political environment, social environment, market environment, technical environment and cultural environment. The mutual guarantee system of SME obtains the necessary resources of information and material from such environment, which is the necessary condition of self-organizing systems. The various components within the mutual credit guarantee system for SMEs itself is an open subsystem, connected with the external environment. The system achieves development by opening up to the external resources and upgrading of the internal level.

3.3 System Far From Equilibrium

Far-from-equilibrium is a type of dynamic equilibrium in which the state of a system is constantly changing with time due to an external energy (or matter) input. Far-from-equilibrium is a state – an ongoing, systemic condition that has been shown to increase innovativeness and performance (Smith & Comer, 1994). Far from equilibrium systems do strange things. They become inordinately sensitive to external or internal influences. Small changes can yield huge, startling effects up to and including reorganization of the entire system. We expect self-organization in slowly driven interaction dominated systems. Far-from-equilibrium conditions represent a culture or business model of high-energy, innovation-based organizing. This state is highly dynamic, and yet it can generate and regenerate itself for long periods of time, i.e. it is a dynamic state that appears to be relatively "stable." What actually causes the emergence of new order? In the far-from-equilibrium approach, the entire system moves into a regime that is away from equilibrium; this "far-from-equilibrium" organizing leads to non-linearity, adaptive tensions, and ultimately to perturbations of novelty. Under continuing far-from equilibrium conditions, new order will emerge (Benyamin, 2007). This is represented in Figure 2.

The mutual guarantee system of SMEs is a system far from equilibrium with the time and the openness. The borrowing capacity between SMEs in the mutual guarantee system is different. It is different between the financing needs of SMEs and actual financing capability of MGS. Far-from-equilibrium also means the different between the scales of development of MGS.

3.4 Non-Linearity and Feedback

Many complex self-organizing systems are complex because of their non-linear nature. When change occurs in complex systems it occurs in a non-linear fashion. Linear change is where there is a sequence of events that affect each other in order as they appear one after the other.

Figure 2. Far-from-equilibrium as the driver of order emergence

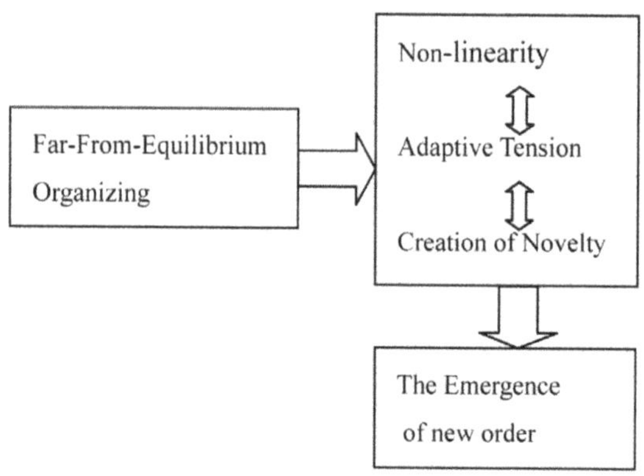

In contrast, in non-linear change, one sees elements being changed by previous elements, but then in turn these changed elements affect the elements that are before it in the sequence. Thus in non-linear analysis, researchers look at how everything in the sequence has the possibility of affecting everything else in the sequence before and after it. Thus often the result ends up being unproportional to the original input. This type of dynamic in a complex system is much closer to how things actually happen in nature. Almost never in nature does a purely linear sequence of events and change occur. Non-linearity can simply be understood through the interplay of positive and negative feedback-cycles that fuel natural selection. This leads to wider ecological processes, which are also continuous and subject to cybernetic principles of regulation. While positive feedback increases the number of configurations, negative feedback controls and stabilizes them. The interaction between them creates intricate and chaotic patterns, which can develop very quickly until they reach a stable configuration, an attractor so to speak. That is, even if the elements themselves are very simple their interaction and relationships can become complex. Small changes can have large effects (through positive feedback), and conversely, large changes can have relatively small affect (through negative feedback). Non-linearity in mutual guarantee system of SMEs means a kind of quantitative characteristics and non-superposition of the interaction of elements within the mutual guarantee system. The change of the elements is not linear with the change of the external environment such as social environment and market environment. When such constant change accumulated to a certain extent--threshold, innovation would occur.

3.5 Fluctuations

Self-organization is normally triggered by internal variation processes, which are usually called "fluctuations" or "noise". Self-organized criticality in a wide variety of systems is seen to arise as a consequence of a singularity in the diffusion coefficient of the hydrodynamic limit. Information is a process that exists as a relationship between specific self-organized units of matter, it is a relationship of reflection between a fluctuation that causes inner-systemic changes and the system's structure. A certain fluctuation causes nonlinear changes in the system, i.e. new order in the system emerges, the system changes its structure due to interactions of its elements. The fluctuating instability is reflected within the system's components, their relationships, and interactions. Reflection means that the system reacts to fluctuations and reproduces fluctuations in its inner structure as self-organized change. Information is not a structure given in advance, it is produced within material relationships. Reflection doesn't mean that an outside reality is mechanically copied or reproduced within the system, it means that a complex, nonlinear relationship between cause and effect is established in a self-organizing social system (Krugman, 1996). All self-organizing systems are information-producing systems. Information is a dynamic process of reflection where certain stimuli, fluctuations, or perturbations are reflected in a complex and non-linear way within the structural interactions and patterns of elements of a system. Knowledge is the manifestation of information in social systems that involves the interpretation, experience, evaluation, and usage of data and can be found in various subsystems of society. In this case fluctuations play an important role, and if fluctuations are large enough then pure power laws in event-size distributions are observed. According to empirical observations, some pattern formation phenomena in driven many-particle systems are more pronounced in the presence of a certain noise level. This phenomenon of fluctuation-driven ordering with a cellular automaton model of interactive motion in space and find optimal noise strength, while order breaks down at higher fluctuation levels. Additionally, we discuss the phenomenon of noise and drift-induced self-organization in systems that would show disorder in the absence of fluctuations (Carlson et al., 1993). The changes of factors such as policy, market, technology, talent and capital will cause the fluctuations of mutual guarantee system for SMEs. When the system is far from equilibrium, fluctuations may be amplified and the mutual guarantee system of SMEs will promote the evolution and development. The direction and process of the mutual guarantee system of SMEs is determined by the fluctuations—the changes of policy, market, technology, talent and capital.

4. DEVELOPMENT SUGGESTIONS OF MGS BASED ON SELF-ORGANIZATION THEORY

4.1 Make Full Use of Market Means To Develop MGS

The implication of self-organization theory is that under some conditions, MGS owns internal tendency of autopoietic, self-evolution and evolution from disorder to order, from low level to high level to solve SME's financing problems in some parts. We cannot find the ideal model of MGS because of the complex financing needs and the different environment throughout the country. The driving forces of MGS are banks and SMEs suffering from the financing problem. So the government is suggested to push development of MGS not by too much intervening but by encouraging innovation, loosening limitation of industrial entrance, making full use of market means to develop MGS from low level to high level.

4.2 Establish the Guidance Mechanism for MGS

Although the mutual guarantee system for SMEs is an autopoietic self-organization formed directly by SMEs, the development of self-organization need induced conditions. Government guide would be important for the development of MGS. The guarantee institutions vary in different operational objectives, marketing position, custom and operating mechanism, but there is no guidance mechanism special for mutual guarantee institutions. There is a negative impact on the development of mutual guarantee institutions because the scale of mutual guarantee institutions is much smaller than commercial guarantee institutions and political guarantee institutions. The guidance mechanism for MGS, such as regulatory system, support policies and performance evaluation, should be established to form the policy environment for MGS.

4.3 Support Non-Linear Development of MGS

The self-organizing systems are complex because of their non-linear nature. The interaction and relationships between the elements are very complex. Dominant power will emerge to lead to a new order after the competition and cooperation between the elements within MGS. The development of the mutual guarantee system is a systematic system. The core part of the mutual guarantee system is the mutual guarantee institution (MGI) formed directly by SMEs, aiming at improve their borrowing capacity. The supporting part is the financial institution such as rural credit cooperative, commercial bank and policy bank and the auxiliary part is the industry association. The government is suggested to strengthen the supporting force to increase public finance investment to the key institution-- mutual guarantee institution, including start bonus, risk compensation, incentive fund and tax preference policy. The capital scale of the mutual guarantee institutions with non-linear feature would accumulate to a certain extent—threshold once they obtain the support of public finance. Fundamental change will occur and the mutual guarantee institutions will develop quickly without the security restrictions.

4.4 Improve the Quality of Employees of MGS

The changes of factors such as policy, market, technology, talent and capital will cause the fluctuations of mutual guarantee system for SMEs. The direction and process of the mutual guarantee system of SMEs is determined by the fluctuations—the changes of policy, market, technology, talent and capital. However the quality of credit guarantee agents is considered as one of the very important induced conditions for improving MGS. Professional employees training system and professional qualification should be set up to improve their professional quality and skills, using various ways such as training exchanges and talent introduction.

5. CONCLUSION

Self-organization theory is a branch of systems theory that relates to the process of order formation in complex dynamic systems. Drawing from the fields of cybernetics and psychology, self organization theory offers a new paradigm for perception and change. From the perspective of system theory this paper illuminated the self-organization mechanisms and the self-organization characteristics of openness, far-from-equilibrium, nonlinearity and fluctuation of mutual guarantee system of SMEs. The implication of self-organization theory is that under some conditions, mutual guarantee system owns internal tendency of autopoietic, self-evolution and evolution from disorder to order, from low level to high level to solve SME's financing problems in some parts. The government is suggested to push development of GMS by encouraging innovation, loosening limitation of industrial entrance, supporting key institutes and improve the financial environment of GMS.

ACKNOWLEDGMENT

This paper is based on research done within the framework of the project "Market structure, spillover effect and supervision mechanism of guarantee industry" funded by Beijing Municipal Education Commission.

REFERENCES

Ashby, W. R. (1947). Principles of the Self-Organizing Dynamic System. *The Journal of General Psychology*, *37*, 125–128. doi:10.1080/00221309.1947.9918144

Ba, J.-S. (2007). The Establishment of China's SME Credit Guarantee Model. *Shanghai Finance*, *8*, 56–63.

Berger, A. N., & Udell, G. F. (1998). The economics of small business finance: the roles of private equity and debt markets in the financial growth cycle. *Journal of Banking & Finance*, *22*, 613–673. doi:10.1016/S0378-4266(98)00038-7

Berger, A. N., & Udell, G. F. (2006). A more complete conceptual framework for SME finance. *Journal of Banking & Finance*, *30*(11), 2945–2966. doi:10.1016/j.jbankfin.2006.05.008

Berger, U. (2002). Small Business Credit Availability and Relationship Lending: The Importance of Bank Organizational Structure. *The Economic Journal*, *112*(447), 32–53. doi:10.1111/1468-0297.00682

Carlson, J. M., Grannan, E. R., Singh, C., & Swindle, G. H. (1993). Fluctuations in self-organizing systems. *The American Physical Society*, *48*(22), 688–698.

Coco, G. (2000). On the use of collateral. *Journal of Economic Surveys*, *14*(2), 191–214. doi:10.1111/1467-6419.00109

Columba, F., Gambacorta, L., & Mistrulli, P. E. (2009). *The effects of mutual guarantee consortia on the quality of bank lending*. Munich, Germany: University Library of Munich.

Haken, H. (1983). *Synergetics: An Introduction. Nonequilibrium Phase Transition and Self-Organization in Physics, Chemistry, and Biology* (3rd ed.). Berlin: Springer-Verlag.

Krugman, P. (1996). *The Self-organizing Economy*. Cambridge, MA: Blackwell Publishers.

Lichtenstein, B. B. (2007). *Moving Far From Far-From-Equilibrium: Opportunity Tension as the Driver of Emergence.* Paper presented at the Organization Science Winter Conference.

Molleman, E. (1998). Variety and the Requisite of Self-Organization, the International. *Journal of Organizational Analysis*, *6*(2), 109–131. doi:10.1108/eb028880

Peng, J.-B. (2008). The establishment of SME credit guarantee system based on mutual guarantee. *Finance Reaserch*, *2*, 75–82.

Petersen, M. A., & Rajan, R. G. (1994). The benefits of lending relationships: evidence from small business data. *The Journal of Finance*, *49*(1), 3–37. doi:10.2307/2329133

Ronald, F. (1971). Entropy reduction in open systems. *Journal of Theoretical Biology*, *31*(1), 43–46. doi:10.1016/0022-5193(71)90119-6

Saadia, M. (2009). Framework for the Role of self-organization in the handling of adaptive challenges. *Complexity & Organization*, *131*(2), 1–14.

Schrödinger, E. (1967). *What is Life?: The Physical Aspect of the Living Cell with Mind and Matter and Autobiographical Sketches*. Cambridge, MA: Cambridge University Press.

Smith, C., & Comer, D. (1994). Self-organization in small groups: a study of group effectiveness within non-equilibrium conditions. *Human Relations*, *47*(5), 553–581. doi:10.1177/001872679404700505

Von Foerster, H. (1960). On Self-Organizing Systems and their Environments. In Yovitts, M. C., & Cameron, S. (Eds.), *Self-Organizing Systems* (pp. 31–50).

Willems, J. C. (1972). Dissipative dynamical systems, part I: General theory; part II: Linear systems with quadratic supply rates. *Archive for Rational Mechanics and Analysis*, *45*, 321–393. doi:10.1007/BF00276493

Zhao, H., & Xia, H. (2009). Study on the constitution, evolution and development suggestions of entrepreneurial finance system: perspective from self-organization theory. *China Soft Science*, *9*, 107–112.

This work was previously published in International Journal of Asian Business and Information Management, Volume 1, Issue 3, edited by Dasho Karma Ura and Patricia Ordóñez de Pablos, pp. 1-9, copyright 2010 by IGI Publishing (an imprint of IGI Global).

Chapter 19
A Trilogy of Unfortunate Events in China:
Reflecting on the Management of Crises

Zhang Long
China University of Geosciences, China

William Crandall
University of North Carolina at Pembroke, USA

John A. Parnell
University of North Carolina at Pembroke, USA

ABSTRACT

In this paper, the authors address three recent organizational crises that have occurred in China; the SK-II cosmetic incident, the Sharon Stone comment on the May 2008 earthquake in China, and the melamine milk contamination crisis. Each held significant notoriety due to the crises involving major companies and an assortment of negative outcomes. After presenting an overview of each case, the authors outline their reflections on the management of these crises in relation to their cultural context.

INTRODUCTION

The management of organizational crises has been a key concern in recent years in both developed and developing nations (Parnell, Köseoglu, & Spillan, 2010). Three events of interest to crisis management researchers have occurred in the People's Republic of China since September 2006. The discovery of contaminants in Proctor & Gamble's cosmetic line-SK-II in 2006, the comments by U.S. actress Sharon Stone regarding the May 12, 2008 earthquake in China, and the melamine milk crises that occurred in the summer of 2008 each generated their own degree of negative outcomes for the organizations involved. The fact that these events occurred in the People's Republic of China is of special interest to crisis researchers, given the widespread globalization that has spread to this country.

This article begins with an overview of these three events and the negative outcomes that caused them to be crises for their organizations. We then

DOI: 10.4018/978-1-4666-0276-2.ch019

offer our reflections on the significance of these events within the framework of crisis management and the cultural context of China.

A TRILOGY OF UNFORTUNATE EVENTS IN CHINA

The SK-II Crisis

The SK-II product line of cosmetics is manufactured by Proctor & Gamble (P&G) in Japan and distributed to stores in Australia, China, Hong Kong, Japan, Korea, Malaysia, Singapore, Taiwan, the United Kingdom, and the United States. In the United States, the product line is available through Saks, with a market comprising high-end consumers.

The SK-II crisis began on September 14, 2006, when authorities in South China's Guangdong Province detected chromium and neodymium in a type of SK-II cosmetic. Because these metals can cause skin irritation and disease, they are banned in all cosmetics in China. Sales of SK-II in China represent less than seven percent of the brand's global sales (Crandall, Parnell, Xihui, & Long, 2007).

Initially, P&G denied there was a problem with the cosmetics, instead stating that it was working with the authorities to verify the validity of the findings. After the types of allegedly contaminated SK-II products increased to nine, the company agreed to offer refunds to consumers. To be eligible for those refunds, consumers had to bring the product back to the store of purchase with no less than one-third remaining, complete and sign a form acknowledging that the product was of good quality, and wait several weeks for a refund to be processed (*China Daily*, 2006a; Guan, 2006). On September 21, hundreds of Shanghai women sought refunds at P&G's specified locations, only to become frustrated when told that their refunds would take three weeks to process. On September 22, P&G announced that it would suspend its refund operations due to security concerns. A few hours later, tempers flared as an angry group of consumers kicked down the front door at P&G's Shanghai office. Media calls to P&G officials in Guangzhou and Shanghai were not returned. The company's China website was reportedly hacked that weekend. Some retailers of the SK-II products began offering immediate cash refunds to customers after P&G suspended its refund program (*China Daily*, 2006b, 2006c).

The Sharon Stone Crisis

A major earthquake measuring 7.8 on the Richter scale jolted Wenchuan County in southwest China's Sichuan Province on May 12, 2008 (Xinhua, 2008a), resulting in almost 70,000 deaths. An outpouring of relief efforts from around the world followed to help the victims of this tragedy. In addition, the three days from May 19 to 21 were declared as national mourning days by the Chinese government for the earthquake victims.

The tragedy took a strange twist during the Cannes Film Festival on May 24, however. During an interview, actress Sharon Stone offered these comments on the earthquake in China:

"Well you know it was very interesting because at first, I'm you know, I'm not happy about the ways the Chinese were treating the Tibetans because I don't think anyone should be unkind to anyone else. And so I have been very concerned about how to think and what to do about that because I don't like that. And then I've been this, you know, concerned about, oh how should we deal with the Olympics because they are not being nice to the Dalai Lama, who is a good friend of mine. And then all this earthquake and all this stuff happened and I thought: Is that karma, when you are not nice that bad things happen to you ..." (Gardner, 2008)

The "Karma" comment created uproar in the news media and a wave of criticism on Internet blogs in China since the actress was suggesting that the country's earthquake was retribution for its policies on Tibet (Passariello & Meichtry, 2008). Chinese citizens left Internet messages, saying she was ignorant of the Tibet issue and had no sympathy for those who were suffering. One online comment, from a woman calling herself Mariah, summed up the reaction of many Chinese citizens: "These kind of remarks deeply hurt Chinese people's feeling and are totally unacceptable" (Roberts, 2008, p. 12).

The media reported that protestors were tearing down billboards featuring the actress in advertisements (McLaughlin & Kaiser, 2008). In addition, many music stores on the Chinese mainland and in Hong Kong cinemas pledged not to show her films again. Ng See-Yuen, the founder of the UME Cineplex chain, said films featuring Stone would be banned from any UME cinema in Hong Kong and the Chinese mainland. Moreover, the Shanghai International Film Festival had also decided to permanently ban Stone and her films (Du, 2008).

The Sharon Stone comment also created a crisis for Dior, the French company that has used Stone in their ads. Dior offers upscale clothing, cosmetics, and apparel, and has a branch in China, appropriately called Dior China. On May 27, 2008, three days after the Stone comment, Dior China issued the following statement through *SINA*, a popular news-oriented website in China: "We absolutely disagree with Stone's remarks and are deeply sorry for them" (*SINA*, 2008a).

On May 28, Dior China issued another statement noting that Stone was "deeply sorry" for the offense and anger caused by her comment and had offered to take part in relief work in the Sichuan disaster area. However, in an interview in the *New York Times*, she insisted her comments in Cannes had been taken out of context (Horyn, 2008). On May 31, Stone seemed to make amends for her misstep, issuing a formal statement of apology through CNN. However, a survey of 250,000 Chinese citizens found that 69 percent did not accept her apology and vowed to never forgive her (Xinhua, 2008b). Many Chinese citizens wrote letters to Dior China to express their opinions indicating they would never buy any Dior products as long as the company was affiliated with Sharon Stone (Passariello & Meichtry, 2008).

Under pressure from its Chinese consumers, Dior China began removing all of its advertisements with Stone's image nationwide (McLaughlin & Kaiser, 2008). On June 2, the company issued an official statement indicating it would stop using Sharon Stone as a spokesperson in China (Roberts, 2008).

The Melamine Milk Crisis

Melamine is composed of nitrogen, carbon and hydrogen, and was invented in the 1830s by a German scientist. In the 1930s, it was used to make plastics and laminates. Ironically, when added to a food product in powdered form, it can boost the protein rating of that food item. Unfortunately, once it is ingested inside the body, it can cause kidney stones and renal failure (Pickert, 2008; Tschang, 2008). This false boosting of protein levels is what led to the now infamous melamine milk crisis that occurred in China during 2008. At the bottom of the crisis was this practice – dairy middlemen were spiking milk with melamine, after they had watered down the milk to extend the product usage. The higher protein content enabled these middlemen to command higher prices on the market. The Sanlu Group is one company that bought melamine spiked milk, at first unknowingly. Later, investigators found that some Sanlu staff were aware that melamine was being added, but that Sanlu continued to produce and distribute baby milk formula, even months after the discovery (DeLaurentis, 2009).

On June 28, 2008, an infant with kidney stones was admitted to the People's Liberation Army No. 1 Hospital in Gansu's provincial capital of Lanzhou. The parents told doctors that they had been feeding the baby milk produced by the Sanlu Group in Shijiazhuang, Hebei province since the baby was born. Within two months, fourteen infants with similar problems had been admitted to the hospital (Cheng, 2008). By September 11, 2008, other cases were reported in provinces of the Ningxia Hui Autonomous region, Shandong, Jiangxi, Hubei, Shanxi, Jiangsu, Shandong, Anhui and Hunan. All of the affected infants had been fed the milk formula produced by the Sanlu Group. The melamine contaminations of the milk products produced by Sanlu were about to become a major scandal, complete with global media coverage.

By September 20, 2008, the World Health Organization (WHO) reported more than 54,000 children in China had sought medical treatment, 12,000 were hospitalized and at least four infants died from melamine-contaminated dairy products (Schlein, 2008). The number of infants affected was likely much higher than the WHO reported, however, as concerned parents were still queuing up at hospitals throughout China to have their children examined.

By September, the controversy had spread outside the Chinese mainland. Five children in Hong Kong had been diagnosed with kidney stones as a result of drinking tainted Chinese milk formula (Liu, 2008). The melamine was even found in a third Chinese-made dairy product, White Rabbit Candy, a popular milk-flavoured toffee, causing its producer to halt domestic and foreign sales of the candy. Public health officials in Taiwan also announced their findings of melamine-tainted instant coffee, milk tea and chicken-and-corn soup. The import of all such products into Taiwan was banned, including instant coffee made by the popular Taiwan brand Mr. Brown, which is made in China (Tan, 2008).

In other locations, Japan's Marudai Food Co. issued a voluntary recall of five China-made products, indicating they may contain the toxin melamine. Brunei ordered a ban on all China-made milk products and dairy items. Malaysia imposed a "level six import ban" on all Chinese dairy products. Bangladesh put a ban on three brands of Chinese-made milk powder. The African nations of Tanzania and Gabon imposed bans on Chinese dairy products, followed shortly thereafter by Burundi (Tan, 2008). Finally, on September 25, the 27-nation European Union banned imports of baby food containing Chinese milk (Liu, 2008). By this time, the milk contamination case had become a worldwide crisis not only for milk companies like the Sanlu Group, but also for the Chinese government.

After receiving complaints about its baby formula milk powder in March, the Sanlu Group initially claimed that its products had repeatedly passed quality tests, met national quality standards, and that sick babies must have been fed counterfeit milk powder that used the Sanlu brand name. Later, the company did confirm the problems by importing melamine-testing equipment before early August, which was validated by Fonterra, a New Zealand based company, which owns 43 percent of Sanlu (Fonterra, 2008),. For a time afterwards, however, Sanlu neither reported the problem to the Chinese government nor revealed the information to the public until September 11, at which time the Ministry of Health suspected melamine contamination (*Economist*, September 20, 2008).

On the morning of September 11, a company spokesperson interviewed by a reporter from the *People Daily Online*, still insisted that its products were in strict accordance with national standards and met quality tests, and that it believed that the quality inspection department would give a clear explanation later (*SINA*, 2008b). Ironically, the Sanlu Group issued an announcement later that day recalling all of its 700 tons of milk because of the melamine contamination.

Faced with increasingly angry criticism from Chinese parents and the public, Zhang Zhenling, Sanlu's Vice President, delivered an apology letter on behalf of the company at a news briefing on September 15, four days after the public became aware of the contamination. The apology expressed regret and included a declaration to recall all the infant milk powder produced prior to August

6, as well as an optional recall for milk produced after that date if consumers have concerns about sick infants. But the late apology and the dismissal of Sanlu president Tian Wenhua—who thought the company was totally innocent and had been disgraced by the fraudulent milk suppliers—did not seem to appease the public sufficiently, as many citizens lost confidence in the Sanlu brand (Xinhua, 2008c).

REFLECTING ON THE MANAGEMENT OF CRISES

The three cases discussed herein represent an eclectic trio. Two of the cases concern multinational corporations (Dior and P&G), while the third case focused on a former Chinese state-owned business (Sanlu Group). Two of the cases centered on a faulty product (P&G and Sanlu), while the other case revolved around a faulty spokesperson (Dior). Each case carries with it, some degree of political overtones. All three cases resulted in the same outcomes, negative publicity for the companies and a public boycott of their products. In the next section, we overview our reflections on this trio of unfortunate events in relation to crisis management in China.

The Issue of Product Quality in China

At the heart of many crises in China is the problem of product quality. This one issue has plagued the country in terms of its reputation and reliability. The problem has appeared on two fronts; products produced in China, and products imported to China via multinational corporations (MNCs). The first front has been well documented, and will not be discussed here. However, the second front reflects a growing consumer activism on the part of the Chinese, based somewhat on strong nationalist sentiments (Tai, 2008).

The SK-II case in particular highlights the problem of a quality issue that emerged from a popular MNC. Herein though, lies an interesting dilemma, was the SK-II product actually defective? What is known is that there were traces of chromium and neodymium in the cosmetic brand, but these amounts were far below the levels considered internationally hazardous. Some feel that P&G might have been a victim of an international trade dispute between Japan and China (Lin, 2007). Regardless of the original source of the crisis, the goal of P&G was to take control of the situation in an effective manner.

Taking responsibility for a crisis does not mean the company is admitting guilt for any wrong doing. It does mean the company is concerned about the event and looking into what possibly went wrong. However, it appears that P&G, although trying to create a sense of good will with its customers by offering a refund, probably created the perception that the brand was slipping in the eyes of the Chinese consumers (Lin, 2007). Closing the SK-II counters in China immediately after the crisis unfolded further eroded the consumer perception of the brand image. Looking back, it can be seen that at the heart of the crisis was the concern over product quality.

The Sanlu Group moved slowly in its response to the crisis it faced. The Sanlu Group blamed illegal dairy dealers for adding melamine to milk, which in fact, was true. But the posture of its comments indicated that as a company, it had nothing to do with the scandal. The public and the Chinese government were not convinced. More than 70 percent of those responding to on an online survey thought the company should take the major responsibility rather than its milk suppliers (Ye, 2008). Even the apology letter became an object to be criticized by the public since it appeared less than forthright in its management of the crisis. The Sanlu Group seemed to avoid addressing the problem head on, taking limited action instead and displacing the blame to other parties.

The Symbolic Power of the Company Spokesperson in China

Actress Sharon Stone's unexpected comments immediately damaged Dior China's reputation. Although Dior did nothing to bring about the crisis, one could argue that Stone might not have been the best spokesperson for the company in the first place. Sometimes spokespersons can carry negative perceptions with consumers, not because of what they say, but because of something they represent. P&G found this out when they employed Taiwanese model Lin Zhiling, as a brand spokesperson. Zhiling's father once campaigned for Chen Shuibian in his bid to become Taiwan's president. Because Shuibian is an ardent supporter of independence for the province, many Mainland Chinese consumers have linked P&G to Taiwanese independence through its affiliation with Zhiling (Crandall, Parnell, Xihui, & Long, 2007). P&G officials, however, have stated that the firm holds no political position on the matter. Nonetheless, in August 2005, a group of "netizens" (i.e., citizens utilizing the Internet) sought to secure 400,000 signatures on a petition opposing Lin Zhiling and P&G products. This group has committed to a boycott of all P&G products as well as those produced by plants her father owns in Mainland China (*Taipei Times*, 2005).

A spokesperson can also generate negative perceptions if they have engaged in controversial film projects. Pond Cosmetic Company hired as its brand spokesperson Tang Wei, a female star in the popular film *Lust, Caution*. The actress drew the criticism of the Chinese government, however, for several pornographic scenes and for the historical retort in the film (Fitzsimmons, 2008). A number of scenes were deleted from the version shown in China, but many consumers were able to access the complete film on the Internet. As with the previous example, Pond advertisements featuring Tang were banned from Chinese television.

As these examples illustrate, P&G and Pond suffered both financial and reputational losses as a result of their decisions to hire these spokespersons. So did Dior China. The key implication is that firms need to consider the long term fit between its products and the spokespersons it employs to represent those products.

The Political Unity Factor that Must be Anticipated in China

China is a country that is very different from its western counterparts. Political issues are treated sensitively and seriously in China, and the media plays an important guidance role under the control of the government. Compared with its western counterparts, China exerts more centralization with political issues, especially on the subject of China's unification. Moreover, the Chinese people have a deep patriotic feeling for China's unification. Therefore, any topics associated with China's unification are very sensitive in the eyes of both the Chinese government and its citizens. China once banned one of Taiwan's top pop stars, A-Mei, after she sang the island's national anthem at the inauguration of President Chen Shuibian in 2000 (*BBC News Online*, 2000).

A company spokesperson with strong political positions can easily be misunderstood as the voice of the firm's position as well. If the spokesperson is controversial in the eyes of the host government, image problems are inevitable. A spokesperson represents the style, image and characteristics of a brand and many people may purchase products because of advertisements featuring that individual. But the spokesperson also represents the political bias of the company, whether the company is aware of that or not. Hence, firms like P&G and Dior need to be especially cognizant of the use of their spokespersons since considerable consumer dissatisfaction may result.

Tibet is a recognized part of China. The Dalai Lama has been campaigning for Tibet's independence for many years since late 1950s. From its perspective, the Chinese government has instituted democratic reforms, abolished the feudal serf system, and allocated a significant number of resources to improve the life of Tibetans. Although Sharon Stone is entitled to her own opinion of the Tibet situation, her attempt to link it to an unrelated disaster such as an earthquake is reckless. From a European perspective, this would be akin to linking human suffering from a natural disaster to one's disagreement with some aspect of an EU nation's foreign policy. One simply does not cause the other, and to suggest a connection is ill-advised at best and callous at worst.

The Growing and More Vocal Middle Class in China

Multinational corporations have known for a number of years that the rising middle class in China represents a large potential market. What should also be remembered is that, like the United States and other developed countries, the middle class in China is becoming more vocal in its demand for truthful advertising and products that deliver what the ads say they will provide (Tai, 2008). As Mao Shoulong, professor of Sociology at Renmin University of China put it, "For all foreign brands, the most important thing is to win the heart of Chinese consumers and to maintain their credibility," (Lan, 2007 quoted from China Economic Net). The SK-II incident illustrated how Proctor & Gamble did not anticipate this new level of consumer advocacy in China.

Likewise, Dior did not anticipate the backlash from more affluent consumers to the Sharon Stone comment. The company took prompt action to address the Sharon Stone crisis, distancing itself from the context of her remarks. The firm's effort to issue a statement on her behalf illustrates its concern, but the extent to which Stone was involved in preparing and refining the statement is unclear. In the end, Dior China found itself in an unenviable position. Battered by Stone's hesitation to address the comments in a responsible manner, the company took what was probably the most appropriate course of action by removing advertisements with Stone's image and cutting ties with Stone in the Chinese market. The firm's reputation was at stake, and prompt, decisive action was justified.

The Growing Power of the Internet in China

The Internet can substantially influence the outcome of a crisis. In China, the growth of the Internet, particularly through blogging, enables news to spread quickly (Betser & Fraser, 2006). The first case of tainted milk resulting in infant kidney stones was disclosed on the Internet on the popular Tianya Forum in June 2008. Only two days after September 11—the day the melamine contamination of the milk was made public—there were more than 2.7 million related webpage discussing the melamine tainted milk incident (Fangjianjiaping, 2008). The "melamine" ranked the first key word for search in Google in 2008 (Wei, 2008). Hence, knowledge of the crisis became widespread in a short period of time. To address such crises, firms must be poised to act quickly and decisively. Sanlu seemed to have overlooked the speed at which such information can travel.

For P&G, the SK-II crisis became a major event on the Internet, with literally thousands of articles about the incident on websites (Tai, 2008). In the case of the Sharon Stone crisis, the "karma" comment created an uproar in the news media and a wave of criticism on Internet blogs in China. Of the three cases in the trilogy, Dior China utilized one of China's most popular websites, *SINA*, to address crisis concerns and clarify the company's stance. On the other hand, the Sanlu Group tried to conceal the truth and postponed presentation to the media, creating even more suspicion (DeLaurentis, 2009).

FUTURE DIRECTIONS

Several opportunities exist for future research. First, the present study examined three prominent crises in China, but did not consider others. A closer examination of all major crises in a given year, for example, could provide greater insight into key crisis management patterns common to Chinese firms.

Second, the generalizability of crisis management practices across cultures and nations remains unclear and requires additional investigation (Parnell, Köseoglu, & Spillan, 2010). For example, in what ways does crisis management practice in emerging economies differ from that in developed nations? To what extent is crisis management practice contingent on specific cultural attributes? An examination of crises in other nations, particularly emerging ones, could shed light on these questions.

Third, additional research that compares and contrasts different approaches to crisis management is needed. Learning from successful firms not only contributes to practical contributions to the field, but can also help scholars rethink and revise their frameworks and models.

Finally, the application of early warning systems is an area that has received relatively little research to date (Crandall, Parnell, & Spillan, 2010). Because crisis avoidance is the most desirable outcome for firms, introducing systems that enable managers to identify prospective crises before they occur is useful.

REFERENCES

BBC News Online (2000). 'China Bans Taiwan's Madonna', http://news.bbc.co.uk/2/hi/asia-pacific/761908.stm (accessed 07 July 2009).

Betser, Y., & Fraser, P. (2006, November). 'Trouble in China'. *Brand Strategy*, *207*, 48–49.

Cheng, C. (2008). 'Pay Attention to the Taxation Problems when Companies Recall the Tainted Milk Formula', http://www.ctaxnews.com.cn/syxw/xwxzt/200809/t20080924_1532169.htm (accessed 26 September 2008).

Daily, C. (2006a, September 19). 'Actress Accused of False Advertising', http://www.chinadaily.com.cn/china/2006-09/19/content_692408.htm (accessed 15 July 2009).

Daily, C. (2006b, September 23). 'Consumers Angry as SK-II Pulled from Shelves', http://www.chinadaily.com.cn/china/2006-09/23/content_695363.htm (accessed 15 July 2009).

Daily, C. (2006c, September 26). 'Consumers Turn Cautious on P&G Skin-care Row'. *China Daily Online*, http://www.chinadaily.com.cn/china/2006-09/26/content_696526.htm (accessed 15 July 2009).

Crandall, W. R., Parnell, J. A., & Spillan, J. E. (2010). *Crisis Management in the New Strategy Landscape*. Los Angeles: Sage.

Crandall, W. R., Parnell, J. A., Xihui, P., & Long, Z. (2007). 'When Crisis Management goes Abroad: The demise of SK-II in China'. *Journal of International Business Research and Practice*, *1*(1), 38–49.

DeLaurentis, T. (2009). 'Ethical Supply Chain Management'. *The China Business Review*, *36*(3), 38–41.

Du, W. (2008, May 30). 'Netizens Annoyed by Stone's Ignorance'. *China Daily Online*, http://www.chinadaily.com.cn/china/2008-05/30/content_6724266.htm (accessed 08 May 2009).

Economist, (2008, September 20). *Formula for Disaster*. Volume 388, pp. 57-58.

Fangjianjiaping (2008). 'Sanlu Affair triggered the Credit Crisis of Baidu', http://hi.baidu.com/fjpl/blog/item/efdb01e9bb1f8339b80e2d94.html (accessed and translated 16 September 2008).

Fitzsimmons, E. (2008, May 16). 'Ban Exposes China's Ad Censorship Anomalies'. *Campaign*, Issue 19, p. 17.

Fonterra. (2008). 'Statement by Fonterra CEO Andrew Ferrier to Media Conference', *Media Release,* http://www.fonterra.com/wps/wcm/connect/fonterracom/fonterra.com/our+business/news/media+releases/statement+by+fonterra+ceo+andrew+ferrier+to+media+conference (accessed 15 September 2008).

Gardner, D. (2008). 'Why Do We Say Silly Things?' *The Ottawa Citizen*, http://www.canada.com/ottawacitizen/columnists/story.html?id=81b2c80c-ddc3-4f4c-94a3-6f9c98f8f608 (accessed 15 July 2009).

Guan, X. (2006, September 18, 2006). 'Refunds Offered on Harmful Cosmetics', *China Daily*. http://www.chinadaily.com.cn/china/2006-09/18/content_691155.htm (accessed 15 July 2009).

Horyn, C. (2008, June 1). 'Actress Stone and Dior Differ Over Apology', *New York Times online,* http://www.nytimes.com/2008/06/01/fashion/01stone.html?_r=1&scp=1&sq=Actress%20Stone%20and%20Dior&st=cse&oref=slogin (06 September 2008).

Lan, X. (2005, May 29). 'Famous Brands Lose Face', *Beijing Review*, http://www.bjreview.cn/EN/En-2005/05-29-e/bus-1.htm (accessed 23 July 2009).

Lin, T. J. (2007). 'Brand Lessons'. *Global Cosmetic Industry*, *175*(Issue 5), 29–32.

Liu, J. (2008) 'Sanlu Group Stop Reimbursing Diagnose Fee for Infants'. *YNET news online,* http://china.ynet.com/view.jsp?oid=43262117 (accessed 17 September 2008).

Liu, M. (2008). 'A Familiar Taint: The Milk Crisis Spotlights Flaws in Beijing's Food System'. *Newsweek Online.* http://www.newsweek.com/id/160895 (25 September 2008).

McLaughlin, K. E., & Kaiser, A. (2008). Dior China Drops Stone after Quake Comments. *Women's Wear Daily*, *195*(Issue 115), 24–25.

Parnell, J. A., Köseoglu, M. A., & Spillan, J. E. (2010). Crisis readiness in Turkey and the United States. *Journal of Contingencies and Crisis Management*, *18*(2), 108–116. doi:10.1111/j.1468-5973.2010.00603.x

Passariello, C., & Meichtry, S. (2008, May 30). 'Dior Pulls Ads with Sharon Stone'. *Wall Street Journal*, B.7.

Pickert, K. (2008). 'Brief History of Melamine'. *Time Online*, http://www.time.com/time/health/article/0,8599,1841757,00.html (accessed 17 September 2008).

Roberts, D. (2008, June 2). 'China: Multinationals Hear it Online'. *Business Week Online*, http://www.businessweek.com/globalbiz/content/may2008/gb20080530_213248.htm (accessed 06 September 2008).

Schlein, L. (2008). 'China's Melamine Milk Crisis Creates Crisis of Confidence'. *Voice of America News Online,* http://www.voanews.com/english/archive/2008-09/2008-09-26-voa45.cfm?CFID=64268891&CFTOKEN=95716309 (accessed 26 September 2008).

SINA (2008a). 'Dior Announcement: Absolutely Disagree with the Comment of Sharon Stone', http://ent.sina.com.cn/s/2008-05-27/16102038967.shtml (accessed 27 May 2008).

SINA (2008b). 'Sanlu Group Insisted Their Products Meet National Standard When Responding to the Milk Accident', http://news.sina.com.cn/c/2008-09-11/105416273760.shtml (accessed 11 September 2008).

Tai, S. (2008). 'Beauty and the Beast: The Brand Crisis of SK-II Cosmetics in China'. *Asian Case Research Journal*, *12*(Issue 1), 57–71. doi:10.1142/S0218927508001023

Times, T. (2005, August 11). 'Chinese Fanatics Urge Boycott over Lin's Green Parents'. *Taipeitimes.com.*, http://www.taipeitimes.com/News/front/archives/2005/08/11/2003267233 (accessed 22 July 2009).

Tan, K. (2008). 'Singapore Finds Melamine in White Rabbit Candies; Chinese Dairy Products now Banned across Asia and Africa'. *Shanghaiist online,* http://shanghaiist.com/2008/09/22/singapore_finds_melamine_in_white_r.php (accessed 22 September 2008).

Tschang, C. (2008, September 29). 'Contaminated Milk Sours China's Dairy Business'. *Business Week Online*, http://www.businessweek.com/globalbiz/content/sep2008/gb20080926_543133.htm (accessed 20 July 2009).

Wei, K. (2008). 'Google Disclosed the Hottest Search in China in 2008'. *China.COM.CN.* http://www.china.com.cn/international/txt/2008-12/11/content_16930439.htm (accessed and translated 11 December 2008).

Xinhua (2008a). 'Earthquake Measuring 7.8 Richter Scale Hits SW China'. *Chinaview online*, http://news.xinhuanet.com/english/2008-05/12/content_8153541.htm (accessed 17 July 2009).

Xinhua (2008b). 'Stone's Quake 'Karma' Apology Doesn't Mollify Many Chinese'. *China Daily Online,* http://www.chinadaily.com.cn/china/2008-05/29/content_6721642.htm (accessed 29 May 2008).

Xinhua (2008c). '4 Officials Sacked Following Baby Milk Scandal'. *China Daily Online*, http://www.chinadaily.com.cn/china/2008-09/17/content_7034236.htm (accessed 08 May 2009).

Ye, M. (2008). 'Sanlu Milk Scandal: Self-reflection of Crisis Public Relations'. *Yahoo online*. http://xk.cn.yahoo.com/articles/080927/1/dpn7_3.html (accessed 27 September 2008).

This work was previously published in International Journal of Asian Business and information Management, Volume 1, Issue 4, edited by Dasho Karma Ura and Patricia Ordóñez de Pablos, pp. 20-29, copyright 2010 by IGI Publishing (an imprint of IGI Global).

Chapter 20
Implementation Concerns of Scholar-Practitioners:
A Pilot Study of the Link between Research and Practice

Claretha Hughes
University of Arkansas, USA

Jai Wang
Texas A&M University, USA

Wei Zheng
Northern Illinois University, USA

Laird McLean
McLean Global Consulting, USA

ABSTRACT

The challenge of combining research and practice in HRD has led to continuing debate on who are scholar-practitioners and how they integrate research and practice in the workplace. This qualitative collective case study provides insights from seven scholar-practitioners to help HRD professionals and researchers understand implementation concerns of scholar-practitioners. The findings reveal scholar-practitioners' perceptions of their roles, the link between research and practice, and actions they took to overcome barriers and challenges related to integrating research and practice. This pilot study can serve as a benchmark for future studies regarding successful integration of research and practice in HRD.

INTRODUCTION

Scholars have been trying to understand how professionals engage themselves in both research and practice to better disseminate research findings in the workplace. In this effort, many terms have been used to describe this group of professionals such as research practitioner (Lynham, 2002), scientist-practitioners (Brewerton & Millward, as cited in Hamlin, 2002; Muchinsky, 2004), scholar-practitioners (Graham & Kormanik, 2004), practitioner-theorists (Lynham, 2002), scholarly practitioners (Ruona, 1999), and reflective practitioners (Jacobs, 1999). Ruona (1999)

DOI: 10.4018/978-1-4666-0276-2.ch020

theorized and differentiated HRD practitioners by categorizing them into four types—atheoretical practitioners, practitioners, reflective practitioners, and scholarly practitioners.

- Atheoretical practitioners were defined as having very little grounding in theory, perhaps no relevant educational qualifications or affiliation with professional associations, lacking the knowledge of sound and credible resources, and lacking the skills to consult and interpret scholarly resources.
- Practitioners were defined as meeting the minimum standards of the profession, having a mastery of the common body of knowledge, continually updating of that knowledge, having a willingness to engage in dialogue, having the ability to strategically and effectively analyze/use scholarly resources and theory in practice, and the ability to "sell" theory to gain support of key stakeholders.
- Reflective practitioners were defined as meeting the standards of the *Practitioner* but also critically reflecting on their practice and consulting scholarly resources as a basis for the improvement of their practice.
- Scholarly practitioners were defined as meeting the standards of both *Practitioner* and *Reflective Practitioner*, and also contributing to theory through research, publication in both refereed and non-refereed journals, involvement in scholarly conferences, and having a goal of further development of the field of HRD (p. 895).

We adopted Short's definition and used it as a guide for the selection of participants because it encompasses Ruona's reflective and scholar practitioners definitions.

HRD scholar-practitioners operate as a bridge between HRD research and HRD practice to improve the understanding and practice of HRD. They ground their practice in research and theory, they are champions of research and theory in the workplace and in professional associations, they conduct research, and they disseminate findings from their own research and practice. They are partners with academics and with other practitioners (p. 261).

HRD is not alone in its quest to understand and define the link between research and practice (Short, 2006b, 2006c; Short, Keefer, & Stone, 2006). In fact, the scholar-practitioner divide has also been noted and explored across many different fields including psychology, nursing, education and business (e.g., Amabile et al., 2001; Argyris, 1985; Burack, 1999; Corwin & Louis, 1982; Dean & Bowen, 1994; Gray & Wood, 1991; Hackman, 1985; Mohrman, Gibson, & Mohrman, 2001; Rynes, Bartunek, & Daft, 2001; Van de Ven & Johnson, 2006).

The gap between research and practice is widely recognized by the HRD community and we are constantly searching for ways to connect the divide. From Estabrooks (1999) we know that "many factors get in the way of using research, and empirically, we know very little about what makes research use happen or not happen" (p. 15). The HRD community remains unclear about the role of HRD scholar-practitioners - how they perceive the relationship between research and practice and more importantly, how they combine these two aspects in the workplace.

Implementation concerns contributes to the gap between theory and practice, because theorists "are relatively unconcerned with how their theories, principles and methods are put into practice in arenas outside of academy study," whereas practitioners are "deeply concerned with matters of implementation because what they do occurs in arenas not created primarily for scientific study" (Muchinsky, 2004, p. 208). Muchinsky further suggested that there be a linkage between organizational change and implementation and that the scientist and practitioner both could benefit from a better understanding of the linkage. Specifically, Muchinsky notes that "issues of implementation are rarely discussed in academic journal articles" (p. 208). This study attempts to enhance our current understanding of scholar-practitioners and add to the much needed literature base regarding the implementation concerns and strategies used to link research to practice. This study explored HRD scholar-practitioners' perceptions and experiences of being and how they are using research to inform HRD practice.

Through our thorough search of multiple databases, we found few articles that specifically address the personal perceptions and experiences of scholar-practitioners and their use of research to inform HRD practice. The extant research on the gap between theory and practice (Banks & Murphy, 1985; Hulin, 2001; Latham, 2001; Muchinsky, 2004; Short, 2006a, b, c; Short et al., 2006; Van de Ven & Johnson, 2006) appears to be a primarily conceptual analysis and very little empirical evidence is available to provide deeper implementation insights. Presented in this section are three prevailing views of the gap between theory and practice: (1) a knowledge transfer problem, (2) two distinct kinds of knowledge, and (3) a knowledge production problem.

Gap between Theory and Practice as a Knowledge Transfer Problem

The gap between theory and practice is typically perceived as a knowledge transfer problem, that is, the problem in translating and transferring or disseminating research knowledge into practical knowledge (Van de Ven & Johnson, 2006). A number of researchers have attempted to explain the difficulty of implementing research knowledge in practice. For example, Short et al. (2006) provided several reasons across disciplines. First, researchers are often disconnected from practice. Second, research questions do not address issues that are important to practitioners. Third, research methodologies do not generate data that help to answer practitioners' questions. Fourth, the findings of research are not disseminated in ways that are likely to make an impact on practice. Furthermore, practitioners have difficulty understanding research and applying research knowledge due to lack of education and training. Finally, practitioners and researchers have limited opportunities to communicate with one another about issues of their common concern.

Reflecting on his professional experiences, Latham (20001) outlined five reasons that practitioners do not read academic journals that may be of use to them in the workplace: (1) lack of time, (2) difficulty in transferring learning from the journals or research findings to practice or a specific application, (3) the difference between the organization and the participant sample from the ones with whom practitioners are working, (4) practitioners' resistance to journals that reject them, and (5) narrow scope in journal topics.

In the similar vein, Hulin (2001) stated that research that was conducted by scientists often does not address immediate problems facing managers in organizations. He further noted, "Practitioners want to solve this problem in this organization at this time [and] their issues of generality are not important because (a) this problem may never come up again, (b) it may come up again but only in my expected (short) tenure in this organization, and (c) if it does come up again, we can try the same intervention that solved it this time" (pp. 227-228).

Through an empirical study on the usefulness of research as perceived by practitioners, Mohrman, Gibson, and Mohrman (2001) found that practitioners in ten companies that were going through organizational change viewed research as useful when they worked collaboratively with researchers to interpret research findings and when they had opportunities to develop actions based on the research findings.

Similarly, Banks, and Murphy (1985) pointed out that there is a divergence in focus indicating that researcher's solutions may not speak to practitioners' problems and that only through "joint effort of researchers and practitioners can useful products be generated and adopted in organizations" (p. 337).

Gap between Theory and Practice as Two Distinct Kinds of Knowledge

Researchers identified theory and practice as two distinct kinds of knowledge (Muchinsky, 2004; Van de Ven & Johnson, 2006). The two distinct kinds of knowledge have been characterized as being scholarly work and managerial work (Van de Ven & Johnson, 2006), scientific and practical (Muchinsky, 2004; Pettigrew, 2001), academic and practitioner (Hulin, 2001; Latham, 2001), or theory and practice (McKelvey, 2006; Van de Ven & Johnson, 2006). Regardless of how they are defined, there appears to be a consensus that theory and practice are two distinct kinds of knowledge that perpetuates the existence of the gap between the two. There are studies that have provided identification and recommended solutions to link the two distinct kinds of knowledge (Johnson, Garzioli, Jamal, & Berryman, 2001; Johnson et al., 2002; Kondrat, 1992).

According to Latham (2001), "the workforce consistently provides thought-provoking questions; the journals consistently provide answers for practice; and the implementation of the answers provides a basis for publication and subsequent research" (p. 202). These thoughts point to a continuous cycle view of research and practice as opposed to the choice of one or the other.

In contrast to Latham's (2001) viewpoint, Hulin (2001) argued,

Journals are the repository of the research we do; if they are not read, our research has no utility. If they are read by other academics, but not by practitioners, the research has no utility for the practitioners ...; [therefore], its utility is not diminished by being unread by practitioners (p. 226).

Practitioners often resolve their issues by using repetitive techniques that have proven successful in the past; allowing them to focus on their singular, organizational goals and not invest in long-term research that may not result in a payoff for the organization. Hulin's (2001) study also alludes to the difference in the time constraints that practitioners focus upon in contrast to that of researchers. Knowledge is often used contextually, contingent upon the situation the researcher and practitioner is involved at the time (Van de Ven & Johnson, 2006).

According to McKelvey (2006), "practitioners need immediate help; they can't wait for scientists' lengthy conception- to – publication time cycle. They need site – and time – specific insights" (p. 826). Van de Ven and Johnson (2006) supported McKelvey's assertion that practitioners are interested in immediate problem-solving while researchers are concerned with generalizable knowledge.

The distinction between the two types of knowledge must be understood by scholar-practitioners as they engage in workplace activities, especially with regards to completion of projects that are time-sensitive. Connecting the divide between theory and practice requires recognition that there are two distinct kinds of knowledge and the relationship that exists between the two.

Gap between Theory and Practice as a Knowledge Production Problem

Another school of thought is that the gap between theory and practice is a knowledge production problem (Van de Ven & Johnson, 2006). Van de Ven and Johnson (2006) pointed out that the knowledge production problem occurs at two levels: institutional and individual. At the institutional level, it becomes an issue of structural reform and its impact on research. In other words, is research enabled or restrained based on institutional arrangements?

At the individual level, there needs to be a deeper form of research that engages the academic and practitioner or engaged scholarship (Pettigrew,

2001). Boyer (1990) defined engaged scholarship as consisting of a set of reforms that minimizes the insular behaviors of academic departments and disciplines which results in a fundamental shift in how scholars define their relationships within their communities. Scholar-practitioners must be able to identify their community of practice.

Van de Ven and Johnson (2006) also suggest four interrelated activities to resolve the knowledge production problem. First, researchers need to ground the research question or problem in a concrete and observable phenomenon in order to appreciate and situate its multiple dimensions and manifestations. Second, researchers must develop plausible concepts and frameworks that represent the core aspects of the observed phenomenon and that thereby provide a base to develop new theories that will likely address the central research question under study. Third, researchers must seek appropriate methods that will guide the design and collection of empirical evidence needed for examining the phenomenon under study. Finally, researchers must find ways to apply and disseminate the research findings from the perspectives of different academic and practitioner users (Aram & Salipante, 2003; Van de Ven, 2000).

In addition to identifying the causes for the gap between research and practice, scholars across academic disciplines have been searching for ways of bridging such gap (Banks & Murphy, 1985; Garland, 1999; Gray & Wood, 1991; Hulin, 2001; Latham, 2001; Muchinsky, 2004; Short, 2006a, b, c; Short et al., 2006; Van de Van & Johnson, 2006). For example, Carr and Kemmis (1986) called for the need to view research and practice as "mutually constitutive, as in a process of interaction which is a continual reconstruction of thought and action" (p. 34). Van de Ven and Johnson (2006) proposed to fill the theory-practice gap through "engaged scholarship in which researchers and practitioners coproduce knowledge that can advance theory and practice in a given domain" (p. 803). The purpose of engaged scholarship is not only to "create practitioner-meaningful research" (Van de Ven & Johnson, 2006, p. 822), but also to create scientifically meaningful research, in other words, to produce knowledge that advances both practice and science. Beer (2001) encouraged researchers to provide practical implications and specific suggestions for how to use the research knowledge they produced.

In the field of HRD, effort has also been made to this end. For example, Short (2006a) conducted a review of the HRD publications between 1995 and 2005 and recommended some activities that are likely to promote the dissemination of research into practice and help develop practitioners into scholarly practitioners. Short (2006a) suggested that practitioners adopt an evidence-based orientation to their practice; consult research and theory on HRD to resolve implementation issues in a timely manner; scientifically test interventions before adopting into practice; provide access to experiences or partner with external researchers to conduct applied research in their organizations; share topics that have a high relevance to the practitioner community with researchers; and participate in conferences and other forums for researchers and practitioners.

Despite the diversity of suggestions provided by researchers across disciplines, what appears to be consistent is that the collaboration between researchers and practitioners is essential for disseminating research knowledge in practice. Understanding the perceptions and experiences of dedicated scholar-practitioners and the mechanisms that facilitate and/or hinder their efforts to combine research and practice may facilitate this effort.

RESEARCH DESIGN

This study was conducted by a research team of four members. Among them, three are graduate faculty members and one is a HRD practitioner/consultant who is also in pursuit of a PhD degree.

Hermeneutic Phenomenology Using Collective Case Studies?

This qualitative study is a hermeneutic phenomenology because it attempts to get at the experiences of scholarly-practitioners as they understand their practice and role. This study also adopted a collective case study approach that is qualitative and exploratory in nature (Stake, 1995). A case study, by definition, is a detailed investigation of a contemporary phenomenon within its realistic context, "especially when the boundaries between phenomenon and context are not clearly evident" (Yin, 2003, p. 13). It is an essential research strategy in applied disciplines such as HRD (Dooley, 2002). This approach is most useful when "little is known about a phenomenon, current perspectives seem to be inadequate because they have little empirical substantiation" (Eisenhardt, 1989, p. 548), or when the researcher "seeks an in-depth understanding of a phenomenon because of its uniqueness" (Ellinger, Watkins, & Marsick, 2005, p. 330). In collective case study, the researchers examine closely multiple cases to gain a solid understanding of a phenomenon of interest (Stake, 1995). In other words, each case becomes a means to an end. In the context of this study, the phenomenon of dynamics between research and practice was investigated from the perspective of a number of scholar-practitioners.

Sampling Strategies

The multiple cases for the study, or individuals, were selected using criterion-based sampling strategies (Patton, 2002). First, the participant had to be a practitioner in the field of HRD. Second, the participants had to have published scholarly works in an AHRD or HRD related refereed journals. In other words, the participant was either actively engaged in scholarly activities and/or using research findings to guide daily practice. Using these criteria, each member of our research team identified two individuals through personal contact resulting in eight participants for the study. The final sample consisted of five male and two female HRD practitioners who had areas of responsibility in areas such as training, strategic planning, project management, leadership development, performance improvement, and research. Among the seven participants, six have earned a PhD degree.

Methods for Data Collection

Data for the study were collected primarily through individual interviews. We used member checks to assist with triangulation of the data. To ensure the consistency of information obtained from different participants (Patton, 2002), we developed an interview guide with 13 semi-structured and open-ended questions. Each member of the research team conducted two interviews with the two participants identified. Three interviews were conducted face to face, and five via telephone. Prior to the actual interviewing, participants were sent a cover letter explaining the purpose and procedures of the study, and their consent to participate was obtained through their submission of signed Informed Consent forms. Each interview lasted between 60 – 90 minutes. All the interviews were tape recorded. For confidentiality, participants were identified as P1 – P7.

The main unit of analysis was the participants interviewed. All the interviews were thoroughly transcribed by the researchers, respectively. Transcripts were analyzed within and across the units using the constant comparative method (Merriam, 2001) at two stages. First, all the transcripts were reviewed and coded for themes by two researchers independently. A table was developed to represent emerging themes coded based on the interview questions. At this stage, thirteen coding categories were used; key phrases and illustrative examples were identified to support themes under each category. Then the two researchers exchanged results of their independent analysis. Both similarities and differences were noted, compared, and discussed until consensus was reached. Through constant comparison and further analysis guided by the researcher questions, the emerging themes were re-defined, merged, refined, and organized under a reduced five coding categories. The constant comparative method was also used by all researchers during their respective interviews and after transcribing to clarify their understanding of participant responses. Member checks were also conducted.

FINDINGS

A number of findings were elicited from the thematic analysis of seven interview transcripts. Only selective and most relevant themes are reported in this section with support of direct quotes and illustrative examples. Despite the variation of professional experiences, what appeared to be commonly agreed among the seven participants was the following characteristics of a scholarly practitioner: (a) is likely to have a higher degree of education and read broadly outside of his/her field of practice, (b) consciously applies research findings to practical problems, (c) is proficient in conducting research, (d) tends to challenge the status quo, and (e) is grounded in practice.

Specifically, we present key findings under five categories: (1) meaning of "scholar-practitioner," (2) relationship between research and practice, (3) driving forces, (4) challenges and barriers, and (5) recommended strategies. For better data representation, long direct quotes are separated, italicized and single spaced.

Meaning of "Scholar-Practitioner"

Participants shared a diversity of understanding of the meaning of a scholar-practitioner. For example, P2, Vice President of Leadership and Development in a publicly traded company, perceived a scholarly practitioner as "someone who can create or develop program content grounded in theory and supported by research, but applicable or useful in whatever area they work in." P6, a training consultant, described a research-oriented practitioner as "being pragmatic, uses tools such as action research or evaluation research, and may even do ethnographic research. P7, a research director in a large Fortune 500 corporation, defined scholarly practitioners in comparison to non-scholarly practitioners.

A practitioner is somebody who does things according to common practice. They are following the rules of the profession and they don't question the rules. In other words, I would describe a practitioner as a single-loop learner. A research oriented practitioner or scholarly practitioner would be a double-loop learner who is always questioning common practice based on research.

P7 went on to share more insights into the differences,

A research oriented practitioner would probably be somebody who would read broadly and read outside of his particular practice, and therefore, look for general principles to guide practice. A practitioner would just be looking for better practice within the same paradigm...would not be seeing ways to improve or question what they are doing...is looking for things that basically reinforce the way that they do things and perhaps help them do the same old things better. A scholarly practitioner would be more likely to challenge the things that are being done.

Relationship between Research and Practice

Speaking of the role of research in practice, the interviewed HRD practitioners listed values such as providing measurement, reliability and validity of practice, informing them of doing things differently, and helping making better decisions. More specifically, P2 said,

Research supports the practice. I think anything done should be grounded with some reliability and validity. Therefore, you know what you're doing makes sense, can be replicated and applied to others, and is not some pie in the sky presence.

As an executive overseeing training and HR issues, P1 pointed out that "the only value of research is if it can substantiate that various approaches work." In P6's experience, research has made her "a conscious researcher," that is, "be conscious of what the opportunities are around me." P5, serving as the head of a research group in a research-based organization, viewed research as an ongoing practice and a requirement for the company to deliver quality products that are evidence and research-based:

It's a little bit easier at XXX than it might be at some other places because we are a research-based organization. XXX takes great pride in that fact,, that even though it isn't as lucrative for us, that our products and service are not fad based. We don't come out with the flavor of the month. So anything we deliver has to be delivered with the proper amount of reliability and validity evidence…we maintain several proactive research projects that involve collecting data from our clients on an ongoing basis, doing analyses that offer insights into our clients.

Echoing other participants' points, P7 felt very strongly about the influence of research in his practice:

I am always looking for new ways to transform the things that I am doing. My eye is more on research than practice, because I look at practices being something we need to change. And I see my role as being a change agent... My interest in practice has always been to change it and to improve it. In my consulting activities, my interest is always going in and shaking things up by bringing in new ideas from research, never trying to create a more stable practice, always to change things. That's why I say I am a research oriented practitioner rather than practitioner oriented researcher.

In terms of the overall relationship between research and practice, P1 stressed that "they have to be together. One is not exclusive of the other." He went on with the following explanation:

As a practitioner, you have to stay current and reading and trying to stay abreast of what's going on and doing ongoing research, and trying to figure out what are the best practices that exist out there. As a researcher, you have to also want to understand what's going on out in the workplace and the workforce. They go hand in hand. They are not mutually exclusive. It is extremely beneficial to have a combination of both.

Unlike other participants, P7 provided a different perspective by describing the relationship as "very difficult" and "uneasy" and by referring to the field of HRD in particular.

People in HRD tend to be very practice oriented… they tend to embrace things that are very humanistic, whether they are research based or not. So I find there is a tendency to accept certain types of research that tend to fit the mental model of the HRD practitioner and ignore the other things. That's one thing.

Another thing is that because HRD people are not very well grounded in research, because they are not very scholarly, they have difficulty explaining and defending the things that they do. One of the things I have observed is the way we embraced ROI as a panacea. We came up with all kinds of weird ways to justify what we are currently doing, rather than going out and find new ways to do things and being experimental. So I think that practitioners, by their very nature, tend to be status oriented, and in HRD, that's definitely the case.

Implementation Strategies

A number of implementation strategies were identified by the participants to drive their integration of research findings into daily practice. Some mentioned implementation issues were legal issues, change orientation, seeking sound practice, passion about research, credibility, fairness, ethics, efficacy, and return on investment. Among these issues, two were made prominent by the participants: (1) strong appreciation for research and (2) environmental and organizational demands.

Appreciation for research. Evidently, all the participants in this study highly valued research. Such appreciation, coupled with the internal drive for sound practice and desire for new ideas and change, motivated the participants to consistently look for research evidence to improve their practice. In this regard, P6 and P7's remarks were most insightful.

Research lends credibility and understanding to our everyday ventures. Research provides us opportunity to understand what we don't know. Research establishes the bar for professional conduct and dialogue. (P6)

The research keeps me focused on an experimental orientation. I am not just a researcher; I am an innovator. I think researchers can just do every conservative research. I tend to be somebody who gets bored with the status quo, so I am a change agent, an innovator who is also a researcher. (P7)

From a learning-oriented perspective, P7 articulated, "*My whole orientation is to be driven by research, new ideas, and changes, as opposed to be driven by practice.*"

Environmental and organizational demands. Another force that was noted by a few participants was the organizational demand for them to justify their practices and provide evidence for return on investment. P2 described his situation under the organizational pressure for demonstrating benefits to the bottom line:

I always have to justify what I'm doing and prove return on investments. The research is the metrics and measurement of what I'm doing that tells whoever's paying for it; it's reasonable and valid to pay for it because they're going to get some return, meaning some desired thing will change or desired efficiencies.

Challenges and Barriers

Challenges and barriers for linking research and practice are different in nature. This is perhaps largely due to the diversity of backgrounds and experiences the seven participants had. Some of the major themes were selectively reported below with support of representative quotes for each.

Theme 1: Lacking knowledge of/disconnection between practice and research

There is a huge disconnect because they are totally different agendas. There are totally different goals, motives, [and] systems. I don't think they necessarily are designed to co-exist... they are just not necessarily compatible. It's like Sushi and Steak. (P3)

The biggest challenge for me as a scholar-practitioner is being a broad enough thinker, because I have one foot in HR and HRD practice, and the other foot in the scholarly pursuits. I think scholar-practitioners have particular difficulty thinking beyond the area of practice. (P7)

Most of the barriers are organization oriented. It's not a lack of knowledge. I become a barrier when being too knowledgeable coming to a training

role, because people do not look at training as a science. So when you approach it as a scientific process, they get real squirrelly. They don't want to look at it that way, because you start to complicate their world. (P3)

Lack of a colloquium for understanding and exchange. In AHRD there are various philosophical camps, we need to have a "convening" session, and agree to that there are a number of schools of thought—we be real practical and brash—researchers would have no topics to investigate if it were not for the practitioners! Too often at the Academy we were asked to define what research role I play, scholar or practitioner. I have a hard time separating these two roles—I believe that the HR practitioner lives both lives! (P6)

The researchers want to do research for research sake, but sometimes don't look for the application of the research in the way that allows a practitioner to use it. (P2)

I think that the term practice, or practitioner, connotes a resistance to change...Changing practice is very difficult because practitioners seem very comfortable with what they are doing (P7)

You get involved with so many other things that you don't have the luxury of perusing through tons of journals... many times you are not drilling down as deep as you could probably go. (P1)

I think what keeps them separated is that a lot of practitioners don't take the time or have the time to listen to the academic jargon and think that it's a lot of fluff. (P 2)

Other barriers, noted that were not dominate themes include passion and personality. As P1 also noted, "some people are more practical minded than practitioner type oriented people."

Recommended Strategies

In addition to sharing their perceptions about the relationship between practice and research and their practice of engaging research in practice, the participants also recommended a few strategies for linking research to practice.

Table 1 provides a summary of how our research findings address implementation concerns as identified by Short (2006):

The most important issue that was emphasized by all the participants was to (1) make sure that research does have practical application, as P2 urged,

always make your research useful and applicable to someone who is willing to pay for it or support it. And always make their practice focus on cost effective measures and efficiencies because that's something that people will pay for and support as well.

P3 expressed a very similar view as follows:

You have to work on something that has practical applications, even if it's theoretical. It has to inform you to do something differently or better; can't do it in a vacuum. It's a waste of time...make sure that the type of research you are engaged in will have some real tangible benefit to the workforce and society in general versus so much of the research that you see in scholarly journals tend to be self-preservation of species. It is research that really doesn't have any application.

Other recommendations that were offered by the participants include (2) building communities of scholars and practitioners through constant dialogues, (3) constantly educating organizations and members about the value of research, (4) being the proponents of the practitioner, and (5) becoming more of an empathetic practitioner. These suggestions are evidenced by the representative remarks below.

Table 1. Research findings and implementation concerns

Short (2006)	Research Findings/Recommended Strategies
1. Researchers are often disconnected from practice.	make sure that research does have practical application
2. Research questions do not address issues that are important to practitioners.	being the proponents of the practitioner
3. Research methodologies do not generate data that help to answer practitioners' questions.	being the proponents of the practitioner
4. The findings of research are not disseminated in ways that are likely to make an impact on practice.	constantly educating organizations and members about the value of research
5. Practitioners have difficulty understanding research and applying research knowledge due to lack of education and training.	becoming more of an empathetic practitioner; constantly educating organizations and members about the value of research
6. Practitioners and researchers have limited opportunities to communicate with one another about issues of their common concern.	building communities of scholars and practitioners through constant dialogues

I think you have to have a dialogue. You have to build a sense of community where people can have a dialogue and learn about what other people do. You need to hear what my needs are. I need to her what your needs are. We have to care.... The conference is a good opportunity to build a community. I think we need to create opportunities. I have to help practitioners discover what is practical in scholarly work. (P3)

In advising those who aspire to become a scholar-practitioner, P3 shared his viewpoint,

Never take yourself out of the arena completely. Don't isolate yourself with books and research articles. It will lose focus on what it is you are trying to do... the advice is the caution. Don't alienate yourself from the field in which you are engaged in. Don't get off the playing field. Be on the playing field... be a coach and a mentor and a player as well.

In the same effort, the participants encouraged those who want to become scholar-practitioners to "*get good training on the scientist side*," "*find a place of work that values the two,*" and "*do a very good job being the proponents of the practitioner*."

DISCUSSIONS AND IMPLICATIONS

Scholar-practitioner participants referred to the role of the scholar-practitioner using similar terms and characteristics as identified by Short (2006a), Ruona (1999), Muchinsky (2004). This study added a layer of depth to those definitions by explicating components of the role of the scholar-practitioner, encompassing operational definitions, implementation issues, alienating forces, and strategies. These components could be a starting point for constructing competency models or performance measures for scholar-practitioners.

We did not include motivational theories within the theoretical framework of this study; however, further research should be conducted to explore its role in scholar-practitioners implementation of research into practice. Elements of internal and external motivations for integrating research and practice was identified by participants. As revealed by this study, the internal motivations include scholar-practitioners' drive for informed decisions, best practices, and new ideas. External motivations involve the organization's demand for practitioners to show evidence, to justify their practices, and to demonstrate return on investment (ROI). These factors could be combined and reinforced by professional organizations such as AHRD, through disseminating HRD findings to the practitioner world, as well as supporting skill development in evidence-based research. Further research is needed to examine the effectiveness of the efforts of professional organizations in addressing and utilizing the motivations for professional development purposes.

Along the same line of thinking, this study points to the need for training/educating the scholar-practitioner that is different than training one or the other alone; for example, skills include combination of influencing skills, consuming and critiquing research, designing and conducting research, translating research findings into useful plans, to name a few. More studies need to be conducted to take a closer look at the formative events in scholar-practitioners' careers. Academic preparation of HRD professionals also need to be re-visited in light of preparing scholar-practitioners, because the graduate school experience usually serves as an important gateway to an HRD career and an HRD identify. More emphasis needs to be placed on understanding the practical concerns in the HRD world, constructing lines of scholarly inquiry out of practical experiences, forging engaged scholarship between scholars and practitioners, as well as navigating between scholarly and practical representations of research.

Additionally, from a career point of view, a scholar-practitioner might be best served to seek out positions in organizations where research is equally valued as practice (Muchinsky, 2004). The organization should be supportive of scholar-practitioners who use research findings to improve their practice. In this study, most of the participants were marginalized in their organizations to a varied degree, because research may not have been valued or understood. As a result, these scholar-practitioners had to be good influencers and persistent in their effort to overcome the resistance to change. Negotiation skills may be a competency to explore in future research with regards to how scholar-practitioners influence change within their organization.

Furthermore, they must be highly productive and efficient due to the lack of time. They must also be comfortable with ambiguity given the tradeoffs they must have to make because of the opposing issues linking research to practice. Such reality poses a question for researchers and HRD professionals, that is, how can we help universities and organizations foster an appreciation for research-based practice and help them build organizational systems that will encourage "engaged scholarship?"

Findings of the study also indicate that, in general, there are opportunities for connecting the linking research to practice within the workplace. The implementation issues identified by study participants included: (a) lacking knowledge of the gap between practice and research or knowledge transfer problem; (b) resistance to change or a knowledge production problem; and (c) lack of time or two distinct kind of knowledge problem, which are similar to what was found by Ruona (1999) and Van de Ven and Johnson (2006). For instance, Muchinsky (2004) noted that there was a link between implementation and organizational change that strongly related to the gap between research and practice. Participants recognized a perceived gap between research and practice; however, they viewed themselves as capable, within the constraints of their particular organization, to bridge the gap. More research is needed on the role of organizational culture and change and implementation strategies scholar-practitioners are employing to effectively link research to practice as opposed to identifying that a gap exists.

Scholar-practitioners, as evidenced by this study, work in various roles within various types of organizations.

It was also evident that the impact of the organization may determine the effectiveness of the HRD scholar-practitioner. As noted by P5, *since the organization valued research, there was not a problem to implement research into operational practice*. However, that may not be the case within other organizations. Further research on change needs to be reviewed to add depth to how scholar-practitioners may be able to implement change initiatives to link research and practice within organizations. According to Moran and Brightman (2000), to manage change within organizations, leaders must understand the three most powerful drivers of work behavior: purpose, identity, and mastery in individuals. Leaders of change efforts must inspire individuals to align work behavior components and the environment effectively with the necessary organizational change effort. More research is needed to explore scholar-practitioners' perception or organizational change and how they may lead research and practice change efforts within their organizations.

The participants recommended ways to improve the relationships of scholars and practitioners that support the literature, specifically, that research has practical application. Hulin (2001) argued that research does not have to have immediately apparent practical application to be sound for practice. Latham (2001) argued that time was the issue of concern. All of these issues and perspectives are important and are supported by the study participants. More empirical data is needed to support the theories and assess why these barriers and challenges continue to exist despite attempts to bridge the gap by organizations such as American Society for Training and Development (ASTD) and the Academy of Human Resource Development (AHRD) within the field of HRD (Short, 2006c). This study provided empirical evidence of how scholar-practitioners identify themselves and are working to link research to practice. The specific case examples are instances where scholar-practitioners are actively involved in trying to incorporate valid and reliable research into their practice. Future research should include a broader survey study to see if other scholar-practitioners agree with the definition, relationship between research and practice, implementation issues and potential strategies to link research to practice in HRD. Van de Ven and Johnson (2006) argued that "scholarship that is engaged with practice not only enhances the practical relevance of research but also contributes significantly to advancing fundamental knowledge in a given domain" (p. 830). The descriptions provided within this study could be used to further develop theories and competencies for HRD scholar-practitioners as they inform HRD practice.

REFERENCES

Amabile, T. M., Patterson, C., Mueller, J., Wojcik, T., Odomirok, P. W., Marsh, M., & Kramer, S. J. (2001). Acaddmic-practitioner collaboration in management research: A case of cross-profession collaboration. *Academy of Management Journal*, *44*, 418–431. doi:10.2307/3069464

Aram, J. D., & Salipante, P. F. Jr. (2003). Bridging scholarship in management: Epistemological reflections. *British Journal of Management*, *14*, 199–205. doi:10.1111/1467-8551.00374

Argyris, C. (1985). Making knowledge more relevant to practice: Maps for action. In Lawler, E. E. III, Mohrman, A. M., Mohrman, S. A., Ledford, G. E. Jr, & Cummings, T. G. (Eds.), *Doing research that is useful for theory and practice* (pp. 29–125). San Francisco, CA: Jossey-Bass.

Banks, C. G., & Murphy, K. R. (1985). Toward narrowing the research-practice gap in performance appraisal. *Personnel Psychology*, *38*, 335–345. doi:10.1111/j.1744-6570.1985.tb00551.x

Beer, M. (2001). Why management research findings are unimplementable: An action science perspective. *Reflections: The SoL Journal*, *2*(3), 58–65. doi:10.1162/152417301570383

Burack, E. H. (1999). Bridging research to corporate application. In Larwood, L., & Gattiker, U. E. (Eds.), *Impact analysis: How reach can enter application and make a difference* (pp. 17–46). Mahwah, NJ: Erlbaum.

Carr, W., & Kemmis, S. (1986). *Becoming critical: Knowing through action research.* Victoria, Australia: Deakin University.

Corwin, R. G., & Louis, K. S. (1982). Organizational barriers to the utilization of research. *Administrative Science Quarterly*, *27*, 623–640. doi:10.2307/2392534

Dean, J. W., & Bowen, D. E. (1994). Management theory and total quality: Improving research and practice through theory development. *Academy of Management Review*, *19*, 392–418. doi:10.2307/258933

Dooley, L. M. (2002). Case study research and theory building. *Advances in Developing Human Resources*, *4*(3), 335–354.

Eisenhardt, K. M. (1989). Building theories from case study research. *Academy of Management Review*, *14*(4), 532–550. doi:10.2307/258557

Ellinger, A., Watkins, K. E., & Marsick, V. J. (2005). Case study research methods. In Swanson, R., & Holton, E. F. III, (Eds.), *Research in organizations: Foundations and methods of inquiry* (pp. 327–350). San Francisco, CA: Berrett-Koehler.

Estabrooks, D. A. (1999). Mapping the research utilization field in nursing. *Canadian Journal of Nursing Research*, *31*(1), 53–72.

Garland, H. (1999). Management research and management practice: Learning from our colleagues in economics. In Larwood, L., & Gattiker, U. E. (Eds.), *Impact analysis: How research can enter application and make a difference* (pp. 129–135). Mahwah, NJ: Erlbaum.

Graham, P., & Kormanik, M. (2004). Bridging the conference gap: A challenge to enhance the research-practice dialogue. *Human Resource Development International*, *7*(3), 391–393. doi:10.1080/1367886042000245996

Gray, B., & Wood, D. J. (1991). Collaborative alliances: Moving from practice to theory. *The Journal of Applied Behavioral Science*, *27*, 3–22. doi:10.1177/0021886391271001

Hackman, J. R. (1985). Doing research that makes a difference. In Lawler, E. E. III, Mohrman, A. M., Mohrman, S. A., Ledford, G. E. Jr, & Cummings, T. G. (Eds.), *Doing research that is useful for theory and practice* (pp. 126–149). San Francisco, CA: Jossey-Bass.

Hamlin, R. G. (2002). In support of evidence-based management and research-informed HRD through HRD professional partnerships: An empirical and comparative study. *Human Resource Development International*, *5*(4), 467–491. doi:10.1080/13678860210122643

Hulin, C. (2001). Applied psychology and science: Differences between research and practice. *International Association for Applied Psychology*, 225-234.

Johnson, P., Grazoli, S., Jamal, K., & Berryman, G. (2001). Detecting deception: Adversarial problem solving in a low base-rate world. *Cognitive Science*, *25*, 355–392. doi:10.1207/s15516709cog2503_2

Johnson, P., Veazie, P., Kochevar, L., O'Connor, P., Potthoff, S., Verma, D., & Dutta, A. (2002). Understanding variation in chronic disease outcomes. *Health Care Management Science*, *5*, 175–189. doi:10.1023/A:1019740401536

Kondrat, M. E. (1992). Reclaiming the practical: Formal substantive rationality in social work practice. *The Social Service Review*, *166*, 237–255. doi:10.1086/603912

Latham, G. P. (2001). The reciprocal transfer of learning from journals to practice. *Applied Psychology: An International Review, 50*(2), 201–211. doi:10.1111/1464-0597.00054

Lynham, S. A. (2002). The general method of theory-building research in applied disciplines. *Advances in Developing Human Resources, 4*(3), 221–241.

McKelvey, B. (2006). Van de Ven and Johnson's "engaged scholarship": Nice try, but.... *Academy of Management Review, 31*(4), 822–829.

Merriam, S. B. (2001). *Qualitative research and case study applications in education*. San Francisco, CA: Jossey-Bass.

Mohrman, S., Gibson, C., & Mohrman, A. (2001). Doing research that is useful to practice: A model and empirical exploration. *Academy of Management Journal, 44*, 357–375. doi:10.2307/3069461

Moran, J. W., & Brightman, B. K. (2000). Leading organizational change. *Journal of Workplace Learning: Employee Counseling Today, 12*(2), 66–74.

Muchinsky, P. M. (2004). When the psychometrics of test development meets organizational realities: A conceptual framework for organizational change, examples and recommendations. *Personnel Psychology, 57*(1), 179–205. doi:10.1111/j.1744-6570.2004.tb02488.x

Patton, M. Q. (2002). *Qualitative research & evaluation methods* (3rd ed.). Thousand Oaks, CA: Sage.

Pettigrew, A. M. (2001). Management research after modernism. *British Journal of Management, 12*, S61–S70. doi:10.1111/1467-8551.12.s1.8

Ruona, W. E. A. (1999). Theory in 'theory to practice': Voices of practitioners. In K. P. Kuchinke (Ed.), *Proceedings of the AHRD Conference* (pp. 888-896). Bowling Green, OH: AHRD.

Rynes, S. L., Bartunek, J. M., & Daft, R. L. (2001). Across the great divide: Knowledge creation and transfer between practitioners and academics. *Academy of Management Journal, 44*, 340–355. doi:10.2307/3069460

Short, D. C. (2006a). HRD scholar-practitioners: A definition and description of core activities. In F. M. Nafukho & H. C. Chin (Eds.), *Proceedings of the AHRD Conference* (pp. 258-264). Bowling Green, OH: AHRD.

Short, D. C. (2006b). The link between research and practice: Experiences of different professions and implications for HRD. In F. M. Nafukho & H. C. Chin (Eds.), *Proceedings of the AHRD Conference* (pp. 265-272). Bowling Green, OH: AHRD.

Short, D. C. (2006c). The gap between research and practice in HRD: A summary of the discussion 1995-2005. In F. M. Nafukho & H. C. Chin (Eds.), *Proceedings of the AHRD Conference* (pp. 250-257). Bowling Green, OH: AHRD.

Short, D. C., Keefer, J. M., & Stone, S. J. (2006). The link between research and practice: Experiences of different professions and implications for HRD. In F. M. Nafukho & H. C. Chin (Eds.), *Proceedings of the AHRD Conference* (pp. 914-921). Bowling Green, OH: AHRD.

Stake, R. E. (1995). *The art of case study research*. Thousand Oaks, CA: Sage.

Van de Ven, A. H. (2000). Professional science for a professional school: Action science and normal science. In Beer, M., & Nohrta, N. (Eds.), *Breaking the code of change* (pp. 393–413). Boston: Harvard Business School Press.

Van de Ven, A. H., & Johnson, P. E. (2006). Knowledge for theory and practice. *Academy of Management Review, 31*(4), 802–821.

Van de Ven, A. H., & Johnson, P. E. (2006). Reply: Nice try, Bill, But.... There you go again. *Academy of Management Review, 31*(4), 802–821.

Yin, R. K. (2003). *Case Study Research: Design and Methods (applied social research methods)* (3rd ed.). Thousand Oaks, CA: Sage.

This work was previously published in International Journal of Asian Business and information Management, Volume 1, Issue 2, edited by Dasho Karma Ura and Patricia Ordóñez de Pablos, pp. 33-47, copyright 2010 by IGI Publishing (an imprint of IGI Global).

Compilation of References

Abendblatt. (2005, April 2). *Herr der Waschmaschinen (Master of Washing Machines)*. Retrieved October 5, 2010, from http://www.abendblatt.de/daten/2005/04/02/416504.html

Accenture. (2005). *China spreads its wings – Chinese companies Go Global*. Retrieved from www.accenture.com

Adorno, T. (1991). *The Culture Industry: Selected Essays on Mass Culture*. London: Routledge Press.

Ajzen, I. (2006). *Constructing a Theory of Planned Behavior Questionnaire: Conceptual & Methodological Considerations*. Retrieved from www.unix.oit.umass.edu/~aizen/pdf/tpb.measurement.pdf

Ajzen, I. (1991). The theory of planned behavior. *Organizational Behavior and Human Decision Processes*, *50*(2), 179–211. doi:10.1016/0749-5978(91)90020-T

Ajzen, I. (2008). Consumer attitudes and behavior. In Haugtvedt, C. P., Herr, P. M., & Kardes, F. R. (Eds.), *Handbook of Consumer Psychology* (pp. 525–548). New York: Lawrence Erlbaum.

Ajzen, I., & Fishbein, M. (2005). The influence of attitudes on behavior. In Albarracín, D., Johnson, B. T., & Zanna, M. P. (Eds.), *The Handbook of Attitudes* (pp. 173–221). Mahwah, NJ: Erlbaum.

Alomaim, N., Zihni, M., Tunca, S., & Zairi, M. (2003). Customer satisfaction @ virtual organizations. *Management Decision*, *41*(7), 666–670. doi:10.1108/00251740310495595

Alon, I., & McIntyre, J. R. (2008). *Globalizing of Chinese Enterprises*. New York: Palgrave Macmillan.

Alzua, A., O'Leary, J. T., & Morrison, A. M. (1998). Cultural and heritage tourism: Identifying niches for international travelers. *Journal of Tourism Studies*, *9*(2), 2–13.

Amabile, T. M., Patterson, C., Mueller, J., Wojcik, T., Odomirok, P. W., Marsh, M., & Kramer, S. J. (2001). Acaddmic-practitioner collaboration in management research: A case of cross-profession collaboration. *Academy of Management Journal*, *44*, 418–431. doi:10.2307/3069464

Amighini, A., & Chiarlone, S. (2003). Rischi e opportunità dell'integrazione commerciale cinese per la competitività internazionale dell'Italia (Working Paper, 149, CESPRI).

Andersson, A. E. (1985). Creativity and Regional Development. *Papers / Regional Science Association. Regional Science Association. Meeting*, *56*(1), 5–20. doi:10.1007/BF01887900

Anholt-GfK Roper. (2008). *2008 Anholt-GfK Roper Nation Brands Index[SM] Report Highlights*. Retrieved October 5, 2010, from http://www.gtai.com/uploads/media/Report_Highlights_Roper_Nation_Brands_Index.pdf

Anzkiewicz, A., & Whalley, J. (2006). Recent Chinese buyout activity and the implications for global architecture (Working Paper, 12072, NBER).

Aram, J. D., & Salipante, P. F. Jr. (2003). Bridging scholarship in management: Epistemological reflections. *British Journal of Management*, *14*, 199–205. doi:10.1111/1467-8551.00374

Ardichvili, A. (2001). Leadership styles of Russian entrepreneurs and manager. *Journal of Development Entrepreneurship*, *6*(2), 169–187.

Argote, L., & Ingram, P. (2000). Knowledge transfer: A basis for competitive advantage in firms. *Journal of Organizational Behaviour and Human Decision Processes*, *82*(1), 150–169. doi:10.1006/obhd.2000.2893

Argyris, C. (1985). Making knowledge more relevant to practice: Maps for action. In Lawler, E. E. III, Mohrman, A. M., Mohrman, S. A., Ledford, G. E. Jr, & Cummings, T. G. (Eds.), *Doing research that is useful for theory and practice* (pp. 29–125). San Francisco, CA: Jossey-Bass.

Ashby, W. R. (1947). Principles of the Self-Organizing Dynamic System. *The Journal of General Psychology*, *37*, 125–128. doi:10.1080/00221309.1947.9918144

Aulakh, P. S. (2007). Emerging multinationals from developing economies: motivations, paths and performance. *Journal of International Management*, *13*(3), 235–402. doi:10.1016/j.intman.2007.05.001

Aydalot, P., & Keeble, D. (Eds.). (1988). *High Technology Industry and Innovative Environments: The European Experience*. London: Routledge Press.

Aykut, D., & Goldstein, A. (2006). Developing countries multinationals: south-south investment comes of age (Working Paper, 257, OECD).

Ba, J.-S. (2007). The Establishment of China's SME Credit Guarantee Model. *Shanghai Finance*, *8*, 56–63.

Baker, J. H., & Baker, G. A. (2001). *Leadership, Culture and Knowledge Management*. Retrieved from http://leadendtoday.org.

Baker, D. (2009). *Plunder and Blunder: The Rise and Fall of The Bubble Economy*. Sausalito, CA: Polipoint Press.

Ballared, C. L., & Medema, S. G. (1993). The Marginal Efficiency Effects of Taxes and Subsidies in the Presence of Externalities. *Journal of Public Economics*, *52*(2), 199–216. doi:10.1016/0047-2727(93)90020-T

Banks, C. G., & Murphy, K. R. (1985). Toward narrowing the research-practice gap in performance appraisal. *Personnel Psychology*, *38*, 335–345. doi:10.1111/j.1744-6570.1985.tb00551.x

Bao J. (2007). Reconstruction of the traditional concept of circular economy. *Research of Nature of Processes*, (4), 91-94

Barber, B. R. (1992, March). Jihad vs. McWorld. *The Atlantic*. Retrieved October 5, 2010, from http://www.theatlantic.com/doc/199203/barber

Barbera, R. J. (2009). *The Cost of Capitalism: Understanding Market Mayhem and Stabilizing Our Economic Future*. New York: McGraw-Hill.

Barkema, H. G., & Vermeulen, F. (1998). International Expansion through Start-Up or Acquisition: A Learning Perspective. *Academy of Management Journal*, *41*(1), 7–26. doi:10.2307/256894

Barth, J. R., Li, T., Lu, W., Phumiwasana, T., & Yago, G. (2009). *The Rise and Fall of the U.S. Mortgage and Credit Markets: A Comprehensive Analysis of the Market Meltdown*. New York: John Wiley.

Bartlett, C. A., & Ghoshal, S. (1989). *Management Across Borders. The Transnational Solution*. Boston, MA: Hutchinson Business Books.

Bartlett, C. A., & Ghoshal, S. (1989). *Managing across borders: The transnational solution*. Boston: Harvard Business School Press.

Bartlett, C. A., & Ghoshal, S. (2000). Going global: Lessons from late movers. *Harvard Business Review*, *78*(2), 133–142.

Bass, B. M. (1985). *Leadership and Performance Beyond Expectation*. New York: Free Press.

Bass, B. M. (1990). From transactional to transformational leadership: learning to share the vision. *Organizational Dynamics*, *18*(3), 19–31. doi:10.1016/0090-2616(90)90061-S

Bass, B. M. (2002). Cognitive, social, and emotional intelligence of transformational leaders. In Riggio, R. E., Murphy, S. E., & Pirozzolo, F. J. (Eds.), *Multiple Intelligences and Leadership* (pp. 105–118). Mahwah, NJ: Lawrence Erlbaum.

Bass, B. M., & Avolio, B. J. (1994). *Improving Organizational Effectiveness through Transformational Leadership*. Thousand Oaks, CA: Sage.

Bass, B. M., Waldman, D. A., Avolio, B. J., & Bebb, M. (1987). Transformational leadership and the falling dominoes effect. *Group and Organization Studies*, *12*(1), 73–87. doi:10.1177/105960118701200106

Bassett, K., Griffiths, R., & Smith, I. (2002). Cultural Industries, Cultural Clusters and the City: the Example of Natural History Film-making in Bristol. *Geoforum*, *33*(2), 165–177. doi:10.1016/S0016-7185(01)00032-X

Battat, J. (2006). *China's Outward Foreign Direct Investment*. Washington: FIAS, World Bank Group.

BBC News Online (2000). 'China Bans Taiwan's Madonna', http://news.bbc.co.uk/2/hi/asia-pacific/761908.stm (accessed 07 July 2009).

Beamer, L., & Varner, I. (2001). *Intercultural Communication in the Global Workplace*. New York: McGraw-Hill Higher Education.

Beamish, P. W. (1993). The Characteristics of Joint Ventures in the People's Republic of China. *Journal of International Marketing*, *1*, 29–48.

Becker, G. (1998). Human Capital and Poverty. *Religion and Liberty*, *8*(1), 5–7.

Beer, M. (2001). Why management research findings are unimplementable: An action science perspective. *Reflections: The SoL Journal*, *2*(3), 58–65. doi:10.1162/152417301570383

Bellabona, P., & Spigarelli, F. (2007). Moving from Open Door to Go Global: China Goes on the World Stage. *International Journal of Chinese Culture and Management*, *1*(1), 93–107. doi:10.1504/IJCCM.2007.016170

Bellabona, P., & Spigarelli, F. (2006). La dimensione transnazionale dell'economia cinese. In Abbiati, M. (Ed.), *Propizio è intraprendere imprese* (pp. 147–165). Venice: Cà Foscarina.

Bell, S. (2008). *International Brand Management of Chinese Companies: Case Studies on the Chinese Household Appliances and Consumer Electronics Industry Entering US and Western European markets*. Heidelberg, Germany: Physica Verlag.

Bennett, R., & Rundle-Thiele, S. (2004). Customer satisfaction should not be the only goal. *Journal of Services Marketing*, *18*(7), 514–523. doi:10.1108/08876040410561848

Berger, A. N., & Udell, G. F. (1998). The economics of small business finance: the roles of private equity and debt markets in the financial growth cycle. *Journal of Banking & Finance*, *22*, 613–673. doi:10.1016/S0378-4266(98)00038-7

Berger, A. N., & Udell, G. F. (2006). A more complete conceptual framework for SME finance. *Journal of Banking & Finance*, *30*(11), 2945–2966. doi:10.1016/j.jbankfin.2006.05.008

Berger, U. (2002). Small Business Credit Availability and Relationship Lending: The Importance of Bank Organizational Structure. *The Economic Journal*, *112*(447), 32–53. doi:10.1111/1468-0297.00682

Bertoni, F., Elia, S., & Rabbiosi, L. (2008, October 9-10). *Drivers of acquisition from Brics to advanced countries: firm level evidence*. Paper presented at Emerging Multinationals: Outward Foreign Direct Investment from Emerging and Developing Economies Conference, Copenhagen, Denmark.

Betser, Y., & Fraser, P. (2006, November). 'Trouble in China'. *Brand Strategy*, *207*, 48–49.

Bhargava, S. (2003). *Transformational leadership: Value-based management for Indian organizations*. New Delhi, India: Response Books.

Bhattacharyya, S. K., & Rahman, Z. (2003). Sources of first mover advantages in emerging markets - an Indian perspective. *European Business Review*, *15*, 359–369. doi:10.1108/09555340310500541

Bi, J-Y. (2006). The economic globalization time international industry shifts new tendency with our country face the opportunity and challenges. *China Price*, 9-18.

Bian, Y. (1997). Bringing strong ties back in: Indirect ties, network bridges, and job searches in China. *American Sociological Review*, *62*, 366–385. doi:10.2307/2657311

Bian, Y. (1999). Getting a job through a web of guanxi in China. In Wellmam, B. (Ed.), *Networks in the global v illage: Life in contemporary communities* (pp. 275–296). Boulder, CO: Westview Press.

Bian, Y., & Ang, S. (1997). Guanxi networks and job mobility in China and Singapore. *Social Forces*, *75*, 981–1005. doi:10.2307/2580527

Bilkey, W. J., & Nes, E. (1982). Country-of-Origin Effects on Product Evaluations. *Journal of International Business Studies*, *13*(1), 89–99. doi:10.1057/palgrave.jibs.8490539

Bilton, C. (2010). *Management and Creativity: From Creative Industries to Creative Management*. Beijing, China: New Word Press.

Black, N. L. (1990). A model and methodology to assess change to heritage buildings. *Journal of Tourism Studies*, *1*(1), 15–23.

Bollinger, A. S., & Smith, R. D. (2001). Managing Organizational Knowledge as a Strategic Asset. *Journal of Knowledge Management*, *5*(1), 8–18. doi:10.1108/13673270110384365

Bonaglia, F., Goldstein, A., & Mathews, J. A. (2007). Accelerated internationalization by emerging markets' multinationals: The case of the white goods sector. *Journal of World Business*, *42*(4), 369–383. doi:10.1016/j.jwb.2007.06.001

Bond, M. H. (2003). Cross-cultural social psychology and the real world of culturally diverse teams and dyads. In D. Tjosvold & K. Leung (Eds.), *Cross-cultural management: Foundation and Future* (pp 45-58). Aldershot, Hampshire: Ashgate Publishing.

Bond, M. H. (1992). *Beyond the Chinese face: Insights from Psychology* (4th ed.). New York: Oxford University Press.

Bonner, W., Incontrera, K., & Wiggin, A. (2009). *Financial Reckoning Day Fallout: Surviving Today's Global Depression*. New York: Wiley.

Bonner, W., & Wiggin, A. (2009). *The New Empire of Debt: The Rise and Fall of an Epic Financial Bubble*. New York: Wiley.

Borsari, S., Ceccagno, A., Gallo, A., & Gaudino, S. (2006). *I cinesi e gli altri. L'imprenditoria extracomunitaria in Italia*. Bologna, Italy: Consorzio Spinner.

Boulding, W., & Christen, M. (2001). Idea - First-mover disadvantage. *Harvard Business Review*, *79*, 20–21.

Bourdieu, P. (1993). *The field of cultural production: Essays on art and literature*. New York: Columbia University Press.

Bovenberg, A. L., & Goulder, L. H. (1996). Optimal environmental taxation in the presence of other taxations: general equilibrium analysis. *The American Economic Review*, *86*(4), 985–1000.

Bradley, F. (2002). *International Marketing Strategy* (4th ed.). Upper Saddle River, NJ: Prentice Hall.

Brandenburgs, D. (2007). *Haier will Higher*. Retrieved October 5, 2010, from http://www.brandenburgs.net/site/DE/int/03_presse/03_02_berichte/44_container_datum.php?we_lv_start_self=23&PulldownListviewCat=

Branstetter, L., & Lardy, N. (2006). *China's embrace of globalization* (Working Paper, 12373, NBER).

Brouthers, K. D., & Bamossy, G. J. (1997). The role of key stakeholders in international joint venture negotiations: Case studies from Eastern Europe. *Journal of International Business Studies*, *28*, 285–308. doi:10.1057/palgrave.jibs.8490102

Buch, C. M., Kleinert, J., & Toubal, F. (2006). Where enterprises lead, people follow? Links between migration and FDI in Germany. *European Economic Review*, *50*(8), 2017–2036. doi:10.1016/j.euroecorev.2005.11.002

Buckley, P. J. (2002). Is the International Business research agenda running out of steam? *Journal of International Business Studies*, *33*(2), 365–373. doi:10.1057/palgrave.jibs.8491021

Buckley, P. J., Clegg, J., Cross, A. R., Liu, X., Voss, H., & Zheng, P. (2007). The Determinants of Chinese Outward Foreign Direct Investment. *Journal of International Business Studies*, *38*(4), 499–518. doi:10.1057/palgrave.jibs.8400277

Buckley, P. J., Clegg, J., Cross, A., Voss, H., Rhodes, M., & Zheng, P. (2008a). Explaining China's Outward FDI. In Sauvant, K. P. (Ed.), *The Rise of Transnational Corporations From Emerging Markets: Threat or Opportunity?* Willinston, VT: Edward Elgar Publishing.

Buckley, P. J., Cross, A. R., Tan, H., Liu, X., & Voss, H. (2008b). Historic and Emergent trends in Chinese Outward Direct Investments. *Management International Review*, *48*(6), 715–748. doi:10.1007/s11575-008-0104-y

Burack, E. H. (1999). Bridging research to corporate application. In Larwood, L., & Gattiker, U. E. (Eds.), *Impact analysis: How reach can enter application and make a difference* (pp. 17–46). Mahwah, NJ: Erlbaum.

Burgess, R. (1987). *In the field: An introduction to field research*. London: George Allen and Unwin.

Burns, J. M. (1978). *Leadership*. New York: Harper.

Burton, R. (1995). *Travel Geography*. Pitman Publishing.

Businessweek. (2008, July 31). *Haier: Taking a Brand Name Higher*. Retrieved October 5, 2010, from http://www.businessweek.com/magazine/content/06_31/b3995009.htm

Buttery, A. E., & Wong, Y. H. (1999). The development of a Guanxi framework. *Marketing Intelligence & Planning*, *17*(3), 147–154. doi:10.1108/02634509910271605

C. B. R. C. (2008). *Annual Report 2008*. Beijing, China: China Banking Regulatory Commission.

Cacioppo, K. (2007, November-December). *Harvard Business Review*.

Calomiris, C. (2009). Financial innovation, regulation, and reform. *The Cato Journal*, *29*(1), 65.

Cantwell, J., & Barnard, H. (2008). Do Firms from Emerging Markets have to Invest Abroad? Outward FDI and the Competitiveness of Firms. In Sauvant, K. P. (Ed.), *The Rise of Transnational Corporations From Emerging Markets: Threat or Opportunity?* Willinston, VT: Edward Elgar Publishing.

Capgemini & Merrill Lynch. (2007, 2008, 2009). *World Wealth Reports 2007, 2008, 2009. Capegemini Consulting Technology & Merrill Lynch Wealth Management.* Retrieved from http://www.capgemini.com

Carabelli, A., Hirsch, G., & Rabellotti, R. (2006). *Italian SMEs and Industrial District on the move: Where are they going?* (Working Paper, 3). Bologna, Italy: University of Bologna, Department of Economic Science.

Carlson, J. M., Grannan, E. R., Singh, C., & Swindle, G. H. (1993). Fluctuations in self-organizing systems. *The American Physical Society*, *48*(22), 688–698.

Carow, K., Heron, R., & Saxton, T. (2004). Do early birds get the returns? An empirical investigation of early-mover advantages in acquisitions. *Strategic Management Journal*, *25*, 563–585. doi:10.1002/smj.404

Carr, W., & Kemmis, S. (1986). *Becoming critical: Knowing through action research.* Victoria, Australia: Deakin University.

Celep, C., & Cetin, B. (2005). Teachers' Perception about the Behaviours of School Leaders with regard to Knowledge Management. *International Journal of Educational Management*, *19*(2), 102–117. doi:10.1108/09513540510582408

Chakraborty, S. K. (1993). *Managerial Transformation by Values: A Corporate Pilgrimage.* New Delhi, India: Sage.

Chakraborty, S. K. (1995). Wisdom leadership: leading self by the SELF. *Journal of Human Values*, *1*, 205–220. doi:10.1177/097168589500100205

Chakraborty, S. K. (1999). *Wisdom Leadership: Dialogues and Reflections.* New Delhi, India: Wheeler Publishing.

Chakraborty, S. K., & Chakraborty, D. (2008). Spirituality. In *Management: Means Or End?* Oxford, UK: Oxford University Press.

Chang, T. C., Milne, S., Fallan, D., & Pohlmann, C. (1996). Urban heritage tourism: The global-local nexus. *Annals of Tourism Research*, *23*(2), 284–305. doi:10.1016/0160-7383(95)00064-X

Chen, N., & Tjsovold, D. (2007). Guanxi and leader member relationships between American managers and Chinese employees: open-minded dialogue as mediator. *Asia Pacific Journal of Management*, *24*(2), 171–189. doi:10.1007/s10490-006-9029-9

Cheng, C. (2008). 'Pay Attention to the Taxation Problems when Companies Recall the Tainted Milk Formula', http://www.ctaxnews.com.cn/syxw/xwxzt/200809/t20080924_1532169.htm (accessed 26 September 2008).

Chen, H., & Chen, T. J. (1998). Network linkages and location choice in foreign direct investment. *Journal of International Business Studies*, *29*, 455–478. doi:10.1057/palgrave.jibs.8490002

Chen, Y. (2008). Meinuo 109 Nian de Zhuanzhu (Miele: Watching 109 Years). *Zhongguo Xin Shidai*, *8*, 53–55.

Cheung, S. C. H. (1999). The meanings of a heritage trail in Hong Kong. *Annals of Tourism Research*, *26*, 570–588. doi:10.1016/S0160-7383(99)00006-7

Child, J., & Pleister, H. (2003). Governance and management in China's private sector. *Management international*, *7*(3), 13-24.

Child, J., & Rodrigues, S. B. (2005). The internationalization of Chinese firms: a case of theoretical extension? *Management and Organization Review*, *1*(3), 381–410. doi:10.1111/j.1740-8784.2005.0020a.x

China Statistical Yearbook. (1982-2010).

Chin, K. F. (2001). *Developing a Knowledge Management Strategy.* Price Water House Coopers.

Chin-loy, C. (2003). *Assessing the Influence of Organizational Culture on Knowledge Management Success.* Unpublished UMI dissertation, Wayne Huizeng School of Business and Entrepreneurship. Nova Southeastern University.

Chittoor, R., & Ray, S. (2007). Internationalization paths of Indian pharmaceutical firms – A strategic group analysis. *Journal of International Management*, *13*(3), 338–355. doi:10.1016/j.intman.2007.05.008

Chorafas, D. N. (2009). *Financial Boom and Gloom: The Credit and Banking Crisis of 2007-2009 and Beyond.* New York: Palgrave Macmillan.

Christopherson, S., & Storper, M. (1986). The City as Studio; the World as Back Lot: the Impact of Vertical Disintegration on the Location of the Motion Picture Industry. *Environment and Planning. D, Society & Space, 4*(3), 305–320. doi:10.1068/d040305

Chung, M. (2006). *The Impact of Cultural Difference: An Australian Business in China.* Unpublished Doctoral Thesis, Monash University, Melbourne.

Chung, M. (2008). *Shanghaied: Why Foster's Could Not Survive China.* Melbourne, Australia: Heidelberg Press.

Chung, M., & Gong, W. Y. (2009). *How Food Is Used as an Effective Business Tool in China?* Paper presented at the Symposium on Food Law & Policy, Melbourne, Australia.

Chung, M., & Smith, W. (2007). Overseas Chinese as Expatriate Managers in China – Is Their Recruitment a Solution to Cross-Cultural Management Problems for Multinationals Operating in China? *International Journal of Diversity in Organisations, Communities and Nations, 7.*

Clow, K. E., & Baack, D. (2010). *Integrated Advertising, Promotion and Marketing Communication.* Upper Saddle River, NJ: Prentice Hall.

Coco, G. (2000). On the use of collateral. *Journal of Economic Surveys, 14*(2), 191–214. doi:10.1111/1467-6419.00109

Cohen, J. (1988). *Statistical Power Analysis for the Behavioral Sciences* (2nd ed.). Hillsdale, NJ: Lawrence Erlbaum.

Collins Cobuild. (2009). *Collins Cobuild Advanced Dictionary.*

Columba, F., Gambacorta, L., & Mistrulli, P. E. (2009). *The effects of mutual guarantee consortia on the quality of bank lending.* Munich, Germany: University Library of Munich.

Corò, G., Volpe, M., & Bonaldo, S. (2005, May 27). *Local production systems in Italy between fragmentation and international integration.* Paper presented at the Economia Internazionale e Sviluppo conference, University of Urbino, Urbino, Italy.

Corwin, R. G., & Louis, K. S. (1982). Organizational barriers to the utilization of research. *Administrative Science Quarterly, 27,* 623–640. doi:10.2307/2392534

Cossons, M. (1989). Heritage tourism: Trends and tribulations. *Annals of Tourism Research, 10,* 192–194.

Council Development Research Center Industry Economical Research Department. (2006, November 9). Continues the international industrial shift: Our country must deal with six aspects the challenge. *China Economical Times Daily,* 11-09.

Cousin, V. (2007). *Banking in China.* New York: Palgrave Macmillan.

Crandall, W. R., Parnell, J. A., & Spillan, J. E. (2010). *Crisis Management in the New Strategy Landscape.* Los Angeles: Sage.

Crandall, W. R., Parnell, J. A., Xihui, P., & Long, Z. (2007). 'When Crisis Management goes Abroad: The demise of SK-II in China'. *Journal of International Business Research and Practice, 1*(1), 38–49.

Crawford, C. B. (2004). *Transformational Leadership, Innovation and Knowledge Management: Empirical Findings and Emergent Conclusions.* Hays, KS: Fort Hays State University.

Cross, A. R., & Voss, H. (2008, February 1-2). *Chinese direct investment in the United Kingdom: An assessment of motivations and competitiveness.* Paper presented at the Corporate Strategies in the New Asia Conference, University of Bremen, Bremen, Germany.

Cuervo-Cazurra, A. (2007). Sequence of value-added activities in the multinationalization of developing country firms. *Journal of International Management, 13*(3), 258–277. doi:10.1016/j.intman.2007.05.009

Cuervo-Cazurra, A., & Genc, M. (2008). Transforming disadvantages into advantages: developing-country MNEs in the least developed countries. *Journal of International Business Studies, 39*(6), 957–979. doi:10.1057/palgrave.jibs.8400390

Cui, D. (2009). About Low-carbon Economy. *Environmental Education,* (7), 19-21.

Customer definition. (n.d.). Retrieved January 24, 2009, from http://www.ichnet.org/glossary.htm

Customer satisfaction. (n.d.). Retrieved January 24, 2009, from http://www.ic.gc.ca/eic/site/stco-levc.nsf/eng/h_qw00037.html

d'Italia, B., & Cambi, U. I. (2004). *Manuale della Bilancia dei Pagamenti e della posizione patrimoniale dell'Italia sull'estero*. Roma. Retrieved from http://www.bancaditalia.it/statistiche/quadro_norma_metodo/metodoc/manuale_bilpag.pdf

Dai, H. (2007). International industrial shift new tendency and to our country's enlightenment. *International trade, 2*, 13-14.

Dai, H. (2006). *The international industrial shift with the Chinese manufacturing industry develops*. Beijing, China: People's Publishing Agency.

Daily, C. (2006a, September 19). 'Actress Accused of False Advertising', http://www.chinadaily.com.cn/china/2006-09/19/content_692408.htm (accessed 15 July 2009).

Daily, C. (2006b, September 23). 'Consumers Angry as SK-II Pulled from Shelves', http://www.chinadaily.com.cn/china/2006-09/23/content_695363.htm (accessed 15 July 2009).

Daily, C. (2006c, September 26). 'Consumers Turn Cautious on P&G Skin-care Row'. *China Daily Online*, http://www.chinadaily.com.cn/china/2006-09/26/content_696526.htm (accessed 15 July 2009).

Danieir, T. (1998). Networking your knowledge. *Management Review.*

Darvill, T. (1987). *Ancient Monuments in the Countryside*. London: English Heritage.

Davenport, T. H. (1997). Ten principles of Knowledge management and four case studies. *Knowledge and Process Management, 4*, 187–208. doi:10.1002/(SICI)1099-1441(199709)4:3<187::AID-KPM99>3.0.CO;2-A

Davenport, T. H., & Prusak, L. (1998). *Working Knowledge: How Organizations Manage What They Know*. Boston: Harvard Business School Press.

Davies, K. (2007). China's Investment Watch. Organisation for Economic Cooperation and Development. *The OECD Observer. Organisation for Economic Co-Operation and Development, 260*, 19.

Davis, K. (2009). While global FDI falls, China's outward FDI doubles. In *Proceedings of the Columbia FDI in Perspectives, Perspectives on topical foreign direct investment issues, Vale Columbia Center on Sustainable Investment*.

Dawes, J., & Rowley, J. (1996). The waiting experience: Towards service quality in the leisure industry. *International Journal of Contemporary Hospitality Management, 8*(1), 16–21. doi:10.1108/09596119610108608

De Felice, A. (2006). Investimenti Diretti Esteri in Romania: nuove prospettive per il Mezzogiorno. In Del Prete, F. (Ed.), *Prossimità e Sviluppo*. Milano, Italy: Franco Angeli.

De Long, D. W., & Fahey, L. (2000). Diagnosing Cultural Barriers to Knowledge Management. *The Academy of Management Executive, 14*(4), 113–127.

Dean, J. W., & Bowen, D. E. (1994). Management theory and total quality: Improving research and practice through theory development. *Academy of Management Review, 19*, 392–418. doi:10.2307/258933

Deguo Meinuo Gongsi. (n.d.). *Miele Zai Zhongguo Miele (Miele in China)*. Retrieved October 5, 2010, from http://www.miele.cn/cn/domestic/ 5727.htm

DeLaurentis, T. (2009). 'Ethical Supply Chain Management'. *The China Business Review, 36*(3), 38–41.

Deng, J. A. (2007). Comparative Study of environmental taxation and its impact on China. *Tax and Economics*, (3), 95-100.

Deng, L. (2006). Deguo – Zou Chuqu de Xin Wotu (Germany – The New Fertile Soil of (Chinese Enterprises') Going Global). *Da Jingmao, 6*, 50–52.

Deng, P. (2003). Foreign direct investment by transnationals from emerging countries: the case of China. *Journal of Leadership & Organizational Studies, 10*(2), 113–124. doi:10.1177/107179190301000211

Deng, P. (2004). Outward investment by Chinese MNCs: motivations and implications. *Business Horizons, 47*(3), 8–16. doi:10.1016/S0007-6813(04)00023-0

Department of Foreign Affairs and Trade. (2007). *People's Republic of China Country Brief - August 2006*. Retrieved February 19, 2008 from http://www.dfat.gov.au/geo/china/cb_index.html

Department of Foreign Affairs and Trade. (2008). *Australia-An Overview*. Retrieved June 27, 2008 from http://www.dfat.gov.au/aib/overview.html

Desatnick, R. L. (1989, April). Long Live the King. *Quality Progress*, 24-26.

DeTienne, K. B., Dyer, G., Hoopes, C., & Harris, S. (2004). Toward a Model of Effective Knowledge Management and Directions for Future Research: Culture, Leadership, and CKOs. *Journal of Leadership & Organizational Studies*, *10*(4), 26–43. doi:10.1177/107179190401000403

Di Minin, A., & Zhang, J. (2008, October 9-10). *Preliminary Evidence on the International R&D strategies of Chinese companies in Europe*. Paper presented at the Emerging Multinationals: Outward Foreign Direct Investment From Emerging and Developing Economies Conference, Copenhagen, Denmark.

Diao, D., & Zhang, L. (2004). Implementation the science and technology trade strategy and development high-tech product to export. *Commercial Research*, *3*, 12–13.

Ding, D. Z. (1997). Control, Conflict, and Performance: A Study of U.S.-Chinese Joint-ventures. *Journal of International Marketing*, *5*(3), 31–45.

Doh, J., & Ramamurti, R. (2003). Reassessing risk in developing country infrastructure. *Long Range Planning*, *36*, 337–353. doi:10.1016/S0024-6301(03)00069-4

Dooley, L. M. (2002). Case study research and theory building. *Advances in Developing Human Resources*, *4*(3), 335–354.

Dradio. (2004). *Das Gewissen der Verbraucher (The Conscience of Consumers)*. Retrieved October 5, 2010, from http://www.dradio.de/dlf/sendungen/hiwi/324537

Drake, G. (2003). This Place Gives Me Space: Place and Creativity in the Creative Industries. *Geoforum*, *34*(4), 511–524. doi:10.1016/S0016-7185(03)00029-0

Du, W. (2008, May 30). 'Netizens Annoyed by Stone's Ignorance'. *China Daily Online*, http://www.chinadaily.com.cn/china/2008-05/30/content_6724266.htm (accessed 08 May 2009).

Dunning, J. H. (1977). Trade, Location of Economic Activity and the MNE: A Search for an Eclectic Approach. In B. Ohin, P. O. Hessalbom, & P. M. Wijkman (Eds.), *Proceedings of the International Allocation for Economic Activity*. London: Macmillan.

Dunning, J. H. (1979). Explaining Changing Patterns of International Production: In Defence of the Eclectic Theory. *Oxford Bulletin of Economics and Statistics*, *41*(4), 269–296.

Dunning, J. H. (1980). Toward an Eclectic Theory of International Production: Some Empirical Tests. *Journal of International Business Studies*, *11*(1), 9–31. doi:10.1057/palgrave.jibs.8490593

Dunning, J. H. (1981). Explaining the International Direct Investment Position od Countries: Toward a Dynamic and Development Approach. *Weltwirtschaftliches Archiv*, *117*(5), 30–64. doi:10.1007/BF02696577

Dunning, J. H. (1988). The Eclectic Paradigm of International Production, a Restatement and Some Possible Extensions. *Journal of International Business Studies*, 1–31. doi:10.1057/palgrave.jibs.8490372

Dunning, J. H. (1993). *Multinational Enterprises and the Global Economy*. Reading, MA: Addison-Wesley.

Dunning, J. H., Kim, C., & Lin, J. (2001). Incorporating trade into the investment development path: A case studyof Korea and Taiwan. *Oxford Development Studies*, *29*(2), 145–154.

Dunning, J. H., Kim, C., & Park, D. (2008). Old Wine in New Bottles: a Comparison of Emerging Market TNCs Today and Developed Country TNCs Thirty Years Ago. In Sauvant, K. P. (Ed.), *The Rise of Transnational Corporations From Emerging Markets: Threat or Opportunity?* Willinsont, VT: Edward Elgar Publishing.

Dunning, J. H., & Narula, R. (1996). The Investment Development Path Revisited: Some Emerging Issues. In Narula, J. H. (Ed.), *Foreign Direct Investment and Governance*. London: Routledge.

Dunning, J., & Lundan, S. M. (2008). *Multinational Enterprises and the Global Economy* (2nd ed.). Cornwall, UK: Edward Elgar.

Durán, J. J., & Úbeda, F. (2001). The investment path: A new methodological approach and some theoretical issues. *Transnational Corporations*, *10*(2), 1–34.

Durán, J. J., & Úbeda, F. (2005). The Investment Development Path of Newly Developed Countries. *International Journal of the Economics of Business*, *12*(1), 123–137. doi:10.1080/1357151042000323076

E. I. U. (2009). New York: Economist Intelligence Unit. *Country Report: Taiwan, 6*, 1–24.

Economist, (2008, September 20). *Formula for Disaster*. Volume 388, pp. 57-58.

Edson, G. (2004). Heritage: Pride or passion, product or service? *International Journal of Heritage Studies, 10*(4), 333–348. doi:10.1080/1352725042000257366

Edwards, J. A., & Coit, J. C. L. (1996). Mines and quarries: Industrial heritage tourism. *Annals of Tourism Research, 23*, 341–363. doi:10.1016/0160-7383(95)00067-4

Eisenhardt, K. M. (1989). Building theories from case study research. *Academy of Management Review, 14*(4), 532–550. doi:10.2307/258557

Eisenstadt, S. N. (1973). *Tradition, Change and Modernity*. New York: John Wiley & Sons Inc.

Ekeledo, I., & Sivakumar, K. (1998). Foreign Market Entry Mode, Choice of Service Firms: A Contingency Perspective. *Journal of the Academy of Marketing Science, 26*(4), 274–292. doi:10.1177/0092070398264002

Electrojournal.at. (2006). *Miele verlängert 'Abo' als Testsieger: 'Hier offenbart sich Güte der Geräte-Konstruktion' (Miele Continues, Subscription' to Test Winner: Displaying the Quality of Construction)*. Retrieved October 5, 2010, from http://www.elektrojournal.at/ireds-44767.html

Ellinger, A., Watkins, K. E., & Marsick, V. J. (2005). Case study research methods. In Swanson, R., & Holton, E. F. III, (Eds.), *Research in organizations: Foundations and methods of inquiry* (pp. 327–350). San Francisco, CA: Berrett-Koehler.

Erdener, C., & Shapiro, D. M. (2005). The internationalization of Chinese family enterprise and Dunning's ecletic MNE paradigm. *Management and Organization Review, 1*(3), 411–436. doi:10.1111/j.1740-8784.2005.00021.x

Esch, F.-R., Tomczak, T., & Kernstock, J. (2004). *Corporate Brand Management – Marken als Anker strategischer Führung von Unternehmen* (Corporate Brand Management – Brands as Anchor of Strategic Corporate Governance). Wiesbaden, Germany: Gabler.

Estabrooks, D. A. (1999). Mapping the research utilization field in nursing. *Canadian Journal of Nursing Research, 31*(1), 53–72.

Falk, J., & Dierking, L. (1992). *The Museum Experience*. Washington, DC: Whalesback Publications.

Fangjianjiaping (2008). 'Sanlu Affair triggered the Credit Crisis of Baidu', http://hi.baidu.com/fjpl/blog/item/efdb01e9bb1f8339b80e2d94.html (accessed and translated 16 September 2008).

Farh, J. L., Tsui, A. S., Xin, K. R., & Cheng, B. S. (1998). The influence of relational demography and guanxi: the Chinese case. *Organization Science, 9*(4), 471–488. doi:10.1287/orsc.9.4.471

Farrell, D., Gersch, U. A., & Stephenson, E. (2006). The Value of China's Emerging Middle Class. *The McKinsey Quarterly*, 60–69.

Fatiguso, R. (2006, November 14). Investire in Italia? Che fatica. *Il Sole24ore*.

Faulkner, B., Opperman, M., & Fredline, E. (1999). Destination Competitiveness: An Exploratory Examination of South Australia's Core Attractions. *Journal of Vacation Marketing, 5*(2), 125–139. doi:10.1177/135676679900500202

Federal Quality Institute. (1991). *Introduction to Total Quality Management in the Federal Government*. Washington, DC: Government Printing Office.

Federico, S., & Minerva, G. A. (2007). Outward FDI and local employment growth in Italy. *Temi di Discussione del Servizio Studi* (Economic Working Papers, 613). Banca d'Italia.

Feigenbaum, A. V. (1986). *Total Quality Control*. New York: McGraw-Hill.

Financial Times Deutschland. (2005, May 25). *Haier and Higher*. Retrieved October 5, 2010, from http://www.karma-news.com/archives/56

Fishbein, M., & Ajzen, I. (1975). *Belief, attitude, intention, and behavior: An introduction to theory and research*. Reading, MA: Addison-Wesley.

Fisher, G., Hughes, R., Griffin, R. W., & Pustay, M. W. (2006). *International Business: Managing in the Asia-Pacific*. Frenchs Forest, NSW: Pearson Education.

Fitzsimmons, E. (2008, May 16). 'Ban Exposes China's Ad Censorship Anomalies'. *Campaign,* Issue 19, p. 17.

Fladmoe-Lindquist, K., & Tallman, S. (1997). In Wortzel, H. V., & Wortzel, L. H. (Eds.), *Strategic Management in a Global Economy* (pp. 149–167).

Fleming, J. H., & Asplund, J. (2007, September 13). Customer Satisfaction: A Flawed Measure; Your customers want more than "satisfying" transactions — they want engaging relationships. *Gallup Management Journal*. Retrieved January 25, 2009, from http://findarticles.com/p/articles/mi_6770/is_2007_Sept_13/ai_n28455398/pg_2?tag=artBody;col1

Flisi, S., & Murat, M. (2009). Immigrants links, Diasporas and FDI. An empirical Investigation on Five Eureopean Countries. *Materiali di discussion* (608). Università degli Studi di Modena e Reggio Emilia. Retrieved from http://merlino.unimo.it/web_dep/materiali_discussione/0608.pdf

Fonterra. (2008). 'Statement by Fonterra CEO Andrew Ferrier to Media Conference', *Media Release,* http://www.fonterra.com/wps/wcm/connect/fonterracom/fonterra.com/our+business/news/media+releases/statement+by+fonterra+ceo+andrew+ferrier+to+media+conference (accessed 15 September 2008).

Fortis, M. (2006). I distretti industriali e la loro rilevanza nell'economia italiana: alcuni profili di analisi. In Quadrio Curzio, A., & Fortis, M. (Eds.), *Società, Stato e Mercato in Italia* (pp. 477–483). Bologna, Italy: Sussidiarietà, Laboratori, Distretti, Il Mulino.

Fowler, S. W., Lawrence, T. B., & Morse, E. A. (2004). *New ventures in the information economy: Exploiting virtual embeddedness and knowledge-based assets.* Paper presented at the annual meeting of the Academy of Management, New Orleans, LA.

Francis, J. J., Johnston, M., Eccles, M. P., Grimshaw, J., & Kaner, E. F. S. (2004). *Measurement Issues in The Theory of Planned Behavior: A Supplement to The Manual for Constructing Questionnaires Based on The Theory of Planned Behavior & Theory of Planned Behavior Questionnaires* (Discussion Paper, pp. 43-76). Newcastle upon Tyne, UK: University of Newcastle, Centre for Health Services Research.

Frankfurter Allgemeine Zeitung. (2008, September 17). *Verseuchte Milch: Die Profiteure gehen über Kinderleichen (Poisoned Milk: Profit-makers Walk Across Children's Bodies)*. Retrieved October 5, 2010, from http://www.faz.net/s/Rub8E1390D3396F422B869A49268EE3F15C/Doc~E9E86EA9FF47D4C1B81C2D818094AE143~ATpl~Ecommon~Scontent.html

Fridgen, J. D. (1992). *Dimensions of Tourism*. The Educational Institute.

Fu, Y. (2008). Low-carbon Economic Development Model, China's population. *Resources and Environment*, (3), 14-23.

Fubini, F. (2006, July 7). Il distretto di Prato? Salvato dai cinesi. *Il Corriere della Sera.*

Fuchs, H. J. (n.d.). Vor einer grossen Schlacht (Before a Great Battle). *Denkpausen – Inspirationen zu Management und Leadership, 5*, 52-58.

Fukuyama, F. (1995). *Trust: The social virtues and the creation of prosperity*. New York: The Free Press.

Fu, Z. (2007). *Cultural Imagination and Humanist Criticism*. Beijing, China: Communication University of China Press.

Gaba, V., Pan, Y., & Ungson, G. R. (2002). Timing of entry in international market: An empirical study of U.S. fortune 500 firms in China. *Journal of International Business Studies, 33*, 39–55. doi:10.1057/palgrave.jibs.8491004

Gallagher, G. J. (2004). Great expectations: Challenge and opportunity for today's wealth manager. *The Journal of Wealth Management, 7*(2), 87–91. doi:10.3905/jwm.2004.434570

Gammeltoft, P. (2008). Emerging multinationals: outward FDI from the BRICS countries. *Int. J. Technology and Globalisation, 4*(1), 5–22. doi:10.1504/IJTG.2008.016184

Gao, L. (2008). *Determinants of China's Outward Foreign Direct Investment*. Paper presented in CEA Conference: China's Three Decades of Economic Reform (1978-2008), Cambridge, University, UK.

Gao, P., Woetzel, J. R., & Wu, Y. (2003). Can Chinese Brands Make it Abroad? *The McKinsey Quarterly*, 55–65.

Gao, T. (2003). Ethnic Chinese networks and international investment: evidence from inward FDI in China. *Journal of Asian Economics, 14*, 611–629. doi:10.1016/S1049-0078(03)00098-8

Gardner, D. (2008). 'Why Do We Say Silly Things?' *The Ottawa Citizen*, http://www.canada.com/ottawacitizen/columnists/story.html?id=81b2c80c-ddc3-4f4c-94a3-6f9c98f8f608 (accessed 15 July 2009).

Garg, M., & Delios, A. (2007). Survival of the foreign subsidiaries of TMNCs: the influence of business group affiliation. *Journal of International Management, 13*(3), 278–395. doi:10.1016/j.intman.2007.05.007

Garland, H. (1999). Management research and management practice: Learning from our colleagues in economics. In Larwood, L., & Gattiker, U. E. (Eds.), *Impact analysis: How research can enter application and make a difference* (pp. 129–135). Mahwah, NJ: Erlbaum.

Garrod, B., & Fyall, A. (2000). Managing heritage tourism. *Annals of Tourism Research, 27*, 682–708. doi:10.1016/S0160-7383(99)00094-8

Garvin, D. A. (1987, November-December). Competing on the Eight Dimensions of Quality. *Harvard Business Review*, 101–109.

Gibson, C., & Hordman, D. (1998). Regenerating urban heritage for tourism. *Managing Leisure, 3*(1), 37–54. doi:10.1080/13606719808559324

Giri, A. K. (1998). *Values Ethics and Business Challenges for Education and Management*. India: Rawat Publications.

Gloet, M., & Berrel, M. (2003). The Dual Paradigm Nature of Knowledge Management: Implications for Achieving Quality Outcomes in Human Resource Management. *Journal of Knowledge Management, 7*(1), 78–89. doi:10.1108/13673270310463635

Goldstein, A. (2007). *Multinational Companies from Emerging Economies. Composition, Conceptualization and Direction in the Global Economy*. New York: Macmillan, Palgrave.

Goldstein, A. (2008). Who's Afraid of Emerging Market TNCs? Or: Are Developing Countries Missing Something in the Globalization Debate? In Sauvant, K. P. (Ed.), *The Rise of Transnational Corporations From Emerging Markets: Threat or Opportunity?* (pp. 183–203). Willinston, VT: Edward Elgar Publishing.

Goldstein, A., & Pananond, P. (2008). Singapore Inc. goes shopping abroad: profits and pitfalls. *Journal of Contemporary Asia, 38*(3), 417–438. doi:10.1080/00472330802078493

Gomez, E. T. (1999). *Chinese business in Malaysia: Accumulation, accommodation and ascendance*. Richmond, Surrey: Curzon Press.

Gong, Y. (2003). The special area of export products is for processing export-orient economy development another position. *Modern management science, 6*, 21-22.

Goodall, K., Li, N., & Warner, M. (2007). Expatriate managers in China: The influence of Chinese culture on cross cultural management. *Journal of General Management, 32*(2), 57–76.

Govindarajan, V., & Gupta, A. (2001). *The Quest for Global Dominance*. San Francisco: Jossey Bass.

Graham, B., & Murray, M. (1997). The spiritual and the profane: The pilgrimage to Santigo de Compostela. *Tourism Management, 9*, 44–62. doi:10.1016/0261-5177(88)90057-X

Graham, P., & Kormanik, M. (2004). Bridging the conference gap: A challenge to enhance the research-practice dialogue. *Human Resource Development International, 7*(3), 391–393. doi:10.1080/1367886042000245996

Gray, B., & Wood, D. J. (1991). Collaborative alliances: Moving from practice to theory. *The Journal of Applied Behavioral Science, 27*, 3–22. doi:10.1177/0021886391271001

Gray, L. (2004). *The New Family Office: Innovative Strategies for Consulting to The affluent*. London: Institutional Investor Books.

Greenhalgh, S. (1994). Re-orientalizing the Chinese family firm. *American Ethnologist, 21*(4), 746–775. doi:10.1525/ae.1994.21.4.02a00050

Guan, X. (2006, September 18, 2006). 'Refunds Offered on Harmful Cosmetics', *China Daily*. http://www.chinadaily.com.cn/china/2006-09/18/content_691155.htm (accessed 15 July 2009).

Guo, K. (2000). The foreign direct investment to Chinese industrial structure influence studies. *Management World, 2*, 37–38.

Guptara, P. (1999). Why Knowledge Management Fails: How to Avoid Common Pitfalls. *Knowledge Management Review, 9*, 26–29.

Hackman, J. R. (1985). Doing research that makes a difference. In Lawler, E. E. III, Mohrman, A. M., Mohrman, S. A., Ledford, G. E. Jr, & Cummings, T. G. (Eds.), *Doing research that is useful for theory and practice* (pp. 126–149). San Francisco, CA: Jossey-Bass.

Haier Jituan (a). (n.d.). *Gongsi jianjie (Company Profile)*. Retrieved October 5, 2010, from http://www.haier.cn/about/about.shtml

Haier Jituan (b). (n.d.). *Fazhan Zhanlüe (Developing Strategy)*. Retrieved October 5, 2010, from http://www.haier.cn/about/vision_index.shtml

Haken, H. (1983). *Synergetics: An Introduction. Non-equilibrium Phase Transition and Self-Organization in Physics, Chemistry, and Biology* (3rd ed.). Berlin: Springer-Verlag.

Haley, G. T., Tan, C. T., & Haley, U. C. V. (1998). *New Asian emperors*. Burlington, MA: Butterworth Heinemann.

Hall, E. T. (1976). How cultures collide. *Psychology Today*, 67–74.

Hall, E. T. (1966). *The Hidden Dimension*. New York: Doubleday.

Hall, P. (2000). Creative Cities and Economic Development. *Urban Studies (Edinburgh, Scotland)*, *37*(4), 639–649. doi:10.1080/00420980050003946

Hamlin, R. G. (2002). In support of evidence-based management and research-informed HRD through HRD professional partnerships: An empirical and comparative study. *Human Resource Development International*, *5*(4), 467–491. doi:10.1080/13678860210122643

Han, M. C., & Terpstra, V. (1988). Country-of-Origin Effects for Uni-National and Bi-National Products. *International Business Studies*, *19*(2), 235–255. doi:10.1057/palgrave.jibs.8490379

Hardy, M. A. (1993). *Regression with Dummy Variables*. Newbury Park, CA: Sage.

Hartley, J. (2007). *Creative Industries*. Beijing, China: Tsinghua University Press.

Heilmann, S. (2008). Policy Experimentation in China's Economic Rise. *Studies in Comparative International Development*, *43*, 1–26. doi:10.1007/s12116-007-9014-4

Hemerling, J., Michael, D. C., & Michaels, H. (2006). *China's Global Challengers - The Strategic Implications of Chinese outbound M&A (BCG Rep.)*. Boston: Boston Consulting Group.

Hennart, J. F. (1988). A transaction costs theory of equity joint ventures. *Strategic Management Journal*, *9*, 361–374. doi:10.1002/smj.4250090406

Hernandez, Z. (2008). Outward foreign direct investment from emerging markets: annotated bibliography. In Sauvant, K. P. (Ed.), *The Rise of Transnational Corporations From Emerging Markets: Threat or Opportunity?* (pp. 339–373). Willinston, VT: Edward Elgar Publishing.

Hess, W. (2006). Going Outside, Round tripping and dollar diplomacy: an introduction to outward direct investment. *World Market Analysis*, January. Retrieved from www.globalinsight.com

Hill, C. W. L., Hwang, P., & Kim, W. C. (1990). An Eclectic Theory of the Choice of International Entry Mode. *Strategic Management Journal*, *2*, 117–128. doi:10.1002/smj.4250110204

Hirn, W. (2005). Kennen Sie Huawei (Do You Know Huawei)? *Manager Magazine*. Retrieved October 5, 2010, from http://www.managermagazin.de/unternehmen/mittelstand/0,2828,344730,00.html

Hofstede, G. (1984). *Cultures Consequences* (abridged ed.). Thousand Oaks, CA: Sage.

Hofstede, G. (1991). *Culture and Organisations: Software of the Mind; Intercultural Cooperation and its Importance for Survival*. London: McGraw-Hill.

Hofstede, G. (2001). *Culture's Consequences, Comparing Values, Behaviors, Institutions, and Organizations across Nations*. Thousand Oaks, CA: Sage Publication.

Holland, D. (1999). Ten Ways to Embed KM into Organizational Culture. *Knowledge Management Review, 7*.

Hollinshead, K. (1997). Heritage tourism under post-modernity: Truth and the past. In Ryan, C. (Ed.), *The Tourist Experience: A New Introduction* (pp. 170–193). London: Cassell.

Holowetzki, A. (2002). *The Relationship between Knowledge Management and Organizational Culture: An Examination of Cultural Factors that Supports the Floe and Management of Knowledge within an Organization*. Eugene, OR: University of Oregon Applied Information Management Program.

Homepage, M. (n.d.). *Das Unternehmen – Zeitreise (The Company –Journey through Time)*. Retrieved October 5, 2010, from http://www.miele.de/de/haushalt/ unternehmen/2335_4859.htm

Hong, E., & Sun, L. (2004). *Go overseas via Direct Investment. Internationalization strategy of Chinese Corporation in a comparative prism* (Discussion Paper 40). London, UK: SOAS University of London, Center for financial and management studies.

Hong, E., & Sun, L. (2006). Dynamics of Internationalization and Outward Investment: Chinese Corporations Strategies. *The China Quarterly*, *187*(September), 610–634. doi:10.1017/S0305741006000403

Horx, M., & Wippermann, P. (1996). *Was ist Trendforschung* (What is Trend Research). Düsseldorf, Germany: Econ Verlag.

Horyn, C. (2008, June 1). 'Actress Stone and Dior Differ Over Apology', *New York Times online*, http://www.nytimes.com/2008/06/01/fashion/01stone.html?_r=1&scp=1&sq=Actress%20Stone%20and%20Dior&st=cse&oref=slogin (06 September 2008).

House, R. J. (2004). *Culture, Leadership, and Organizations: The GLOBE Study of 62 Societies*. Thousand Oaks, CA: Sage.

Howard, P. (2003). *Heritage: Management, Interpretation, Identity*. London: Continuum.

Hu, J. (2007). International industrial shift basic rule and tendency. *International Trade Question research, 8*, 46-47.

Hu, X. (2004). The Thought of China Continues international industrial shift to promote regional economies development. *Economical latitude and longitude, 5*, 33-34.

Huang, Q., Davison, R., & Gu, J. (2008). Impact of personal and cultural factors on knowledge sharing in China. *Asia Pacific Journal of Management*, *25*, 451–471. doi:10.1007/s10490-008-9095-2

Huang, Y. (2000). The banking market system. In Norton, J. J., Li, C. J., & Huang, Y. (Eds.), *Financial Regulation in the Greater China Area: Mainland China, Taiwan and Hong Kong SAR* (pp. 25–63). The Hague, The Netherlands: Kluwer Law International.

Hudson, I. (2009). From deregulation to crisis. In Guard, J., & Antony, W. A. (Eds.), *Bankruptcies and Bailouts*. Halifax, Nova Scotia: Fernwood.

Hui, C., & Graen, G. (1997). Guanxi and professional leadership in contemporary Sino-American joint ventures in mainland China. *The Leadership Quarterly*, *8*(4), 451–465. doi:10.1016/S1048-9843(97)90024-2

Hulin, C. (2001). Applied psychology and science: Differences between research and practice. *International Association for Applied Psychology*, 225-234.

Humphreys, P. K., Shiu, W. K., & Chan, F. T. S. (2001). Collaborative buyer-supplier relationships in Hong Kong. *Supply Chain Management: An International Journal*, *6*(4), 152–162. doi:10.1108/EUM0000000005708

Huntington, S. (1996). *The Clash of Civilizations and the Remaking of World Order*. New York: Simon & Schuster.

Hu, S. (2006). The international industrial shift optimizes with our country industrial structure. *Exploration*, *1*, 27–28.

Hu, T., & Yu, Q. (2002). Zhongguo Qiye Hexin Jingzhengli Peiyang de Liang Zhong Guoji Qudao - 'Haier Moshi' he 'Huali Moshi' de Bijiao Yanjiu. *Hubei Shangye Gaodeng Zhuanke Xuexiao Xuebao*, *14*(4), 30–32.

Hu, Z. (2007). Miele - Baise Jiadian zhong de Benchi Miele (Miele – Mercedes-Benz among the White Goods). *Qiye Wenhua*, *6*, 55–57.

Hwang, K. (1987). Face and favour: The Chinese power game. *American Journal of Sociology*, *92*(4), 944–974. doi:10.1086/228588

ICE. (2008). Rapporto ICE 2007-2008 – L'Italia nell'economia internazionale. Retrieved from www.ice.it/statistiche/rapporto.htm

Ilkka, T. (1999). Data is more than knowledge: implications of reserved knowledge hierarchy for knowledge management and organizational theory. *Journal of Management Information Systems*, *16*(3), 103–117.

IMF-International Monetary Fund. (2010). *World Economic Outlook*. Retrieved from www.imf.org

Immergluck, D. (2009). *Foreclosed: High-risk Lending, Deregulation and The Undermining of America's Mortgage Market*. Ithaca, NY: Cornell University Press.

International Monetary Fund. (1993). *Balance of Payments Manual* (5th ed.). Washington, DC. Retrieved from http://www.imf.org/external/pubs/ft/bopman/bopman.pdf

Isdale, M. H. (2006). Strategies for simplifying and increasing the effectiveness of wealth management. *The Journal of Wealth Management*, *9*(2), 14–18. doi:10.3905/jwm.2006.644215

Isobe, T., Makino, S., & Montgomery, D. B. (2000). Resource commitment, entry timing, and market performance of foreign direct investments in emerging economies: The case of Japanese international joint ventures in China. *Academy of Management Journal*, *43*, 468–484. doi:10.2307/1556405

ISTAT. (2007). Cittadini Stranieri, Bilancio demografico anni 2005-7 e popolazione residente al 31 Dicembre. Retrieved from http://demo.istat.it/

Jansen-Verbeke, M. (1999). Industrial heritage: A nexus for sustainable tourism development. *Tourism Geographies*, *1*(1), 70–85. doi:10.1080/14616689908721295

Jaques, E., & Clement, S. D. (1994). *Executive Leadership: a practical guide to managing complexity*. Cambridge, MA: Carson-Hall & Co. Publishers.

Jiang, T. (2007). Promote Ecological Civilization Construction. *Environmental Protection*, (11), 46-48.

Jiang, W. (2006, June 14). Survey: Firms achieve goals abroad. *China Daily*.

Jiang, P. (2007). *Management on Intellectual Property Rights*. Beijing, China: Intellectual Property Press.

Ji, J. (2007). Multinational Corporation direct investment new characteristic and countermeasure. *The Economic Study*, *2*, 9–11.

Jin, Q. (2009). China's Development Strategy for Low-carbon Economy. *China's National Conditions*, (10), 7-9.

Johanson, J., & Vahlne, J. E. (1977). The internationalization process of the firm - Four Swedish case studies. *Journal of Management Studies*, *12*(3), 305–322. doi:10.1111/j.1467-6486.1975.tb00514.x

Johanson, J., & Vahlne, J. E. (1992). Management of foreign market entry. *Scandinavian International Business Review*, *1*(3), 9–27. doi:10.1016/0962-9262(92)90008-T

Johnson, N. C. (1996). Where geography and history meet: Heritage tourism and the big house in Ireland. *Annals of the Association of American Geographers. Association of American Geographers*, *86*(3), 551–566. doi:10.1111/j.1467-8306.1996.tb01766.x

Johnson, P., Grazoli, S., Jamal, K., & Berryman, G. (2001). Detecting deception: Adversarial problem solving in a low base-rate world. *Cognitive Science*, *25*, 355–392. doi:10.1207/s15516709cog2503_2

Johnson, P., Veazie, P., Kochevar, L., O'Connor, P., Potthoff, S., Verma, D., & Dutta, A. (2002). Understanding variation in chronic disease outcomes. *Health Care Management Science*, *5*, 175–189. doi:10.1023/A:1019740401536

Johnston, D., Lowe, R., & Bell, M. (2005). An Exploration of the Technical Feasibility of Achieving CO_2 Emission Reduction in Excess of 60%Within the UK Housing Stock by the Year 2005. *Energy Policy*, (33): 1643–1659. doi:10.1016/j.enpol.2004.02.003

Jolliffe, L., & Smith, R. (2001). Heritage, Tourism and Museums: the case of the North Atlantic Islands of Skye, Scotland and Prince Edward Island, Canada. *International Journal of Heritage Studies*, *7*(2), 149–172. doi:10.1080/13527250120060187

Jornal Tribuna de macau. (2009). Retrieved from www.jtm.com.mo

Kalotay, K. (2008). Russian transnational and international investments paradigms. *Research in International Business and Finance*, *22*(2), 85–107. doi:10.1016/j.ribaf.2006.12.002

Kawase, R., Matsuoka, Y., & Fujino, J. (2006). Decomposition Analysis of CO2 Emission in Long-term Climate Stabilization Scenarios. *Energy Policy*, *34*(15), 2113–2122. doi:10.1016/j.enpol.2005.02.005

Kejriwal, A., & Krishnan, V. R. (2004). The Impact of Vedic Worldview and Gunas on Transformational Leadership. *Vikalpa*, *29*(1), 29–40.

Keller, A. (1999). Jobs and the natural heritage in Scotland. *Countryside Recreation*, *7*(1), 6.

Keller, K. L. (2003). *Strategic Brand Management – Building, Measuring, and Managing Brand Equity*. Upper Saddle River, NJ: Pearson Education.

Kerstetter, D., Confer, J., & Bricker, K. (1998). Industrial heritage attractions: Types and tourists. *Journal of Travel & Tourism Marketing*, *7*(2), 91–104. doi:10.1300/J073v07n02_05

Ketchen, D. J. Jr, Snow, C. C., & Hoover, V. L. (2004). Research on competitive dynamics: Recent accomplishments and future challenges. *Journal of Management, 30*, 779–804. doi:10.1016/j.jm.2004.06.002

Kim, W. C. & Mauborgne. (2005). *Blue Ocean Strategy: How to Create Uncontested Market space and Make Competition Irrelevant*. Boston, MA: Harvard Business School Press.

Kim, C. K. (1995). Brand Popularity and Country Image in Global Competition: Managerial Implications. *Journal of Product and Brand Management, 4*(5), 21–33. doi:10.1108/10610429510103818

Klein, S., & Wocke, A. (2007). Emerging global contenders: the South African Experience. *Journal of International Management, 13*(3), 319–337. doi:10.1016/j.intman.2007.05.002

Knapp, E., & Yu, D. (1999). Understanding Organizational Culture: How Culture Helps or Hinders the Flow of Knowledge. *Knowledge Management Review, 7*, 16–21.

Kogut, B., & Singh, H. (1988). The effect of national culture on the choice of entry mode. *Journal of International Business Studies, 19*(3), 411–432. doi:10.1057/palgrave.jibs.8490394

Koji, S., Yoshitaka, T., Kei, G., & Yuzuru, M. (2007). Developing a long-term local society design methodology towards a low-carbon economy: An application to Shiga Prefecture in Japan. *Energy Policy*, (35): 4688–4703.

Kolter, P. (2003). *Marketing Management* (11th ed.). Upper Saddle River, NJ: Prentice-Hall International Inc.

Kondrat, M. E. (1992). Reclaiming the practical: Formal substantive rationality in social work practice. *The Social Service Review, 166*, 237–255. doi:10.1086/603912

Kreitner, R. (1995). *Management* (6th ed.). Boston: Houghton Mifflin Company.

Krishnan, A., Muniapan, B., Lew, T. Y., & Kong, E. E. F. (2004, December 13-15). *Exploring the Extent of Transformational Leadership in the Context of Miri Entrepreneurs*. Paper presented at the ASEMAL 4 Conference, Penang, Malaysia.

Krishnan, V. R. (2000, September). Training Programs on Leadership: Do They Really makes a Difference? In *Proceedings of the Seminar on "Role of HR: A New Agenda"*, Delhi, India. Retrieved from http://www.geocities.com/rkvenkat/

Krishnan, V. R. (2003). Modernization without demolishing cultural roots: the role of transformational leadership. In J. Gifford & G. Zezulka-Mailoux (Eds.), *Culture and theState, Volume IV (Alternative Interventions)* (pp. 164–173). Edmonton, Alberta, Canada: Canada Research Chairs Humanities Studio, University of Alberta.

Krishnan, V.R. (1990, January 11). Transformational Leadership and Vedanta Philosophy. *Economics Times*, p. 7.

Krishnan, V.R., & Mulla, R. Z. (2009). Do Karma-Yogis make better leaders? Exploring the relationship between the leader's Karma-Yoga and transformational leadership *Journal of Human Values, 15* (No 2, 2009), 167-183.

Krishnan, V. R. (2001). Can Indian worldview facilitate the emergence of transformational leaders? *Management and Labour Studies, 26*, 237–244.

Krugman, P. (1996). *The Self-organizing Economy*. Cambridge, MA: Blackwell Publishers.

Kruse, R. J. (2005). The Beatles as place makers: Narrated landscapes in Liverpool, England. *Journal of Cultural Geography, 22*(2), 87–114. doi:10.1080/08873630509478240

Kumar, K., & McLeod, M. G. (1981). *Multinationals from developing countries*. Lexington, MA: Lexington Books.

Kynge, J. (2006). *China Shakes the World*. Boston: Houghton Mifflin Company.

Lahtl, R. K., & Beyerlein, M. M. (2000). Knowledge transfer and management consulting: A look at the firm. *Business Horizons, 43*(1), 65–74. doi:10.1016/S0007-6813(00)87389-9

Lai, Y. (2007). Our country continues the international industrial shift basic trend of development. *Foreign economy and trade practice, 2*, 9-10.

Lall, S. (1983). *The new multinationals: The spread of third world enterprises*. Chichester, UK: Wiley.

Lambkin, M. (1988). Order of entry and performance in new markets. *Strategic Management Journal, 9*, 127–140. doi:10.1002/smj.4250090713

Lan, X. (2005, May 29). 'Famous Brands Lose Face', *Beijing Review*, http://www.bjreview.cn/EN/En-2005/05-29-e/bus-1.htm (accessed 23 July 2009).

Landry, C. (2000). *The creative city: A toolkit for urban innovators*. London: Earthscan Press.

Lane, K., St-Maurice, I., & Süssmuth Dyckerhoff, C. (2006). Building Brands in China. *The McKinsey Quarterly*, 34–41.

Lane, S. (2009). Financial services in Europe before, during and after the crisis. *Accountancy Ireland, 41*(2), 16.

Lardy, N. R. (1998). *China's unfinished economic revolution*. Washington, DC: Brookings Institution.

Latham, G. P. (2001). The reciprocal transfer of learning from journals to practice. *Applied Psychology: An International Review, 50*(2), 201–211. doi:10.1111/1464-0597.00054

Lau, L. (2009). Low-carbon Economy and Development of Human Society. *Xinhua Digest*, (1), 26-27.

Lau, H. F. (2003). Industry evolution and internationalization process of firms from a newly industrialized economy. *Journal of Business Research, 56*(10), 847–852. doi:10.1016/S0148-2963(02)00472-1

Lawrence, S. V. (2002). Going global. *Far Eastern Economic Review, 165*(12), 30–33.

Lecraw, D. (1977). Direct Investment by Firms from Less-Developed Countries. *Oxford Economic Papers, 29*(3), 442–457.

Lee, D. J., Pae, J. H., & Wong, Y. H. (2001). A model of close business relationships in China (guanxi). *European Journal of Marketing, 35*(1), 51–69. doi:10.1108/03090560110363346

Lee, D. Y., & Dawes, P. L. (2005). Guanxi, trust, and long-term orientation in Chinese business markets. *Journal of International Marketing, 13*(2), 28–56. doi:10.1509/jimk.13.2.28.64860

Lee, J. (1996). Culture and management – A study of small Chinese family business in Singapore. *Journal of Small Business Management, 34*(3), 63–67.

Lee, K. C. P., & Humphreys, K. P. (2006). The Role of Guanxi in Supply Chain Practises. *International Journal of Production Economics*, 450–467.

Leggett, S. (2009). A single global regulatory framework: Solution, or problem? *Corporate Finance Review, 14*(1), 11.

Leopold, L. (2009). *The Looting of America: How Wall Street's Game of Fantasy Finance Destroyed Our Jobs, Pensions, and Prosperity, and What We Can Do About It*. White River Junction, VT: Chelsea Green Publications.

Li, F., & Shooshtari, N. H. (2003). Brand Naming in China - Sociolinguistic Implications. *Multinational Business Review,* 1-17.

Li, H., et al. (2007). Using of international practice on the environment tax system in the building of China's pollution tax system. *Circulation Economy*, (5), 34-36.

Li, J. (2008). Strict and Tenacious German Style – German Top Brand for Electronic Appliances Miele. *Jiadian Keji, 4,* 26-27, 30-31.

Li, C. J. (2000). China's financial reform and its implications. In Norton, J. J., Li, C. J., & Huang, Y. (Eds.), *Financial Regulation in the Greater China Area: Mainland China, Taiwan and Hong Kong SAR* (pp. 319–344). The Hague, The Netherlands: Kluwer Law International.

Lichtenstein, B. B. (2007). *Moving Far From Far-From-Equilibrium: Opportunity Tension as the Driver of Emergence.* Paper presented at the Organization Science Winter Conference.

Lieberman, M. B., & Asaba, S. (2006). Why do firms imitate each Other? *Academy of Management Review, 31*, 366–385.

Lieberman, M. B., & Montgomery, D. B. (1988). First-mover advantages. *Strategic Management Journal, 9,* 41–58. doi:10.1002/smj.4250090706

Lieberman, M. B., & Montgomery, D. B. (1998). First-mover (dis)advantages: Retrospective and link with the resource-based view. *Strategic Management Journal, 19,* 1111–1126. doi:10.1002/(SICI)1097-0266(1998120)19:12<1111::AID-SMJ21>3.0.CO;2-W

Light, D. (2000). Gazing on communism: Heritage tourism and post-communist identities in Germany, Hungary and Romania. *Tourism Geographies, 2*(2), 157–176. doi:10.1080/14616680050027879

Linder, B. S. (1961). *An Essay on Trade and Transformation*. Stockholm, Sweden: Almqvist & Wiksell.

Linong, Z. (2006). *China Business: Environment, Momentum, Strategies, Prospects*. Singapore: Pearson Prentice Hall.

Lin, T. J. (2007). 'Brand Lessons'. *Global Cosmetic Industry*, *175*(Issue 5), 29–32.

Lin, X., & Germain, R. (1998). Sustaining Satisfactory Joint-Venture Relationship: The Role of Conflict Resolution Strategy. *Journal of International Business Studies*, *29*, 179–196. doi:10.1057/palgrave.jibs.8490031

Li, P. P. (2007). Toward an Integrated Theory of Multinational Evolution: the Evidence of Chinese Multinational Enterprises as Latecomers. *Journal of International Management*, *13*(3), 296–318. doi:10.1016/j.intman.2007.05.004

Lipsey, R. E. (2001). Foreign Direct Investment and the Operations of multinational firms: concepts, history and data (Workinng Paper, 8665, NBER).

Liu, J. (2008) 'Sanlu Group Stop Reimbursing Diagnose Fee for Infants'. *YNET news online,* http://china.ynet.com/view.jsp?oid=43262117 (accessed 17 September 2008).

Liu, M. (2008). 'A Familiar Taint: The Milk Crisis Spotlights Flaws in Beijing's Food System'. *Newsweek Online.* http://www.newsweek.com/id/160895 (25 September 2008).

Liu, H. (2009). *Chinese Business; Landscapes and Strategies*. New York: Routledge.

Liu, H., & Li, K. (2002). Strategic implications of emerging Chinese multinational: the Haier case study. *European Management Journal*, *20*(6), 699–706. doi:10.1016/S0263-2373(02)00119-6

Liu, L. (2005). *China's industrial policies and the global business revolution: the case of the domestic appliance industry*. London: Routledge Curzon.

Liu, L., & Tian, Y. (2008). The internationalization of Chinese enterprises: the analysis of the UK case. *Int. J. Technology and Globalisation*, *4*(1), 87–102. doi:10.1504/IJTG.2008.016189

Liu, M. Y. (2007). *Probing into Theory and Practice for Development of Cultural Creativity Industry of Beijing*. Beijing, China: China Economic Press.

Liu, X., Buck, T., & Shu, C. (2005). Chinese Economic Development, the Next Stage: Outward FDI? *International Business Review*, *14*(1), 97–115. doi:10.1016/j.ibusrev.2004.12.003

Li, Y. H. (2008). A Perspective Analysis into New Cultural Phenomena of Cultural and Creative Age. *China-Today Forum*, *6*, 53.

Li, Y. H., & Bai, Q. X. (2008). *Tutorial of Excellent Cultural Creativity Cases*. Beijing, China: China Economic Press.

Li, Y. H., & Lu, H. (2007). The Nature of Cultural Economy and Value of Intellectual Property Rights. *China-Today Forum*, *5*, 42.

Long, G. (2005). China's Policies on FDI: Review and Evaluation. In Moran, T. H., Graham, E. M., & Blomstrom, M. (Eds.), *Does Foreign Direct Investment Promote Development?* (pp. 315–336). Washington, DC: Peterson Institute for International Economics.

Lovett, S., Simmons, L. C., & Kali, R. (1999). Guanxi versus the market: Ethics and efficiency. *Journal of International Business Studies*, *30*, 231. doi:10.1057/palgrave.jibs.8490068

Low, P. (2008). Confucian ethics & social responsibility – The Golden Rule & responsibility to the stakeholders. *Ethics & critical thinking journal, 4*, 46-54.

Lowenthal, D. (1994). Identity, heritage and history. In Gillis, J. R. (Ed.), *Commemorations: The Politics of National Identity* (pp. 41–57). Princeton, NJ: Princeton University Press.

Lowenthal, D. (1998). *The Heritage Crusades and the Spoils of History*. Cambridge, UK: Cambridge University Press. doi:10.1017/CBO9780511523809

Lu, B. (2004). Brief analysis our country continues new round international industrial shift strategy. *Yunnan Finance and Trade Institute Journal*, *3*, 38–40.

Luo, M. (2005). On the Paradigm of Knowledge Management. *Journal of Fiscal Science, 3*.

Luo, Y. (1997). Guanxi and performance of foreign-invested enterprises in China: An empirical inquiry. *Management International Review*, *37*, 51–70.

Luo, Y. (1997). Guanxi: principles, philosophies and implications. *Human Systems Management*, *16*(1), 43–51.

Luo, Y. (1998). Timing of Investment and International Expansion Performance in China. *Journal of International Business Studies*, *29*, 391–407. doi:10.1057/palgrave.jibs.8490042

Luo, Y. (2002a). Contract, cooperation, and performance in international joint ventures. *Strategic Management Journal*, *23*, 903–919. doi:10.1002/smj.261

Luo, Y. (2002b). Product Diversification in International Joint Ventures: Performance Implications in an Emerging Market. *Strategic Management Journal*, *23*, 1–20. doi:10.1002/smj.203

Luo, Y. (2002c). Stimulating exchange in international joint ventures: An attachment-based view. *Journal of International Business Studies*, *33*, 169–181. doi:10.1057/palgrave.jibs.8491010

Luo, Y., & Park, S. H. (2001). Strategic alignment and performance of market-seeking MNCs in China. *Strategic Management Journal*, *22*, 141–155. doi:10.1002/1097-0266(200101)22:2<141::AID-SMJ151>3.0.CO;2-O

Luo, Y., & Peng, M. W. (1998). First mover advantages in investing in transitional economies. *Thunderbird International Business Review*, *40*, 141–163. doi:10.1002/tie.4270400205

Luo, Y., Shenkar, O., & Nyaw, M.-K. (2001). A dual parent perspective on control and performance in international joint ventures: Lessons from a developing economy. *Journal of International Business Studies*, *32*, 41–58. doi:10.1057/palgrave.jibs.8490937

Luo, Y., Xue, Q., & Han, B. (2010). How emerging market governments promote outward FDI: Experience from China. *Journal of World Business*, *45*, 68–79. doi:10.1016/j.jwb.2009.04.003

Lynham, S. A. (2002). The general method of theory-building research in applied disciplines. *Advances in Developing Human Resources*, *4*(3), 221–241.

MacDonald, R., & Jolliffe, L. (2003). Cultural rural tourism: Evidence from Canada. *Annals of Tourism Research*, *30*(2), 307–322. doi:10.1016/S0160-7383(02)00061-0

Magazine, M. (2007). *Gefährliche Exporte: EU kritisiert China (Dangerous Exports: EU Criticizes China)*. Retrieved October 5, 2010, from http://www. manager-magazin.de/unternehmen/artikel/0,2828,496155,00.html

Magi, A. W. (2008). Share of Wallet in Retailing: The Effects of Customer Satisfaction. *Loyalty Cards and Shopper Characteristics, 79*. Retrieved January 25, 2009, from http://www3.babson.edu/Publications/JR/PastIssues/Volume79Issue2/ShareWalletRetailing.cfm

Maher, K.J. (1997). *Gender – Related Stereotypes of Transformational and Transactional Leadership.*

Makadok, R. (1998). Can first-mover and early-mover advantages be sustained in an industry with low barriers to entry/imitation? *Strategic Management Journal*, *19*, 683–696. doi:10.1002/(SICI)1097-0266(199807)19:7<683::AID-SMJ965>3.0.CO;2-T

Mansur, M. (2003). *Entrepreneurship and Innovation in the Knowledge-based Economy: Challenges and Strategies.* Paper presented at the APO Symposium on Entrepreneurship in Knowledge-based Industry, Tokyo.

Marktforschung, G. (2007). *Best Brands - Die Gewinner (Best Brands – The Winners)*. Retrieved October 5, 2010, from http://www.bestbrands.de/index_fs.html

Marr, B. (2003). Consider the Culture when Benchmarking KM Processes. *KM Review*, *5*(5), 6–7.

Maruyama, M. (1994). *Mindscapes in Management: Use of Individual Differences in Multi-Cultural Management.*

Mason, D., & Pauleen, D. J. (2003). Perceptions of Knowledge Management: A Qualitative Analysis. *Journal of Knowledge Management*, *7*(4), 38–48. doi:10.1108/13673270310492930

Mason, P. (2009). *Meltdown: The End of the Age of Greed.* New York: Verso.

Mathews, J. A. (2002). *Dragon multinationals: A new model of global growth.* New York: Oxford University Press.

Mathews, J. A. (2006). Dragon multinationals: New players in 21st century globalization. *Asia Pacific Journal of Management*, *23*(1), 5–27. doi:10.1007/s10490-006-6113-0

Maude, D. (2006). *Global Private Banking and Wealth Management: The New Realities*. New York: Wiley.

McDermott, R., & O'Dell, C. (2001). Overcoming Cultural Barriers to Sharing Knowledge. *Journal of Knowledge Management*, *5*(1), 76–85. doi:10.1108/13673270110384428

McDevitt, R. (2007). Confucianism – Understanding and Applying The *Analects* of Confucius. *Education About Asia, 12*(1).

McKelvey, B. (2006). Van de Ven and Johnson's "engaged scholarship": Nice try, but.... *Academy of Management Review*, *31*(4), 822–829.

McKercher, B., & Du Cros, H. (2002). *Cultural Tourism: The Partnership between Tourism and Cultural Heritage Management*. New York: The Haworth Press.

McLaughlin, K. E., & Kaiser, A. (2008). Dior China Drops Stone after Quake Comments. *Women's Wear Daily*, *195*(Issue 115), 24–25.

Mendonça, A., Passos, J., & Fonseca, M. (2009). *The Investment Path Hypothesis: A panel data Approach for Portugal and the Cohesion Countries, 1990-2007*. Paper presented at the 3rd National Congress of Portuguese Economists in Funchal, Madeira.

Menzies, J., Chung, M., & Orr, S. (2008). *Doing Business with China: How Australian Companies Make their Decisions when Entering the Chinese Market*. Melbourne, Australia: Deakin Prime.

Merriam, S. B. (2001). *Qualitative research and case study applications in education*. San Francisco, CA: Jossey-Bass.

Meyer, K. (1998). *Direct Investment in Economies in Transition*. Cheltenham, UK: Edward Elgar.

Meyer, K. (2004). Perspectives on Multinational Enterprises in emerging economies. *Journal of International Business Studies*, *35*, 259–276. doi:10.1057/palgrave.jibs.8400084

Meyer, K. (2008). China's second economic transition: building national markets. *Management and Organization Review*, (1): 3–15. doi:10.1111/j.1740-8784.2008.00094.x

Meyer, K., & Nguyen, H. V. (2005). Foreign investment strategies and sub-national institutions in emerging markets: evidence from Vietnam. *Journal of Management Studies*, *42*(1), 63–93. doi:10.1111/j.1467-6486.2005.00489.x

Michailova, S., & Worm, V. (2003). Personal networking in Russia and China: Blat and Guanxi. *European Management Journal*, *21*(4), 509–519. doi:10.1016/S0263-2373(03)00077-X

Michale, P. T. (1993). *Economic Development in 3rd world*. Hyderabad, India: Orient Longman Limited.

Micklethwait, J. (1996). The limits of family values. *Economist*, *338*(7956), S10–S18.

Miles, L. (2006). The Application of Anglo-American Corporate Practices in Societies Influenced by Confucian Values. *Business and Society Review*, *111*(3), 305–321. doi:10.1111/j.1467-8594.2006.00274.x

Millar, S. (1989). Heritage Management for Heritage Tourism. *Tourism Management*, *10*(1), 9–14. doi:10.1016/0261-5177(89)90030-7

Mitchell, W. (1989). Whether and when? Probability and timing of incumbents' entry into emerging industrial subfields. *Administrative Science Quarterly*, *34*, 208–230. doi:10.2307/2989896

Mitchell, W. (1991). Dual clocks: Entry order influences on incumbent and newcomer market share and survival when specialized assets retain their value. *Strategic Management Journal*, *12*, 85–100. doi:10.1002/smj.4250120202

Mjoen, H., & Tallman, S. (1997). Control and performance in international joint ventures. *Organization Science*, *8*, 257–274. doi:10.1287/orsc.8.3.257

MOFCOM. (2008). File (2008) 529 issued by the Department of Commerce National Bureau of Statistics State Administration of Foreign Exchange. Retrieved from http://tradeinservices.mofcom.gov.cn/upload/2009/04/10/1239346168141_20005314.pdf

MOFCOM. (2009). Statistical Bulletin of China's Outward Foreign Direct Investments. Retrieved from http://tradeinservices.mofcom.gov.cn/upload/2009/04/10/1239346168141_20005314.pdf

MOFCOM. (2010). *Ministry of Commerce the People's Republic of China -Foreign Trade and Investment Statistics*. Retrieved from www.english.mofcom.gov.cn

Mohrman, S., Gibson, C., & Mohrman, A. (2001). Doing research that is useful to practice: A model and empirical exploration. *Academy of Management Journal*, *44*, 357–375. doi:10.2307/3069461

Molleman, E. (1998). Variety and the Requisite of Self-Organization, the International. *Journal of Organizational Analysis*, *6*(2), 109–131. doi:10.1108/eb028880

Moran, J. W., & Brightman, B. K. (2000). Leading organizational change. *Journal of Workplace Learning: Employee Counseling Today*, *12*(2), 66–74.

Morck, R., Yeung, B., & Zhao, M. (2008). Perspectives on China's outward foreign direct investment. *Journal of International Business Studies*, *39*(3), 337–350. doi:10.1057/palgrave.jibs.8400366

Morelli, A. (2004). Cina alla conquista degli oceani. *Altraeconomia*. Retrieved from www.altraeconomia.it

Mori, A., & Rolli, V. (1998). Investimenti diretti all'estero e commercio: complementi o sostituti? *Temi di discussione*, 337.

Moscardo, G., & Pearce, P. L. (1999). Understanding ethnic tourists. *Annals of Tourism Research*, *26*(2), 416–434. doi:10.1016/S0160-7383(98)00101-7

Mu, D. P. (1995). Culture and Business: Interacting Effectively to Achieve Mutual Goals. *East Asian Executive Reports*, *17*(6), 6.

Muchinsky, P. M. (2004). When the psychometrics of test development meets organizational realities: A conceptual framework for organizational change, examples and recommendations. *Personnel Psychology*, *57*(1), 179–205. doi:10.1111/j.1744-6570.2004.tb02488.x

Muniapan, B. (2006, December 19-21). *Can the Bhagavad-Gita be Used as a Manual for Management Development of Indian Managers Worldwide?* Paper presented at the 5th Asia Academy of Management Conference, Asian Management: Convergence and Divergence, Tokyo.

Muniapan, B. (2009, July 5-8). *The Bhagavad-Gita on Leadership for Good Governance and Sustainable Development.* Paper presented at the 15th Annual International Sustainable Development Research Conference. "Taking up the Global Challenge: Analyzing the Implementation of Innovations and Governance for Sustainable Development", Utrecht, The Netherlands.

Muniapan, B. (2007). Transformational Leadership Style Demonstrated by Sri Rama in Valmiki Ramayana. *International Journal of Indian Culture and Business Management*, *1*(1), 104–115. doi:10.1504/IJICBM.2007.014473

Muniapan, B. (2008). Perspectives and Reflections on Management Education in Malaysia. *International Journal of Management in Education*, *2*(1), 77–87. doi:10.1504/IJMIE.2008.016232

Muniapan, B. (2010). Perplexity, Management and Business in India. In Lowe, S. (Ed.), *Managing in a Changing Times: A Guide to Perplexed Manager* (pp. 317–346). Thousand Oaks, CA: Sage.

Muniapan, B., & Dass, M. (2008). Corporate Social Responsibility: A Philosophical Approach from an Ancient Indian Perspective. *International Journal of Indian Culture and Business Management*, *1*(4), 408–420. doi:10.1504/IJICBM.2008.018622

Muniapan, B., & Dass, M. (2009). An Indian Leadership Perspective from Literature Works of Poet Kannadasan. *International Journal of Indian Culture and Business Management*, *2*(4), 326–340. doi:10.1504/IJICBM.2009.023552

Muolo, P., & Padilla, M. (2008). *Chain of Blame: How Wall Street Caused the Mortgage and Credit Crisis*. New York: Wiley.

Murat, M., & Pistoresi, B. (2006). Emigrants and immigrants networks in FDI. forthcoming in *Applied Economics Letter.* Retrieved from http://merlino.unimo.it/web_dep/materiali_discussione/0546.pdf

National Development and Planning Commission Macro Economic Research Institute Foreign Economics Department. (2007). The international industrial shift and the Chinese countermeasure. *China Economical Times Daily*, 06-08.

Naughton, B. (2000b). *How much can regional integration do to unify China's markets?* (Working Paper No. 58). Palo Alto, CA: Stanford University Center for Research on Economic Development and Policy Reform.

Naughton, B. (2000a). China's trade regime at the end of the 1990s: Achievements, limitations, and impact on the United States. In Carpenter, T. G., & Dorn, J. A. (Eds.), *China's Future: Constructive Partner or Emerging Threat?* Washington, DC: Cato Institute.

Neuborne, E. (2003). Bridging the culture gap. *Sales & Marketing Management, 155*(7).

Nolan, P. (2001). *China and the Global Economy: National Champions, Industrial Policy and the Big Business Revolution*. New York: Palgrave.

Nolan, P., & Zhang, J. (2002). The challenge of globalization for large Chinese firms. *World Development*, *30*(12), 2089–2107. doi:10.1016/S0305-750X(02)00148-1

Nonaka, I., & Noboru, K. (1998). The concept of "Ba": Building a foundation for knowledge creation. *California Management Review*, *40*(3), 40–54.

Nonaka, I., & Takeuchi, H. (1995). *The knowledge – creating company: How Japanese companies create the dynamics of innovation*. Oxford, UK: Oxford University Press.

Nonaka, I., Toyama, R., & Konno, N. (2002). SECI, Ba and Leadership: a Unified Model of Dynami Knowledge Creation. In Little, S., Quitas, P., & Ray, T. (Eds.), *Managing Knowledge: An Essential Reader*. London: Sage.

Noorderhaven, N. (2000). *Hermeneutic Methodology and International Management Research* (Occasional Paper 2000/2). Uppsala, Sweden: Department of Business Studies, Uppsala University.

Nordstrom, K. A., & Vahlne, J. (1992, April 22-25). *Is the globe shrinking? Psychic distance and the establishment of Swedish sales subsidiaries during the last 100 years*. Paper presented at the International Trade and Finance Association's Annual Conference, Laredo, Texas.

North, D. C. (1990). *Institutions, Institutional change and Economic Performance*. Cambridge, MA: Cambridge University Press.

Oberg, K. (1979). Culture Shock and the problem of adjustment in new cultural environments. In E. C. Smith & L. F. Luce (Eds.), *Towards internationalism: Readings in cross-cultural communication* (pp. 185–186). New York: Newbury House Publishers.

OECD. (2006). *Benchmark Definition of Foreign Direct Investments* (3rd ed.). Retrieved from http://www.oecd.org/dataoecd/10/16/2090148.pdf

OECD. (2008). *Benchmark Definition of Foreign Direct Investment* (4th ed.). Retrieved from www.oecd.org

OECD. (2008). *China encouraging responsible business conduct*.

OECD. (2008). Investment Policy Reviews China. *China Outward Direct Investment*. Retrieved from www.oecd.org

O'Grady, S., & Lane, H. (1996). The psychic distance paradox. *Journal of International Business Studies*, *27*(2), 309–333. doi:10.1057/palgrave.jibs.8490137

Ohmae, K. (1989). The global logic of strategic alliances. *Harvard Business Review*, *67*(2), 143–154.

Oldenburg, R. (1999). *The Great Good Place: Cafes, Coffee Shops, Bookstores, Bars, Hair Salons, and Other Hangouts at the Heart of a Community* (3rd ed.). New York: Marlowe & Company.

Orru, M., Biggart, N. W., & Hamilton, G. G. (1997). *The Economic Organization of East Asian Capitalism*. Thousand Oaks, CA: Sage Publications.

Osenton, T. (2009). *Boomer Destiny: Leading the U.S. Through The Worst Crisis Since The Great Depression*. Westport, CT: Praeger.

Pablos, P. (2006). Western and Eastern views on social networks. *The Learning Organization*, *12*(5), 436–456. doi:10.1108/09696470510611393

Pananond, P. (2007). The changing dynamics of Thai multinationals after the Asian economic crisis. *Journal of International Management*, *13*(3), 356–375. doi:10.1016/j.intman.2007.05.005

Pan, Y., & Chi, P. S. K. (1999). Financial performance and survival of multinational corporations in China. *Strategic Management Journal*, *20*, 359–374. doi:10.1002/(SICI)1097-0266(199904)20:4<359::AID-SMJ31>3.0.CO;2-9

Pan, Y., Li, S., & Tse, D. K. (1999). The Impact of Order and Mode of Market Entry on Profitability and Market Share. *Journal of International Business Studies*, *30*, 81–82. doi:10.1057/palgrave.jibs.8490061

Parashar, S. P. (2008). Winning over equals: insights from Bhagwad Geeta (Krishna–Arjuna framework). *International Journal of Indian Culture and Business Management*, *1*(3), 354–359. doi:10.1504/IJICBM.2008.017790

Park, S. H., & Luo, Y. (2001). Guanxi and organizational dynamics: Organizational networking in Chinese firms. *Strategic Management Journal*, *22*, 455–477. doi:10.1002/smj.167

Park, S., & Luo, D. (2001). Guanxi and organizational dynamics: Organizational networking in Chinese firms. *Strategic Management Journal*, *22*(5), 455–477. doi:10.1002/smj.167

Parnell, J. A., Köseoglu, M. A., & Spillan, J. E. (2010). Crisis readiness in Turkey and the United States. *Journal of Contingencies and Crisis Management*, *18*(2), 108–116. doi:10.1111/j.1468-5973.2010.00603.x

Passariello, C., & Meichtry, S. (2008, May 30). 'Dior Pulls Ads with Sharon Stone'. *Wall Street Journal*, B.7.

Patton, M. Q. (2002). *Qualitative research & evaluation methods* (3rd ed.). Thousand Oaks, CA: Sage.

Pearcea, J. D. (2001). *Unique source of information, training and more*. Retrieved January 13, 2009, from www.jpsalestraining.com

Pearce, D. W. (1991). The role of Carbon Taxes in Adjusting to Global Warming. *The Economic Journal*, *101*, 938–948. doi:10.2307/2233865

Peng, J.-B. (2008). The establishment of SME credit guarantee system based on mutual guarantee. *Finance Reaserch*, *2*, 75–82.

Peng, M. W. (2002). Towards an insitution-based view of business strategy. *Asia Pacific Journal of Management*, *19*(2), 251–267. doi:10.1023/A:1016291702714

Peng, M. W., & Luo, Y. (2000). Managerial ties and firm performance in a transition economy: The nature of a micro-macro link. *Academy of Management Journal*, *43*, 486–501. doi:10.2307/1556406

Perlmutter, H. V. (1969). The tortuous evolution of the multinational corporation. *The Columbia Journal of World Business*, *4*(1), 9–18.

Petersen, M. A., & Rajan, R. G. (1994). The benefits of lending relationships: evidence from small business data. *The Journal of Finance*, *49*(1), 3–37. doi:10.2307/2329133

Pettigrew, A. M. (2001). Management research after modernism. *British Journal of Management*, *12*, S61–S70. doi:10.1111/1467-8551.12.s1.8

Pickert, K. (2008). 'Brief History of Melamine'. *Time Online*, http://www.time.com/time/health/article/0,8599,1841757,00.html (accessed 17 September 2008).

Pigott, C. A. (2002). *China in the world economy: The domestic policy challenges*. Paris: Organisation for Economic Co-operation and Development.

Polanyi, M. (1958). *Personal knowledge: toward to post-critical philosophy* (1st ed.). Chicago: University of Chicago Press.

Polanyi, M. (1966). *The tacit dimension*. London: Routledge.

Politis, J. D. (2001). The Relationships of Various Leadership Styles to Knowledge Management. *Leadership and Organization Development Journal*, *22*(8), 354–364. doi:10.1108/01437730110410071

Pompian, M. M., & Longo, J. M. (2004). A new paradigm for practical application of behavioral finance: Creating investment programs based on personality type and gender to produce better investment outcomes. *The Journal of Wealth Management*, *7*(2), 9–15. doi:10.3905/jwm.2004.434561

Poncet, S. (2005). A fragmented China: Measure and determinants of Chinese domestic market disintegration. *Review of International Economics*, *13*(3), 409–430. doi:10.1111/j.1467-9396.2005.00514.x

Poria, Y. (2001). Challenging the present approach to heritage tourism: Is tourism to heritage places heritage tourism? *Tourist Review*, *56*(1/2), 51–53. doi:10.1108/eb058358

Poria, Y. (2001). The show must NOT go on. *Tourism and Hospitality Research*, *3*(2), 115–119.

Poria, Y., Butler, R., & Airey, D. (2000). Clarifying heritage tourism: A distinction between heritage tourism and tourism in historic place. *Annals of Tourism Research*, *28*(4), 1047–1049. doi:10.1016/S0160-7383(00)00069-4

Poria, Y., Butler, R., & Airey, D. (2001). Tourism subgroups: Do they exist? *Tourism Today*, *1*(1), 14–22.

Poria, Y., Butler, R., & Airey, D. (2003). The core of heritage tourism: Distinguishing heritage tourists from tourists in heritage places. *Annals of Tourism Research*, *30*(1), 238–254. doi:10.1016/S0160-7383(02)00064-6

Poria, Y., Butler, R., & Airey, D. (2004). How tourists decide which heritage site to visit? *Tourism Review*, *59*(2), 12–16. doi:10.1108/eb058431

Poria, Y., Butler, R., & Airey, D. (2004). Links between tourists, heritage, and reasons for visiting heritage sites. *Journal of Travel Research*, *43*(8), 19–28. doi:10.1177/0047287504265508

Porter, M., & Scott, S. (2001). Innovation: Location Matters. *MIT Sloan Management Review*, *42*(2), 28–36.

Posner, R. A. (2009). *A Failure of Capitalism: The Crisis of '08 and The Descent into Depression*. Boston: Harvard University Press.

Posner, V. (1961). International Trade and Technical Change. *Oxford Economic Papers*, *13*(3), 323–341.

Prahalad, C. K., & Hammond, A. (2002, September). What Works: Serving the Poor Profitably. A private sector strategy for global digital opportunity. *Harvard business review*.

Preffer, J., & Robert, I. S. (1999). Knowing "what" to do is not enough: turning knowledge into action. *California Management Review*, *42*(1), 83–108.

Prentice, R. (1993). *Tourism and Heritage Attractions*. London: Rutledge.

Pressedienst, M. (2008). *Miele im Geschäftsjahr 2007/2008 erfolgreich bei der Internationalisierung vorangekommen (Miele advanced successfully in its internationalization in 2007/2008)*. Retrieved October 5, 2010, from http://www.miele.de/de/presse/artikel/artikel_184_2008.aspx

Prideaux, B. R., & Kininmont, L. J. (1999). Tourism and heritage are not strangers: A study of opportunities for rural heritage museums to maximise tourism visitation. *Journal of Travel Research*, *37*(3), 299–303. doi:10.1177/004728759903700312

Quintas, P. (2002). Managing Knowledge in a New Century. In Little, S., Quitas, P., & Ray, T. (Eds.), *Managing Knowledge: An Essential Reader*. London: Sage.

Qu, R. (2005). China becomes the strong trade nation strategic way. *Economic Theory and Economy Management*, *9*, 28–19.

Rabellotti, R., & Sanfilippo, M. (2008, October 9-10). *Chinese FDI in Italy.* Paper presented at Emerging Multinationals: Outward Foreign Direct Investment from Emerging and Developing Economies Conference, Copenhagen, Denmark.

Ragner, N. (1973). *Problem of Capital Formation in Underdeveloped countries*. Delhi: Oxford Universities Press.

Raivo, P. J. (2002). The peculiar touch of the East: Reading the post-war landscapes of the Finnish Orthodox Church. *Social & Cultural Geography*, *3*(1), 11–24. doi:10.1080/14649360120114116

Ralston, D. A., Yu, K.-C., Wang, X., Terpstra, R. H., Gustafson, D. J., & Wei, H. (1996). The cosmopolitan Chinese manager: Findings of a study on managerial values across the six regions of China. *Journal of International Management*, *2*, 79–109.

Ramamurti, R. (2000). Risks and rewards in the globalization of telecommunications in emerging economies. *Journal of World Business*, *35*, 149–170. doi:10.1016/S1090-9516(00)00030-4

Ramamurti, R. (2008). What Have We Learned About Emerging-Market MNEs? In Ramamurti, R., & Singh, J. V. (Eds.), *Emerging Multinationals from Emerging Markets. Cambridge, UK: Cambridge University Press*. Cambridge.

Rauch, J. (2001). Business and social networks in international trade. *Journal of Economic Literature*, *39*, 1177–1203.

Reader's Digest. (2010). *Comparison of winning brands 2006 - 2010: Results for: Kitchen Appliances*. Retrieved October 5, 2010, from http://www.rdtrustedbrands.com/trustedbrands/results/tables/Comparison%2520of%2520winning%2520brands%25202006%2520to%25202010.category.Kitchen%2520appliances.shtml

Reuvid, J. (2008). China and the WTO. In Reuvid, J. (Ed.), *Business Insights: China: Practical Advice on Entry Strategy and Engagement*. Philadelphia: Kogan Page.

Reynolds, P. (1994). Culinary heritage in the face of tourism. *Progress in Tourism. Recreation and Hospitality Management*, *6*, 189–194.

Ribiere, V.M., & Sitar, A.S. (2003). Critical Role of Leadership in Nurturing a Knowledge-supporting Culture. *Knowledge Management Research & Practice*, 39-48.

Ridella, R. (2007, November 8). *La presenza diretta delle aziende cinesi in Italia*. Paper presented at China meets Italy Conference, Osservatorio, Turin.

Ritholtz, B., & Task, A. (2009). *Bailout Nation: How Greed and Easy Money Corrupted Wall Street and Shook The World Economy*. New York: Wiley.

Robbins, R.F. (2003). Harnessing Group Memory to Build a Knowledge-Sharing Culture. *Of Counsel*, *22*(6), 7-11.

Roberts, D. (2008, June 2). 'China: Multinationals Hear it Online'. *Business Week Online*, http://www.businessweek.com/globalbiz/content/may2008/gb20080530_213248.htm (accessed 06 September 2008).

Robinson, K. M. (2001). Unsolicited narratives from the Internet: A rich source of qualitative data. *Qualitative Health Research*, *11*(5), 706–714. doi:10.1177/104973201129119398

Roka, P. (2006). *Bhagavad Gita on Effective Leadership: Timeless Wisdom for Leaders*. Bloomington, IN: iUniverse.

Romagnoli, G. F. (1995). Rome's Jewish quarter: A rich culinary heritage. *Gourmet*, *55*(8), 90–94.

Ronald, F. (1971). Entropy reduction in open systems. *Journal of Theoretical Biology*, *31*(1), 43–46. doi:10.1016/0022-5193(71)90119-6

Rong, Y. (2006). Our country's countermeasure in new round international industrial shift. *Shanxi finance and economics college journal*, *1*, 17-18.

Rooney, J. A. (1995). Branding: A Trend for Today and Tomorrow. *Journal of Product and Brand Management*, *4*(4), 48–55. doi:10.1108/10610429510097690

Ross, S. M., & Offermann, L. R. (1997). Transformational leaders: Measurement of personality attributes and work group performance. *Personality and Social Psychology Bulletin*, *23*(10), 1078–1086. doi:10.1177/01461672972310008

Roth, M. S., & Romeo, J. B. (1992). Matching Product Category and Country Image Perceptions: A Framework for Managing Country-Of-Origin Effects. *Journal of International Business Studies*, *23*(3), 477–497. doi:10.1057/palgrave.jibs.8490276

Rowley, J. (1999). What is Knowledge Management? *Library Management*, *20*(8), 416–420. doi:10.1108/01435129910291175

Rui, H., & Yip, G. S. (2008). Foreign acquisitions by Chinese firms: a strategic intent perspective. *Journal of World Business*, *43*(2), 213–226. doi:10.1016/j.jwb.2007.11.006

Rullani, E. (2006). L'internazionalizzazione invisibile. La nuova geografia dei distretti e delle filiere produttive. *Sinergie*, *24*(69), 3–32.

Ruona, W. E. A. (1999). Theory in 'theory to practice': Voices of practitioners. In K. P. Kuchinke (Ed.), *Proceedings of the AHRD Conference* (pp. 888-896). Bowling Green, OH: AHRD.

Ryan, C. (1995). *Researching Tourist Satisfaction*. London: Routledge.

Rynes, S. L., Bartunek, J. M., & Daft, R. L. (2001). Across the great divide: Knowledge creation and transfer between practitioners and academics. *Academy of Management Journal*, *44*, 340–355. doi:10.2307/3069460

Saadia, M. (2009). Framework for the Role of self-organization in the handling of adaptive challenges. *Complexity & Organization*, *131*(2), 1–14.

Salleh, Y., & Goh, W. K. (2002). Managing Human Resources towards Achieving Knowledge Management. *Journal of Knowledge Management*, *6*(5), 457–468. doi:10.1108/13673270210450414

Santos, E. D. (2008). *The Creative Economy Report 2008*. Beijing, China: People's Education Press.

Sarkar, M. B., Cavusgil, S. T., & Aulakh, P. S. (1999). International expansion of telecommunication carriers: The influence of market structure, network characteristics, and entry imperfections. *Journal of International Business Studies*, *30*, 361–381. doi:10.1057/palgrave.jibs.8490074

Satpathy, B. (2006). Transformational Leadership in the Bhagavad-Gita. *The Journal of Indian Management & Strategy.*

Satpathy, B. (2007). Hierarchy of Values in the Bhagavad-Gita. *Journal of SETU*, 15-27.

Satpathy, B., & Muniapan, B. (2008). The Knowledge of "Self" from the Bhagavad-Gita and its Significance for Human Capital Development. *Asian Social Science*, *4*(10), 143–150.

Sauvant, K. P. (2005). New sources of FDI: the BRICs. Outward FDI from Brazil, Russia, India and China. *Journal of World Investment & Trade*, *6*(5), 639–709.

Sauvant, K. P. (2006). Inward and outward FDI and the BRICs. In Subhash, C. J. (Ed.), *Emerging Economies and the Transformation of International Business: Brazil, Russia, India and China* (pp. 313–409). Cheltenham, UK: Edward Elgar.

Sauvant, K. P. (2008). The Rise of TNCs from emerging markets: the issues. In Sauvant, K. P. (Ed.), *The Rise of Transnational Corporations From Emerging Markets: Threat or Opportunity?* (pp. 3–14). Willinston, VT: Edward Elgar Publishing.

Schein, E. H. (1983). The role of founder in creating organizational culture. *Journal of Organizational Dynamics*, *12*(1), 13–28. doi:10.1016/0090-2616(83)90023-2

Schein, E. H. (1992). *Organizational Culture and Leadership* (2nd ed.). San Francisco: Jossey-Bass.

Schlein, L. (2008). 'China's Melamine Milk Crisis Creates Crisis of Confidence'. *Voice of America News Online,* http://www.voanews.com/english/archive/2008-09/2008-09-26-voa45.cfm?CFID=64268891&CFTOKEN=95716309 (accessed 26 September 2008).

Schoenecker, T. S., & Cooper, A. C. (1998). The role of firm resources and organizational attributes in determining entry timing: A cross-industry study. *Strategic Management Journal, 19,* 1127–1243. doi:10.1002/(SICI)1097-0266(1998120)19:12<1127::AID-SMJ7>3.0.CO;2-4

Schramm, M., & Staack, T. (2005). *Evolving Brand Preferences: The Case of China - An Empirical Analysis.* Retrieved October 5, 2010, from www.thinkdesk.biz/news-english/images/IAMA2005_THINKDESK-Paper.pdf

Schrödinger, E. (1967). *What is Life?: The Physical Aspect of the Living Cell with Mind and Matter and Autobiographical Sketches.* Cambridge, MA: Cambridge University Press.

Schwartz, H. M. (2009). *Subprime Nation: American Power, Global Capital, and The Housing Bubble.* Ithaca, NY: Cornell University Press.

Scott, A. (2000). *The Cultural Economy of Cities: Essays on the Geography of Image-Producing Industries.* London: SAGE Publications.

Scott, A. (2005). *On Hollywood: The Place, The Industry.* Princeton: Princeton University Press.

Scott, W. R. (2002). The changing world of Chinese enterprises: an institutional perspective. In Tsui, A. S., & Lau, M. (Eds.), *Management of Enterprises in People's Republic of China* (pp. 59–78). Boston: Kluwer Academic Press.

Seiwert, M. (2003). Geopolitische Spannungen: Müssen die Marken in Zukunft ihre Herkunft verleugnen? (Geo-political tensions: Will Brands Have to Deny their Origins in the Future). *Absatzwirtschaft, 5,* 14–20.

Sen, A. (2000). *Development as freedom.* Oxford, UK: Oxford University Press.

Senge, P. (1997). *The fifth Discipline, the art and practice of the learning organization.* New York: Currency Doubleday.

Shackely, M. (2001). Sacred world heritage sites: Balancing meaning with management. *Tourism Recreation Research, 26*(1), 5–10.

Shackley, M. (1998). Introduction. In Shackley, M. (Ed.), *Visitor Management: Case Studies from World Heritage Sites* (pp. 1–9). Oxford, UK: Butterworth-Heinemann.

Shackley, M. (1999). Visitor management. In Leask, A., & Yeoman, I. (Eds.), *Heritage Visitor.*

Shang, Y. (2006). International industrial shift to our country industrial development influence and countermeasure. *Economic Performance, 10,* 15–16.

Shapiro, R. (1986). *Towards effective supplier management: international comparisons* (Working Paper No. 9-785-062). Boston, MA: Harvard Business School.

Sharma, V. (2002, February). Visionary Enterprise. In *Proceedings of the First International Conference on Vedic Values and Corporate Excellence,* Shantikunj, Haridwar, Uttarakhand (pp. 22-24).

Sharma, G. D. (2001). *Management and the Indian Ethos.* New Delhi, India: Rupa & Company.

Sharma, S. (1996). *Management in New Age: Western Windows Eastern Doors.* New Delhi, India: New Age International Publishers.

Sharma, S. (1998). Enlightened leadership in Indian ethos: the way of the theory K. *Management and Change, 2,* 93–104.

Sharma, S. (1999). Corporate Gita: lessons for management, administration and leadership. *Journal of Human Values, 5,* 103–123. doi:10.1177/097168589900500203

Sharma, S. (2002). Corporate leadership model: an Indian model for corporate development and ethical leadership. In Pareek, U., Osman-Gani, A. M., & Rao, T. V. (Eds.), *Human ResourceDevelopment in Asia: Trends and Challenges* (pp. 291–296). New Delhi, India: Oxford and IBH.

Sharma, S. (2003). Towards corporate VEDA: Indian ethos and corporate development. *Journal of Human Values, 9,* 163–172. doi:10.1177/097168580300900207

Shenkar, O. (1990). International joint ventures' problems in China: Risks and remedies. *Long Range Planning, 23,* 82–90. doi:10.1016/0024-6301(90)90056-A

Shettel-Neubar, J., & O'Reilly, P. (1981). *Now Where? A Study of Visitor Orientation and Circulation in the Arizona-Sonora Desert Museum*. Jacksonville, FL: Jacksonville State University.

Shi, G. (2002, November 13). Speech in the 16th National Congress of the Communist Party of China. *People's Daily*.

Short, D. C. (2006a). HRD scholar-practitioners: A definition and description of core activities. In F. M. Nafukho & H. C. Chin (Eds.), *Proceedings of the AHRD Conference* (pp. 258-264). Bowling Green, OH: AHRD.

Short, D. C. (2006b). The link between research and practice: Experiences of different professions and implications for HRD. In F. M. Nafukho & H. C. Chin (Eds.), *Proceedings of the AHRD Conference* (pp. 265-272). Bowling Green, OH: AHRD.

Short, D. C. (2006c). The gap between research and practice in HRD: A summary of the discussion 1995-2005. In F. M. Nafukho & H. C. Chin (Eds.), *Proceedings of the AHRD Conference* (pp. 250-257). Bowling Green, OH: AHRD.

Short, D. C., Keefer, J. M., & Stone, S. J. (2006). The link between research and practice: Experiences of different professions and implications for HRD. In F. M. Nafukho & H. C. Chin (Eds.), *Proceedings of the AHRD Conference* (pp. 914-921). Bowling Green, OH: AHRD.

Silberg, T. (1995). Cultural Tourism and Business Opportunities for Museums and Heritage Sites. *Tourism Management*, *16*(5), 361–365. doi:10.1016/0261-5177(95)00039-Q

Simic, I. (1998). Transformational Leadership – The key to successful management of transformational organizational changes. *The Scientific Journal FACTA UNIVERSITATIS*, *1*(6), 49–55.

SINA (2008a). 'Dior Announcement: Absolutely Disagree with the Comment of Sharon Stone', http://ent.sina.com.cn/s/2008-05-27/16102038967.shtml (accessed 27 May 2008).

SINA (2008b). 'Sanlu Group Insisted Their Products Meet National Standard When Responding to the Milk Accident', http://news.sina.com.cn/c/2008-09-11/105416273760.shtml (accessed 11 September 2008).

Singh, I. (2004a). Language and ethnicity. In Singh, I., & Peccei, J. S. (Eds.), *Language, Society and Power an Introduction* (2nd ed., pp. 94–111). New York: Routledge.

Singh, I. (2004b). Language, thought and representation. In Singh, I., & Peccei, J. S. (Eds.), *Language, Society and Power an Introduction* (2nd ed., pp. 18–34). New York: Routledge.

Skinner, G. W., & Baker, H. D. R. (1977). *The city in late imperial China*. Palo Alto, CA: Stanford University Press.

Slote, W. H., & DeVos, G. A. (Eds.). (1998). *Confucianism and the family.*

Smith, H. D. (2001). *The British Maritime Heritage*. Paper presented at the RGS-IBG Annual Conference.

Smith, C., & Comer, D. (1994). Self-organization in small groups: a study of group effectiveness within non-equilibrium conditions. *Human Relations*, *47*(5), 553–581. doi:10.1177/001872679404700505

Song, H. (2004). International industrial shift new tendency. *Economics Daily*, 06-16.

Song, J. L., & Werbel, J. D. (2007). Guanxi as impetus? Career exploration in China and the United States. *Career Development International*, *12*(1), 51–67. doi:10.1108/13620430710724820

Song, X. M., DiBenedetto, C. A., & Zhao, Y. Z. L. (1999). Pioneering advantages in manufacturing and service industries: Empirical evidence from nine countries. *Strategic Management Journal*, 20.

Song, Y., & Niu, C. (2005). International industrial shift tendency and our country's essential countermeasure. *Industrial Technology Economy*, *8*, 27–27.

Soros, G. (2008). *The New Paradigm for Financial Markets: The Credit Crisis of 2008 and What It Means* (1st ed.). New York: Public Affairs.

Sowell, T. (2009). *The Housing Boom and Bust*. New York: Basic Books.

So, Y. L., & Walker, A. (2006). *Explaining Guanxi: The Chinese Business Network*. London: Routledge.

Spicer, S. (2008). *Think global, act local: International Employee Benefits.* Retrieved from http://cendoc.esan.edu.pe/paginas/infoalerta/odp/InfoAlertaODP.html

Spigarelli, F. (2008). *Nuovi investitori globali: le imprese cinesi in Italia* (Working Paper, 12). Macerata, Italy: University of Macerata, Department of Studies on Economic Development.

Spigarelli, F. (2009). Multinazionali dai paesi emergenti: il caso delle imprese cinesi. *Economia e Politica Industriale*, *2*, 131–159.

Spigarelli, F., Alon, I., & Wei, W. (2009). *A speed race: Benelli and QJ competing on the motorbike international arena. Richard School of Business*. London, Ontario: Ivey Publishing.

Stake, R. E. (1995). *The art of case study research*. Thousand Oaks, CA: Sage.

Standifird, S. S., & Marshall, R. S. (2000). The transaction cost advantage of Guanxi-based business practices. *Journal of World Business*, *35*(1), 21–42. doi:10.1016/S1090-9516(99)00032-2

Statistical Bulletin of China's Outward Foreign Direct Investment. (2008). *In*. MOFCOM.

Statistical Yearbook (1982-2010). In MOFCOM.

Stauffer, D. (1999). Why people hoard knowledge. *Across the Board*, *36*(8), 16–21.

Steil, B. (2009). *Lessons of the Financial Crisis*. New York: Council on Foreign Relations, Center for Geo-economic Studies.

Strange, C., & Kempa, M. (2003). Shades of dark tourism: Alcatraz and Robben Island. *Annals of Tourism Research*, *30*(2), 386–405. doi:10.1016/S0160-7383(02)00102-0

Styles, C., & Ambler, T. (2003). The coexistence of transaction and relational marketing: Insights from the Chinese business context. *Industrial Marketing Management*, *17*(6), 539–557.

Sun, Y. (2006). The international industrial shift's new tendency and our country's strategy chooses. *Contemporary management of economy*, *5*, 35-36.

Sun, Y., Von Zedtwitz, M., & Simon, D. F. (2007). Globalization of R&D and China: An Introduction. *Asia Pacific Business Review*, *13*(3), 311–319. doi:10.1080/13602380701291867

SureVista Solutions. (2007). *Benefits of Customer Satisfaction*. Retrieved January 25, 2007, from http://www.surevista.com/benefits-customer-satisfaction.html

Swarbrooke, J. (1993). The future of heritage attractions. *Insights*, *4*, D15–D20.

Swarbrooke, J. (1995). *The Development and Management of Visitors Attraction*. Oxford, UK: Butterworth-Heinemann.

Synovate. (2006). *Western Consumers still Skeptical about Quality of 'Made in China' and 'Made in Korea' Labels*. Retrieved October 5, 2010, from http://www.synovate.com/insights/infact/issues/200601/

Szulanski, G. (1996). Exploring internal stickiness: impediments to the transfer of best practice with the firm. *Strategic Management Journal*, 17.

Szutanski, G. (2000). The process of knowledge transfer: A diachronic analysis of stickiness. *Organizational Behavior and Human Decision Processes*, *82*(1), 9–27. doi:10.1006/obhd.2000.2884

Szymanski, D. M., Troy, L. C., & Bharadwaj, S. G. (1995). Order of entry and business performance: An empirical synthesis and reexamination. *Journal of Marketing*, *59*, 17–33. doi:10.2307/1252325

Tai, S. (2008). 'Beauty and the Beast: The Brand Crisis of SK-II Cosmetics in China'. *Asian Case Research Journal*, *12*(Issue 1), 57–71. doi:10.1142/S0218927508001023

Tan, E. K.-B. (2000). Success amidst prejudice: Guanxi networks in Chinese businesses in Indonesia and Malaysia. *Journal of Asian Business*, *16*(1), 65–84.

Tan, K. (2008). 'Singapore Finds Melamine in White Rabbit Candies; Chinese Dairy Products now Banned across Asia and Africa'. *Shanghaiist online*, http://shanghaiist.com/2008/09/22/singapore_finds_melamine_in_white_r.php (accessed 22 September 2008).

Tan, W. (2005, June 23). Culture Conflicts in Sino-Foreign Ventures. *Beijing Review*.

Tan, C. Z., & Yeung, H. W. (2000). The regionalization of Chinese business networks: A study of Singaporean firms in Hainan, China. *The Professional Geographer*, *52*, 437–454. doi:10.1111/0033-0124.00237

Tarmidi, L., & Gammeltoft, P. (2008, October 9-10). *The characteristics of China's multinationals in the manufacturing sector operating in Indonesia*. Paper presented at Emerging Multinationals: Outward Foreign Direct Investment From Emerging and Developing Economies Conference, Copenhagen, Denmark.

Taub, J. (2009). *Enablers of Exuberance: Legal Acts and Omissions That Facilitated the Global Financial Crisis* (SSRN Working Paper Series). SSRN.

Taylor, J. B. (2009). *Getting Off Track: How Government Actions and Interventions Caused, Prolonged, and Worsened the Financial Crisis*. New York: Hoover Institution Press.

Taylor, R. (2002). Globalization strategies of Chinese companies: current development and future prospects. *Asian Business & Management*, *1*(2), 209–225. doi:10.1057/palgrave.abm.9200011

Teo, P., & Yeoh, B. S. A. (1997). Remaking local heritage for tourism. *Annals of Tourism Research*, *24*(1), 192–213. doi:10.1016/S0160-7383(96)00054-0

Terpstra, V., & Sarathy, R. (1994). *International Marketing* (6th ed.). Orlando, FL: The Dryden Press.

Theodore de Bary, W. (1960). *Sources of Chinese Tradition* (1st ed.). New York: Columbia University Press.

Theodore de Bary, W., & Bloom, I. (1999). *Sources of Chinese Tradition* (2nd ed.). New York: Columbia University Press.

Tian, X. (2006). *Managing international business in China*. New York: Cambridge University Press.

Tian, Z. (2005). New tendency of International industrial adjustment and shift - international outsourcing. *Commercial Research*, *17*, 17–19.

Tichy, N., & Devanna, M. A. (1986). *The Transformational Leader*. New York: John Wiley.

Times, K. (2009, January 21). *Korean Brand Value Up, Closer to US, Japan*. Retrieved October 5, 2010, from http://www.koreatimes.co.kr/www/news/biz/2009/01/123_38278.html

Times, T. (2005, August 11). 'Chinese Fanatics Urge Boycott over Lin's Green Parents'. *Taipeitimes.com.*, http://www.taipeitimes.com/News/front/archives/2005/08/11/2003267233 (accessed 22 July 2009).

Timothy, D., & Boyd, S. (2003). *Heritage Tourism*. London: Prentice Hall.

Tong, S. Y. (2003). *Ethnic Chinese Networking in cross-border investment: the impact of economic and institutional development* (Working Paper). Hong Kong, China: University of Honk Kong, Hong Kong Institute of Economics and Business Strategy.

Tong, S. J. (2005). Ethnic Networks in FDI and the Impact of Institutional Development. *Review of Development Economics*, *9*(4), 563–580. doi:10.1111/j.1467-9361.2005.00294.x

Törnqvist, G. (1983). Creativity and the Renewal of Regional Life. In Buttimer, A. (Ed.), *Creativity and Context: A Seminar Report* (pp. 91–112). Lund, Sweden: Gleerup Press.

Törnqvist, G. (2004). Creativity in Time and Space. *Geografiska Annaler*, *86*(4), 227–243. doi:10.1111/j.0435-3684.2004.00165.x

Treffers, T., Faaij, A. P. E., Sparkman, J., & Seebregts, A. (2005). Exploring the Possibilities for setting up Sustainable Energy Systems for the Long Term:Two Visions for the Dutch Energy System in 2050. *Energy Policy*, (33): 1723–1743. doi:10.1016/j.enpol.2004.02.011

Trompenaars, F. (1993). *Riding the Waves of Culture: Understanding Cultural Diversity in Business*. London: Nicholas Brealey Publishing Ltd.

Tsai, K. S. (2002). *Back Alley Banking: Private Entrepreneurs in China*. Ithaca, NY: Cornell University Press.

Tsang, W. K. (1998). Can guanxi be a source of sustained competitive advantage for doing business in China? *The Academy of Management Executive*, *12*, 64.

Tschang, C. (2008, September 29). 'Contaminated Milk Sours China's Dairy Business'. *Business Week Online*, http://www.businessweek.com/globalbiz/content/sep2008/gb20080926_543133.htm (accessed 20 July 2009).

Tu, W. (1998b). Probing the "three bonds" and "five relationships" in Confucian humanism.

Tullk, G. (1967). Excess Benefit. *Water Resources Research*, (3): 643–644. doi:10.1029/WR003i002p00643

Tu, W. (1998a). Joining east and west. *Harvard International Review*, *20*(3), 44–49.

UNCTAD. (2005). *Prospects for Foreign Direct Investment and the Strategies of Transnational Corporations. 2005-2008*. New York: United Nations.

UNCTAD. (2006). World Investment Report. *Overview FDI from Developing and Transition Economies, United Nations*. New York: Geneva.

UNCTAD. (2007). World Investment Report. *Transnational Corporations, Extractive Industries and development, United Nations*. New York: Geneva.

UNCTAD. (2009). *World Investment Report.*

Unterman, A. (2009). Perverse incentives: Risk taking and reform. *Banking and Financial Services Policy Report, 28*(6), 11.

Valdani, E., & Bertoli, G. (2006). *Mercati internazionali e marketing*. Milano, Italy: Egea.

Van de Ven, A. H. (2000). Professional science for a professional school: Action science and normal science. In Beer, M., & Nohrta, N. (Eds.), *Breaking the code of change* (pp. 393–413). Boston: Harvard Business School Press.

Van de Ven, A. H., & Johnson, P. E. (2006). Knowledge for theory and practice. *Academy of Management Review, 31*(4), 802–821.

Van de Ven, A. H., & Johnson, P. E. (2006). Reply: Nice try, Bill, But.... There you go again. *Academy of Management Review, 31*(4), 802–821.

VanderWerf, P. A., & Mahon, J. F. (1997). Meta-analysis of the impact of research methods on findings of first-mover advantage. *Management Science, 43*, 1510–1520. doi:10.1287/mnsc.43.11.1510

Vang, J., & Chaminade, C. (2007, March). Global-local Linkages, Spillovers and Cultural Clusters: Theoretical and Empirical Insights from an Exploratory Study of Toronto's Film Cluster. *CIRCLE Electronic* (Working Paper Series). Retrieved from http://www.circle.lu.se

Vanhonaker, W., & Pan, Y. (1997). The Impact of National Culture, Business Scope, and Geographic Location on Joint -Venture Operations in China. *Journal of International Marketing, 5*(3), 11–30.

Vernon, R. (1966). International Investment and International Trade in the Product Cycle. *The Quarterly Journal of Economics, 82*(2), 190–207. doi:10.2307/1880689

Viitala, R. (2004). Towards Knowledge Leadership. *Leadership and Organization Development Journal, 25*(26), 528–544. doi:10.1108/01437730410556761

Vinciguerra, L. (2007, September 28). Sergio Tacchini, la missione salvataggio arriva dalla Cina. *Il Sole24ore.*

Vines, S. (2009). *Market Panic: Wild Gyrations, Risks, and Opportunities in Stock Markets* (2nd ed.). New York: Wiley.

Voase, R. (1999). 'Consuming' tourist sites/sights: A note on York. *Leisure Studies, 18*(4), 289–296. doi:10.1080/026143699374862

Von Foerster, H. (1960). On Self-Organizing Systems and their Environments. In Yovitts, M. C., & Cameron, S. (Eds.), *Self-Organizing Systems* (pp. 31–50).

Wagner, D., Head, K. C., & Reis, J. C. (2002). Immigration and the Trade of provinces. *Scottish Journal of Political Economy, 49*(5), 507–525. doi:10.1111/1467-9485.00245

Walder, A. G. (1986). *Communist Neo-Traditionalism: Work and Authority in Chinese Industry*. Berkeley, CA: University of California Press.

Walder, A. G. (1995). Local governments as local firms: An organizational analysis of China's transitional economy. *American Journal of Sociology, 101*, 263–301. doi:10.1086/230725

Wang, J. (2004). Environmental taxes in OECD countries and the Enlightenment to China. *Comprehensive Utilization of Resources in China*, (5), 33-36.

Wang, M. (2004). Discussion on Tax Reform and Environmental Protection. *Henan Social Science*, (3), 121-123.

Wang, R. (2008, August 13). *Zhang Ruimin: Dachui zachu shijie mingpai (Zhang Ruimin: A Big Hammer Helps a World-class Brand to Come into Existence)*. Retrieved October 5, 2010, from http://finance.people.com.cn/GB/8215/126457/7661094.html

Wang, R., & Chen, W. (2006). The international industrial shift's new tendency with deals with the mentality. *Scientific and Technical Information Development and Economic development, 3*, 21-22.

Wang, X. (2002, September 24). *Deguo zhizao de xinyu conghe er lai (Where Does the Reputation of, made In Germany' come from)*? Retrieved October 5, 2010, from http://news.xinhuanet.com/fortune/2002-09/24/content_572868.htm

Wang, X. (2008, September 12). *Sanlu pai ying you'er naifen shijian gundong baodao (Reports on the Issue of Sanlu Milk Powder for Children and Infants)*. Retrieved October 5, 2010, from http://news.xinhuanet.com/newscenter/2008-09/12/content_9935963.htm

Wang, Y., & Qin, J. (2002). Empowerment and Knowledge-Sharing. *Economic Forum, 15*.

Wang, Y., Zhang, X. S., & Goodfellow, R. (1998). *Business culture in China*. Singapore: Butterworth Heinemen.

Wang, Z. (2001). *Zhonghua Renmin Gongheguo Guomin Jingji he Shehui Fazhan di shi ge Wunian Jihua gangyao (Outline of the 10. Five-Year-Plan on Economic and Social Development of the People's Republic)*. Retrieved October 5, 2010, from http://www.china.com.cn/ch-15/plan8.htm

Wang, J. (2008). *Brand New China: Advertising, Media, and Commercial Culture*. Cambridge, MA: Harvard University Press.

Wang, L. C. (2005). Guanxi vs. relationship marketing: Exploring underlying differences. *Industrial Marketing Management, 36*(1), 81–86. doi:10.1016/j.indmarman.2005.08.002

Wang, Z. (2003). *The Multinational Corporation develops the new tendency in China*. Beijing, China: New China Press.

Warentest, S. (2008). Die Langsamspüler (The Slow Washers). *Test, 4*, 64–68.

Warner, M., Hong, N. S., & Xu, S. (2004). Late developments experience and the evolution of transnational firms in the People's Republic of China. *Asia Pacific Business Review, 10*(3-4), 324–345. doi:10.1080/1360238042000264397

Webber, R. H. (1969). Convergence or divergence. *The Columbia Journal of World Business, 4*(3), 75–83.

Weber, J. (2009). Using exemplary business practices to identify Buddhist and Confucian ethical value systems. *Business and Society Review, 114*(4), 511–540. doi:10.1111/j.1467-8594.2009.00352.x

Wedow, S. (1977). Review of the tourist: A new theory of the leisure class. *Contemporary Sociology, 6*(2), 200–202. doi:10.2307/2065807

Wei, J. (2008, April 23). Non-financial investment rises 6%. *China Daily*.

Wei, K. (2008). 'Google Disclosed the Hottest Search in China in 2008'. *China.COM.CN*. http://www.china.com.cn/international/txt/2008-12/11/content_16930439.htm (accessed and translated 11 December 2008).

Weidenbaum, M. (1996a). The Chinese family business enterprise. *California Management Review, 38*(4), 141–156.

Weidenbaum, M. (1996b). The bamboo network. *Executive Speeches, 11*(2), 5–6.

Wells, L. (1983). *Third world multinationals: The role of foreign direct investment from developing countries*. Cambridge, MA: MIT Press.

Wessel, D. (2009). *In Fed We Trust: Ben Bernanke's War On The Great Panic*. New York: Crown Business.

Wilg, K. M. (1994). *Knowledge Management: The central management focus for intelligent-acting organizations*. Arlington, TX: Schema Press.

Willems, J. C. (1972). Dissipative dynamical systems, part I: General theory; part II: Linear systems with quadratic supply rates. *Archive for Rational Mechanics and Analysis, 45*, 321–393. doi:10.1007/BF00276493

Wilson, D., & Purushothaman, R. (2003). *Dreaming With BRICs: The Path to 2050* (Global Economics Paper 99). Retrieved from http://www2.goldmansachs.com/ideas/brics/book/99-dreaming.pdf

Wirtschaftswoche. (2006). Werte vernichtet (Values destroyed). *Special Issue on China*, 68-71.

WIR–World Investment Report (2009). Transnational Corporations, Agriculture Production and Development. In UNCTAD.

Wong, J., & Chan, S. (2003). China's outward direct investment: expanding worldwide. *China. International Journal (Toronto, Ont.), 1*(2), 273–301.

Wong, M. (2007). Guanxi and its role in business. *Chinese Management Studies, 1*(4), 257–276. doi:10.1108/17506140710828532

Wong, S. L. (1985). The Chinese family firm: A model. *The British Journal of Sociology, 36*(1), 58–72. doi:10.2307/590402

Wong, Y. H., & Chan, R. Y.-K. (1999). Relationship marketing in China: Guanxi, favouritism and adaptation. *Journal of Business Ethics, 22*, 107–118. doi:10.1023/A:1006077210425

Woodcock, C. P., Beamish, P. W., & Makino, S. (1994). Ownership-Based Entry Mode Strategies and International Performance. *Journal of International Business Studies*, (2): 253–273. doi:10.1057/palgrave.jibs.8490200

Worden, N. (2003). National identity and heritage tourism in Melaka. *Indonesia and the Malay World, 31*(89), 31–43. doi:10.1080/13639810304447

Wright, M., Filatotchev, I., Hoskisson, R. E., & Peng, M. W. (2005). Strategy Research in emerging economies: challenging the conventional wisdom. *Journal of Management Studies*, *42*(1), 1–33. doi:10.1111/j.1467-6486.2005.00487.x

Wu, F. (2005). The Globalization of Corporate China. *NBR Analysis, 16*(3). Retrieved from http://www.nbr.org/publications/analysis/pdf/vol16no3.pdf

Wu, D. H. (2007). *Study into International Protection System of Intellectual Property Rights*. Beijing, China: Intellectual Property Press.

Wu, Y. (2003). The Refrigerator Magnat. *The McKinsey Quarterly*, *3*, 106–111.

Xia, G., et al. (2008). A review of researches on environmental taxes and charges. *Industrial Technology and Economy*, (5), 156-157.

Xia, L. (2008). Develop Low-carbon Economy And Achieve Sustainable Development of Urban. *Environmental Protection*, (2), 33-35.

Xiang, X. P., & Wang, L. J. (2001). Core Competitiveness of Enterprises in the Perspective of Knowledge. *Journal of Economy and Management of Foreign Countries, 3*.

Xing, F. (1995). The Chinese cultural system: Implications for cross-cultural management. *Sam Advanced Management Journal*, *60*(1), 14–20.

Xinhua (2008a). 'Earthquake Measuring 7.8 Richter Scale Hits SW China'. *Chinaview online*, http://news.xinhuanet.com/english/2008-05/12/content_8153541.htm (accessed 17 July 2009).

Xinhua(2008b). 'Stone's Quake 'Karma' Apology Doesn't Mollify Many Chinese'. *China Daily Online,* http://www.chinadaily.com.cn/china/2008-05/29/content_6721642.htm (accessed 29 May 2008).

Xinhua (2008c). '4 Officials Sacked Following Baby Milk Scandal'. *China Daily Online*, http://www.chinadaily.com.cn/china/2008-09/17/content_7034236.htm (accessed 08 May 2009).

Xu, B. T., Bi, L. Y., & Qi, M. (2007). *Protection on Intellectual Property Rights of Modern Enterprises*. Guangdong, China: Zhongshan University Press.

Xu, S. (1996). Management and Leadership Style in Knowledge Society. *World Manager's Digest*, *12*, 44–54.

Yan, A., & Gray, B. (1994). Bargaining power, management control, and performance in United States-China joint ventures: A comparative case study. *Academy of Management Journal*, *37*, 1478–1517. doi:10.2307/256796

Yang, F. (2007). Foreign direct investment with our country advance in technology. *Fujian Forum*, *5*, 24–25.

Yang, G., Xin, R., Liu, S., & Dai, H. (2008). *Xianhua yu jingji: Tanxun Zhongguo qiye quanqiuhua zhi lu* (Fresh Flowers and Sharp Thorns: In Search of a Road to Globalization for Chinese Enterprises). Beijing, China: Shangwu Yinshuguan.

Yang, Z. (2005). International industrial shift with China's countermeasure. *Asking Journal*, *3*, 15–18.

Yan, Y. (2005). *Foreign Investment and Cooperate Governance in China*. New York: Palgrave Macmillan.

Yan, Y., & Child, J. (2004). Investors' resources and management participation in international joint ventures: a control perspective. *Asia Pacific Journal of Management*, *21*, 287–304. doi:10.1023/B:APJM.0000036464.38236.34

Yao, S. (2006). On economic growth, FDI and exports in China. *Applied Economics*, *38*(3), 339–351. doi:10.1080/00036840500368730

Ye, M. (2008). 'Sanlu Milk Scandal: Self-reflection of Crisis Public Relations'. *Yahoo online*. http://xk.cn.yahoo.com/articles/080927/1/dpn7_3.html (accessed 27 September 2008).

Yeung, H. W. C. (2000). Local politics and foreign ventures in China's transitional economy: the political economy of Singaporean investments in China. *Political Geography*, *19*(7), 809–840. doi:10.1016/S0962-6298(00)00027-5

Yeung, H. W. (1997). Cooperative strategies and Chinese business networks: A study of Hong Kong transnational corporations in the ASEAN region. In Beamish, P. W., & Killing, J. P. (Eds.), *Cooperative Strategies: Asian Pacific Perspectives* (pp. 22–56). San Francisco: The New Lexington Press.

Yeung, H. W., & Liu, W. (2008). Globalizing China: the rise of mainland firms in the global economy. *Eurasian Geography and Economics*, *49*(1), 57–86. doi:10.2747/1539-7216.49.1.57

Yijia, J. (2008). *Meinuo (Miele) Zhongguo gongsi xuanchuan chengli (Announcement of the Foundation of Miele China)*. Retrieved October 5, 2010, from http://www.e-jjj.com/Article/HTML/0063/63484.html

Yin, G. (2009). Low-carbon Economic Development Experience and Implications for China. *Reform and Strategy*, (10), 176-179.

Yin, R. K. (2003). *Case Study Research: Design and Methods (applied social research methods)* (3rd ed.). Thousand Oaks, CA: Sage.

Young, A. (2000). The razor's edge: Distortions and incremental reform in the People's Republic of China. *The Quarterly Journal of Economics*, *154*, 1091–1135. doi:10.1162/003355300555024

Youzhou, Z., Jiesheng, F., Wong, M. Y., & Stewart, S. (1996). The 'catfish effect' of the private sector on the economy of The People's Republic of China. *Journal of Enterprising Culture*, *4*, 331–349. doi:10.1142/S0218495896000198

Yum, J. O. (1988). The impact of Confucianism on interpersonal relationships and communication patterns in East Asia. *Communication Monographs*, *55*(4), 374–388. doi:10.1080/03637758809376178

Zanier, V. (2006). Il sistema economico. In Abbiati, M. (Ed.), *Propizio è intraprendere imprese* (pp. 79–105). Venice, Italy: Cà Foscarina.

Zhang, K. (2005). *Going Global. The Why, When, Where and How of Chinese Companies' Outward Investment intention*. Asia Pacific Foundation of Canada, Vancouver. Retrieved from www.asiapacific.ca

Zhang, M., et al. (2003). Current analysis of the introduction of green taxes in China. *China Environmental Management*, (4), 4-5.

Zhang, J. C. (2007). *Report on the Development of Creative Industries in China*. Beijing, China: China Economic Press.

Zhang, X. M., Yin, C. L., & Li, P. (2007). *Report on the Development of International Cultural Industries*. Beijing, China: Social Science Academic Press.

Zhang, Y. (2003). *China's Emerging Global Business: Political Economy and Institutional Investigations*. Basingtoke, UK: Palgrave Macmillan.

Zhao, W., & Liu, L. (2008, October 9-10). *Outward Direct Investment and R&D spillovers: the China's case*. Paper presented at Emerging Multinationals: Outward Foreign Direct Investment From Emerging and Developing Economies Conference, Copenhagen, Denmark.

Zhao, H., & Luo, Y. (2002). Product diversification, ownership structure, and subsidiary performance in China's dynamic market. *Management International Review*, *42*, 27–48.

Zhao, H., & Xia, H. (2009). Study on the constitution, evolution and development suggestions of entrepreneurial finance system: perspective from self-organization theory. *China Soft Science*, *9*, 107–112.

Zhao, J. (2009). A survey on Beijing's cultural creative industries. *International Journal of Chinese Culture and Management*, *2*(2), 146–166. doi:10.1504/IJCCM.2009.029251

Zhao, J. (2010). *Management of Cultural Industries*. Beijing, China: Tsinghua University Press.

Zhou, M. (2006). The Chinese Diaspora and International Migration. In Bian, Y., Kwok-bun, C., & Tak-sing, C. (Eds.), *Social Transformations in Chinese Societies*. Leiden, The Netherlands: Brill Academic Publishers.

Zhou, Z. (2000). The regulation of foreign banks in China. In Norton, J. J., Li, C. J., & Huang, Y. (Eds.), *Financial Regulation in the Greater China Area: Mainland China, Taiwan and Hong Kong SAR* (pp. 65–97). The Hague, The Netherlands: Kluwer Law International.

Zhuang, G. (2005). The Way of Chinese Low Carbon Economy. *International Technical and Economic Study*, (3), 79-87.

Zou, G. W., & Xu, Q. W. (2006). *Globalization and Development of China's Cultural Industries*. Beijing, China: Central Compilation and Translation Press.

Zweig, D. (2002). *Internationalizing China: Domestic Interests and Global Linkages*. Ithaca, NY: Cornell University Press.

About the Contributors

Dasho Karma Ura earned his undergraduate degree at Oxford University and his Masters in Philosophy of Economics from the University of Edinburgh. He worked for the Ministry of Planning for 12 years before becoming the Director of the Centre for Bhutan Studies (CBS) from its founding in 1999 until 2008 when he became its President. Under Dasho Karma Ura, the CBS has been at the forefront in promoting and deepening national and global understanding of Bhutan's home-grown development philosophy of Gross National Happiness. He was a member of the Drafting Committee of Bhutan's first Constitution, enacted in July 2008. He was awarded the Red Scarf and the ancient title of Dasho (knighthood) by His Majesty the Fourth King in December 2006, and in 2010, he was bestowed the honour of Druk Khorlo (Wheel of Dragon Kingdom) by His Majesty the King for his contributions to literature and fine arts. He is also a painter. His paintings can be seen in the renowned iconic Dochula Temple near Thimphu built under the generous patronage of HM the Queen Mother Ashi Dorji Wangmo Wangchuck. Further, he has written several books, including a novel, The Hero with a Thousand Eyes. He holds numerous international positions, including; 2010~ Executive Committee Member of the School of Well-being, Chulalongkorn University, and San Nagarprada Foundation, Thailand; 2010~ Member of the Reflection Group on Global Development Perspectives, Global Policy Forum, Europe, Bonn; and Member of the Chief Economist's Advisory Panel, South Asia Region, World Bank. In Bhutan, he is a member of several bodies, such as the Board of the Royal Monetary Authority (Central Bank) of Bhutan.

Patricia Ordóñez de Pablos is a Professor in the Department of Business Administration in the Faculty of Economics and Business of the University of Oviedo, Spain. She received her PhD from The University of Oviedo (Best PhD Award, 2001) and completed her studies in the London School of Economics (UK). Her teaching and research interests focus on the areas of strategic management, knowledge management, intellectual capital, human resources management and IT. She serves as Editor in Chief of the *International Journal of Asian Business and Information Management,* Associate Editor of *Behaviour and Information Technology* and Executive Editor of the *International Journal of Learning and Intellectual.*

* * *

Mona Chung who lectures in International Business a Deakin University is a bi-cultural expert in cross-cultural communication whose work addresses the cultural gaps between Australians and Chinese in the fields of commerce and education. She specialises in strategic planning, management and marketing practice for international organisations (both commercial and educational) to understand the Chinese market. Dr Chung is a frequent guest speaker at public forums and tertiary institutions and is the author of an extensive list of publications in cross-cultural business studies. She is the author of "Shanghaied: Why Foster's could not Survive China" by Heidelberg Press. She can be contacted on +61 414271678 or mchung@deakin.edu.au.

William "Rick" Crandall serves as Professor of Management at the University of North Carolina at Pembroke. He earned his Ph.D. in Business Administration from the University of Memphis. His primary research interest is in the area of crisis management, helping organizations cope with catastrophic events. He is co-author of the book, *Crisis Management in the New Strategy Landscape* (with John Parnell and John Spillan, also of the University of North Carolina at Pembroke) through Sage Publications. Prior to entering academe, Dr. Crandall worked in management for ARA Services (now ARAMARK), a service management firm based in Philadelphia.

Claretha Hughes, Assistant Professor, teaches Human Resource Development in the Department of Rehabilitation, Human Resources and Communication Disorders, University of Arkansas, Fayetteville, Arkansas. She has extensive professional experience in business and industry and continues to serve as a consultant to international, national, and state organizations. her research focuses on value creation through the use of human resource development and technology development. She is interested in the impact of values (1) on teaching and learning processes and motivation; (2) on organizational culture, change strategies, and leadership; and (3) on technology in the workplace environment and employee behavior.

Fernanda Ilhéu is Professor of International Marketing, and Management at Lisbon's School of Economics and Management (ISEG). She received her PhD degree from Seville University in Spain. Her research focuses International Business and Marketing. She is also coordinates the ChinaLogus – Business Knowledge & Relationship with China, a research and consultancy Centre in CEGE/SEG.

Cora F. Jungbluth, MA, is a member of the research project "Rethinking Trends – Transcultural Flows in Global Public(s)" at the Cluster of Excellence "Asia and Europe in a Global Context: Shifting Asymmetries in Cultural Flows" at Heidelberg University. Her current research focuses on trends in Chinese economic policy, particularly with regard to the framework for outward foreign direct investment of Chinese Companies. Former research includes intercultural management in Chinese companies, China's rural development and the status of migrant workers in Chinese society.

Aliana M. Leong is Professor of Tourism Management, Dean of Faculty of International Tourism, Macau University of Science and Technology, Macau, Special Administrative Region of China.

Huifeng Li: Master of Economics, Ph.D. candidate in China University of Geosciences, Associate Professor in Beijing Union University, research interest is in economic analysis and evaluation on environmental pollution. Long being engaged in research and teaching on economic theory, familiar with Western economics, industrial economics, economic theory on resources and environment. Hosted and/or participated as the main members several research Programs, granted by Beijing Municipal Education Commission, Social Science Fund, the National Natural Science Foundation, the National Social Science Fund, etc., on environmental issues related to resources as well as a number of cross-organizational research projects, published a number of research papers on core journals and 2 relevant books.

Jacky Xi Li is Associate Professor of Tourism Management, Faculty of International Tourism, Macau University of Science and Technology, Macau, Special Administrative Region of China.

Yuhong Li is a professor of Administration Science, she is the Master of administration science, MBA of City University, USA. At present she serves as the director of Business Administration Department and Culture Administration Research Office, Research Institute, Apply Science and Technology College, Beijing Union University, she also lectures in Motorola University, Advisor of Small- & Middle-Scale Enterprise Consultation.

Zhang Long earned a Ph.D. degree from China University of Geosciences-Beijing, where he currently serves as an Associate Professor. His recent work on crisis management appears in the *Journal of International Journal of Business and Emerging Markets,* the *Journal of International Business Research and Practice,* and the *International Journal of Business & Globalisation.*

Laird McLean is a PhD candidate in Human Resource Development, University of Minnesota, St. Paul, Minnesota, and Owner/Senior Consultant at McLean Global Consulting, St. Paul, Minnesota.

Jane Menzies is a Lecturer in International Business at the Deakin Business School at Deakin University. Dr Menzies was awarded her PhD at Monash University in 2005. Her PhD focused on the participation of women in international assignments in multinational enterprises. Jane co-authored a "Selected Paper' presentation at the International Human Resource Management Conference, Cairns (2005). She has published her work at both national and international conferences, and has received media attention for her research and studies in Australia.

Balakrishnan Muniapan is a Senior Lecturer in Human Resource Management at the School of Business & Design, Swinburne University of Technology (Sarawak Campus), Malaysia. He has more than 50 publications in journals, books and conference proceedings. He has presented on HRM at several seminars and academic conferences within Asia, Australia and Europe.

Shujauddin Niyazi: A research scholar in the field of Operations Research and Statistics, teaches Operations Research and Management, Research Methodology, Business Statistics and Business Mathematics in the college of Sahara Arts and Management Academy. The area of his research is fertility modeling and their stochastic processes.

John A. Parnell is the William Henry Belk Distinguished Professor in Management at the University of North Carolina at Pembroke. He is the author of numerous academic papers in strategic management, crisis management, and related areas. He earned the Ed.D. degree from Campbell University, and the Ph.D. degree in Strategic Management from The University of Memphis. Dr. Parnell was the recipient of a Fulbright research award in Egypt and has also lectured in a number of countries, including extensive recent activity in China, Peru and Mexico. His current research focuses on business strategy, crisis management, and related topics. His recent work on crisis management appears in the *Journal of Contingencies and Crisis Management* and the *Journal of International Business Research and Practice.* He is co-author of the book, *Crisis Management in the New Strategy Landscape* (with William "Rick" Crandall and John Spillan, also of the University of North Carolina at Pembroke).

Nattavud Pimpa is a senior lecturer in international business in the School of Management, Royal Melbourne Institute of Technology (RMIT University, Australia). His research interests span marketing for international education, international leadership in South East Asia, and culture and management of business education in Australia and South East Asia.

Ee Yaw Seng is a Lecturer in Management and Marketing with the School of Business and Design, Swinburne University of Technology (Sarawak Campus), Malaysia. His current research interest focuses on the relationship of traditional teachings and contemporary management style.

Piyush Seth: An MBA from Uttar Pradesh Technical University Lucknow. He is doing Ph.D from Gurukul Kangri Vishwavidyalaya Haridwar,UttaraKhand. He is an Lecturer of Marketing in the College of Sahara Arts and Management Academy and his core area of interest is rural marketing and its association with Amartya Sen's Capability Approach. He is a member of Human Development Capability Association (HDCA).

Vinay Sharma PhD from U.P.Technical University, Uttar Pradesh. MBA from Philippine Christian University. Has around 16 years of Experience, in the areas of Business Opportunity, Development, Market Development, Brand Development, IT enabled Services and Teaching for past seven years. He has a considerable experience of working with various organizations in the fields of Social Development, Media and Information Technology along with having worked with the largest read newspaper of India. His areas of interests include Poverty Alleviation through Business Development, Market Development and Technology wherein he has designed and proposed a specialized model recognized by the name "Affordability for the Poor and Profitability for the Provider" for the provision of health to the rural and the poor population in India through an extensive research which, has been acknowledged at various platforms. The model and his research concentrates on development of health leading to business opportunity development which leads to co-creation of value and that subsequently leads to the capability development of the people at the bottom of the pyramid. He is an associate and a member of the founding group of the Network of Asia Pacific Schools and Institutes of Public Administration and Governance constituted by Asian Development Bank in December 2004 and the Member of Editorial Board of the E-Journal of NAPSIPAG called Journal of Administration and Governance. Vinay Sharma teaches Marketing and the allied subjects and has been a founding HOD of Department of Management Studies of an Institution at Lucknow and also has been an Associate Dean of College of Management and Economic Studies, of a University at Dehradun where he was also the member of the prestigious Academic Council of the University. He has published and presented papers, Chaired sessions at national and international platforms especially in his areas of interest. He has edited an internationally circulated journal and is a member of Editorial Board of GBR and Chinese Public Administration Review and is also the member of academic and advisory councils of prestigious institutions.

Francesca Spigarelli is Assistant Professor at the University of Macerata (Italy), Faculty of Law. She teaches economics and economics of innovation. She is author of several publications on firm internationalization strategies. She published peer-reviewed articles and chapters focused both on Italian SMEs and on MNEs from emerging countries, with specific reference to Chinese Go Global policy.

Lin'yan Sun is Professor, PhD supervisor and vice dean of the School of Management of Xi'an Jiaotong University. He has a wide range of research interest, but mainly focus on: advanced manufacturing management and manufacturing strategy in China, as well as logistics & supply chain management.

Jici Wang was born in 1946. She graduated in 1968 from Department of Geology and Geography at Peking University. In 1978 when the Cultural Revolution ended she pursued further nondegree education, major in economic geography. Like many other Chinese senior scholars, she hasn't got any degree because China has no Degree System until 1980s. She joined the faculty of Department of Geography of Peking University in 1980 and became a professor of economic geography in 1991. She was pointed as a full member of the Commission on the Organization of Industrial Space, International Geographical Union (IGU) for the period 1992-2000. Drawing on her good experience of China research and active work in the Commission, she has been included in the Mentor Panel of the commission since 2000. Prof. Wang is a member of the Steering Committee, IGU Commission on the Dynamics of Economic Spaces. http://uac.utoledo.edu/igu_commission/about_us.htm Her recent research interests are regional innovation system and industrial clusters (new industrial districts). Her book Enterprise Clusters and Regional Development and the web in Chinese (www.clusterstudy.com) effected changes in related research fields for innovation-led economic development of China. Her research project sponsored by the National Natural Science Foundation of China (2006-2009) systematically analyzed the spatial dynamics and economic performance of China's clusters and found the way of their upgrading and innovation.

Jia Wang is an assistant professor in the Department of Human Resource Development, Texas A & M University, College Station, TX.

Han Weixi is Master of Science on Project Management, she got the M.S.c degree in Northumbria University at Newcastle upon Tyne. Han's expertise is on Project Management, Partnership and Collaboration and Cultural Trade.

Hong-yan Yan is lecturer of finance of Beijing Union University, Beijing, P.R. China. She is a Ph.D student of Tianjin University of Finance & Economics, Tianjin. Her research interest focuses on the management of capital structure and the financial problems of high technology enterprises, especially the small and medium-sized enterprises.

Yi Yang is professor of finance and dean of the School of Management, Beijing Union University, Beijing, P.R. China. Professor Yang Yi focuses scientific researches on financial markets and commercial banks for a long time. She is one of the scholars who most early proposed that financial innovation must carry on in China to help the small and medium-sized enterprises which lack collateral or credit history.

Ze-yun Yang is lecturer of finance of Beijing Union University, Beijing, P.R. China. He focuses scientific researches on financial markets and financial service for the small and medium-sized enterprises.

Chun Zhang was born in 1983. She is a Phd student in the Department of Regional and Urban Planning, College of Urban and Environmental Sciences at Peking University. And she is currently a Cooperative Training Phd student in the Department of City and Regional Planning at the University of North Carolina at Chapel Hill between 2009 and 2010. Her research interests are community development, community planning and urban transition in Chinese cities. Her email is: zici0723@gmail.com

Jilong Zhang is an undergraduate student in the international business program at the School of Management, Royal Melbourne Institute of Technology(RMIT University, Australia).

Xuanwei Zhang: Master of Management, PhD candidate of Peking University, Senior lecturer of Beijing Union University, research interests are: evaluation on informatization, cultural differences between the East and the West and their implications for social performance, the Chinese economic model, marketing strategies and their cultural processes, etc.. Long being engaged in teaching on economics, marketing, information management, business English, hosted and participated the research program "China's Indicator System for Enterprise Informatization Evaluation", was awarded and completed the 863 research program "Ïnformatization Index of the Manufacturing Industry of China", granted by The Ministry of Science and Technology, and a number of cross-organizational research programs, published over 10 journal articles, co-published two textbooks.

Jingyuan Zhao is a post-doctoral researcher in School of Management, Harbin Institute of Technology (China). Her Ph.D. is in Management Science and Engineering in Chinese Academy of Sciences (CAS) & University Science and Technology of China (USTC). Dr. Zhao's expertise is on regional innovation, knowledge management, organizational learning, technology diffusion, industry cluster, R&D internationalization of MNCs, and management of cultural industry. She serves as a Guest Editor for several international journals and is an Invited Reviewer for China' state-run newspaper *West Times* to provide comments on the economy.

Wei Zheng is an assistant professor of human resource development at Northern Illinois University. Her research interests include social dynamics of innovation, knowledge management, and scholar-practitioner issues. Her practical experiences include consulting, training, and marketing for various companies in the US and in China.

Valerie (C. Y.) Zhu is a PhD graduate from the School of Management of Xi'an Jiaotong University. She is now working as associate professor in Xi'an University of Science & Technology with research interests mainly falling into Globalization and Internationalization; Cross-Cultural Research and knowledge management, as well as Chinese manufacturing strategies comparing that in western countries.

Index

www.ingramcontent.com/pod-product-compliance
Lightning Source LLC
LaVergne TN
LVHW062145200926
848484LV00007B/1863
* 9 7 8 1 4 6 6 6 0 2 7 6 2 *